Systems Engineering: Theoretical and Practical Approaches

Systems Engineering: Theoretical and Practical Approaches

Edited by **Chester Mann**

NY RESEARCH PRESS
New York

Published by NY Research Press,
23 West, 55th Street, Suite 816,
New York, NY 10019, USA
www.nyresearchpress.com

Systems Engineering: Theoretical and Practical Approaches
Edited by Chester Mann

International Standard Book Number: 978-1-63238-433-1 (Hardback)

Printed in China.

Contents

	Preface	**VII**
Introductory Chapter	**A Few Words About Systems Engineering** Boris Cogan	**1**
Part 1	**Systems Engineering Practice**	**11**
Chapter 1	**Methodology for an Integrated Definition of a System and Its Subsystems: The Case-Study of an Airplane and Its Subsystems** Sergio Chiesa, Marco Fioriti and Nicole Viola	**13**
Chapter 2	**Functional Analysis in Systems Engineering: Methodology and Applications** Nicole Viola, Sabrina Corpino, Marco Fioriti and Fabrizio Stesina	**39**
Chapter 3	**Complex-Systems Design Methodology for Systems-Engineering Collaborative Environment** Guido Ridolfi, Erwin Mooij and Sabrina Corpino	**65**
Chapter 4	**Life Cycle Cost Considerations for Complex Systems** John V. Farr	**97**
Chapter 5	**Integrated Product Service Engineering – Factors Influencing Environmental Performance** Sofia Lingegård, Tomohiko Sakao and Mattias Lindahl	**117**
Chapter 6	**A Safety Engineering Perspective** Derek Fowler and Ronald Pierce	**135**
Chapter 7	**Leveraging Neural Engineering in the Post-Factum Analysis of Complex Systems** Jason Sherwin and Dimitri Mavris	**165**

Chapter 8	**System Engineering Method for System Design** Guillaume Auriol, Claude Baron, Vikas Shukla and Jean-Yves Fourniols	183
Chapter 9	**Assessing the Capacity for Engineering Systems Thinking (CEST) and Other Competencies of Systems Engineers** Moti Frank and Joseph Kasser	199
Chapter 10	**An Abstracted and Effective Capabilities Portfolio Management Methodology Using Enterprise or System of Systems Level Architecture** Joongyoon Lee and Youngwon Park	213
Part 2	**New Systems Engineering Theories**	231
Chapter 11	**Augmented Human Engineering: A Theoretical and Experimental Approach to Human Systems Integration** Didier Fass	233
Chapter 12	**Usage of Process Capability Indices During Development Cycle of Mobile Radio Product** Marko E. Leinonen	253
Chapter 13	**Systems Engineering and Subcontract Management Issues** Alper Pahsa	277
Chapter 14	**A System Engineering Approach to e-Infrastructure** Marcel J. Simonette and Edison Spina	289
Chapter 15	**System Engineering Approach in Tactical Wireless RF Network Analysis** Philip Chan, Hong Man, David Nowicki and Mo Mansouri	309
Chapter 16	**Creating Synergies for Systems Engineering: Bridging Cross-Disciplinary Standards** Oroitz Elgezabal and Holger Schumann	329
	Permissions	
	List of Contributors	

Preface

This book aims to highlight the current researches and provides a platform to further the scope of innovations in this area. This book is a product of the combined efforts of many researchers and scientists, after going through thorough studies and analysis from different parts of the world. The objective of this book is to provide the readers with the latest information of the field.

A descriptive account on systems engineering has been provided in this book, elucidating the theoretical as well as practical approaches. It is a compilation of research works contributed by researchers and developers from all over the world. A majority of these works elucidate methodologies for distinct systems engineering procedures while others consider matters of parallel knowledge areas and sub-areas which essentially contribute to the enhancement, maintenance and operation of systems. This book consists of data on spacecraft, aircraft and space systems development, post-analysis of information gathered during working of huge systems, etc. Important matters regarding "bottlenecks" of systems engineering, like security, complexity and authenticity of various types of systems, creation, operation and maintenance of services, system-human communication and management tasks done during system projects have been addressed in this book. It is very useful for those who are interested in the modern state of the systems engineering information field and for systems engineers engaged in various activities of the field. The aim of this book is to serve as a valuable source of reference for students and university lecturers. It can be directly used in systems engineering courses as descriptive materials.

I would like to express my sincere thanks to the authors for their dedicated efforts in the completion of this book. I acknowledge the efforts of the publisher for providing constant support. Lastly, I would like to thank my family for their support in all academic endeavors.

Editor

Introductory Chapter

A Few Words About Systems Engineering

Boris Cogan
Faculty of Computing,
London Metropolitan University,
UK

1. Introduction

First of all, why 'Practice and Theory', not vice versa what probably could be more traditional? - We will see later on…

Systems Engineering is a relatively young area of knowledge. Its main elements got a significant development during World War II (mainly for aircraft development and maintenance) because project managers found that considering product components as 'separate entities' with their own attributes gives additional views to ones for separate elements. Nowadays it is generally accepted that any 'complex engineering product', 'a system', is analysed/viewed as a hierarchy of layers of 'simpler sub-systems'. (This is referred only to the way of how systems analysis is done and has nothing in common with the architecture of a system.)

After the WW II, numerous military applications, spacecrafts of any kind, nuclear power stations etc. (i.e. products with higher requirements to reliability and safety) required separation of Systems Engineering in a branch of engineering with its own methods, techniques, tools, rules etc. that distinguished it from other engineering knowledge areas. It is considered that the evolution of Systems Engineering began during the late 1950's [INCOSE Handbook].

Since the late 1960's, Systems Engineering Standards were recognised as a separate group and corresponding ISO and IEEE standards were labelled as 'Systems Engineering' ones. (It is worth to note that nowadays more and more Systems Engineering standards are combined with Software Engineering ones because modern systems are 'software-intensive' or 'software-based', Systems Engineering processes and Software Engineering ones are similar and practically no technical system exists without massive amount of software.) However, some 'older' IEEE standards are related to the 'Information Technology' category.

In 1990 the *International Council on Systems Engineering* (INCOSE) was founded as a not-for-profit membership organisation to develop and disseminate the interdisciplinary principles and practices that enable the realisation of successful systems [INCOSE]. As its mission, the organisation declared: '*Share, promote and advance the best of systems engineering from across the globe for the benefit of humanity and the planet*'.

Older and newer ISO and IEEE standards and INCOSE materials may give a bit different definitions of what Systems Engineering is. However, actually, they are about the same. '***Systems Engineering:*** *An interdisciplinary approach and means to enable the realisation of successful systems*' [INCOSE Handbook]. Then we need to clarify what kinds of systems are meant: '***System:*** *An interacting combination of elements to accomplish a defined objective. These include hardware, software, firmware, people, information, techniques, facilities, services, and other support elements*' [INCOSE Handbook]. It is actually important that a system under consideration is 'engineered'. As it is said in the same INCOSE materials: '*The term "Systems Engineering" is only used with respect to the act of engineering a specific system*'. It is not good and not bad; it is just as it is.

2. My route in systems engineering

My own acquaintance with Systems Engineering took place in the second part of 1960's at the *Institute of Control Sciences* (ICS), the Soviet Union's leading research institution in the automatic control knowledge area. Main sub-areas of research of the Institute were automatic control theory and development of elements for automatic equipment [ICS]. My University (actually, Moscow Institute of Physics and Technology - MIPT) department was situated at the ICS because, according to the mode of the education at MIPT, last four years (out of six) students were (and are now) taught by acting leading researches in corresponding knowledge areas, and the students worked with the researchers at their work place, not vice versa [MIPT]. Just one example: during one of the semester at my fifth year, one module (lectures) of a day was taught by a Vice President of the International Federation of Automatic Control - IFAC [IFAC], the next module lectures were read by a leading Soviet Union specialist in control issues of air and ballistic missile defence, the following one was given by a leading specialist in air-to-air missile control. It was an extremely wonderful time! Definitely, no military terms were used in the lectures; they were about automatic control, not military applications.

At the fourth-sixth years, each MIPT student worked at an ICS laboratory as a researcher. The sixth year was completely dedicated to the Final Project. According to the mission of ICS, it was involved in all 'big' military and civil Systems Engineering projects of the Soviet Union for solving control problems of the systems. Because of the severe secrecy, usually students did not know what their projects were for. We developed 'theories' and functioning prototypes. Whether they were used or not in real systems, we, of course, had no idea (at least, officially). However, meetings with specialists from various system development organisations allowed us (according to their interest or absence of the interest) to understand our contribution to Systems Engineering. I was involved (definitely, together with staff researchers of ICS) in developing electronic elements based on magnetic cores with a rather complex configuration to be used in different technical facilities, including elements of multi-stage rockets (for manned space flights). Actually, I do not know so far whether they were used in real space programmes or not.

So, it was my first involvement in Systems Engineering. This term was not used in our environment because (1) it did not yet exist and (2) we all were too far from real (and *technological)* Systems Engineering processes (in terms of ISO/IEC 15288:2008 / IEEE Std 15288™-2008 [ISO 15288]). However, it was invaluable experience for my better understanding of engineering that have been used in my following 'adult' life. I am deeply grateful to all my teachers and colleagues from ICS for the years spend at the Institute (I was a PhD student at ICS later on but specialised in Software Engineering).

During my long employment at the Institute for Automation and Control Processes of the Russian Academy of Science in Vladivostok (the Russian Far East) [IACP], I was mainly involved in developing software for different 'computer-based' systems and did research in the Software Engineering knowledge area creating new methods, techniques and tools for increasing software productivity. However, I also took part in Systems Engineering projects, as well.

In 1980's-90's, as Head of Testing, I was involved in a large project to develop a prototype of a distributed military system (a legacy of the 'Cold War'). The 'core' of the system was 'artificial intelligence' (that may be treated in different ways in this context) developed at the Expert Systems Department of IACP. The knowledge base containing validated knowledge of the best specialist in the application area together with an extremely efficient inference engine allowed monitoring corresponding activity in a very large geographical area, practically in real-time. The bottle-necks were sensors (I reminder that it was only a prototype); aircrafts and ships played the role of sensors. Nowadays, spacecrafts are very common 'sensors' for those kinds of applications, and there is no need to move them from one orbit to another but we did not have access to spacecrafts that time. As members of my testing team, I had specialists who developed, tested and maintained software for the 'Buran' system, the Russian analogue of the US's space shuttle, who took part in launching and landing of that extraordinary automatic craft. It was a great experience for me.

In mid 1990's, after a long and interesting discussion during my work on another project in Scotland, Professor Richard Thayer [R. Thayer], invited me to be a member of his team to finish development of *IEEE Std 1362 IEEE Guide for Information Technology – System Definition – Concept of Operations (ConOps) Document*. This standard was published in 1998 as IEEE Std 1362TM-1998 and reaffirmed in 2008 for the next 10 years without changing a letter. '*This guide prescribes the format and contents of the concept of operations (ConOps) document. A ConOps is a user oriented document that describes system characteristics of the to-be-delivered system from the user's viewpoint. The ConOps document is used to communicate overall quantitative and qualitative system characteristics to the user, buyer, developer, and other organisational elements (e.g., training, facilities, staffing, and maintenance). It describes the user organisation(s), mission(s), and organisational objectives from an integrated systems point of view*' [IEEE 1362]. As a matter of fact, this kind of user oriented document should be developed for any system planned to be build.

Thus, in retrospect, those three decades of my professional life actually gave me great practical experience in development of systems, without any deep knowledge of any 'theory' of the development (in Systems Engineering). Really, that time there were no 'validated' 'theory' (process standards) available for the projects I was involved in (I am not saying that there were no theory in the country at all). However, as a matter of fact, I got my understanding if not Systems Engineering processes but at least what Systems Engineering is – from practice.

The next period of my life is lecturing at the London Metropolitan University [LMU]. In my Software Engineering modules for MSc students I needed (and need now) to clarify the place of Software Engineering in the context of Systems Engineering. For that I had to study at least some 'theory' of systems development and be familiar in detail with many Systems Engineering standards related to processes of Systems Engineering projects. So, I would name this period of my life as 'familiarisation with Systems Engineering theory'. Two 'stages' of my professional career: 'practice' and 'theory', are the first reason for the title of the book.

The second reason (and really important one) is a bit 'philosophical'. People develop systems using existing knowledge ('theory'). During a system's development, then operation and maintenance, they better understand the processes that they have applied and used, their merits and demerits; their own mistakes and 'holes' in the theories applied. They need new theories or at least improved old ones. Reasonable theories are always based on practice. It is why (at least, in engineering) theories are following practice, not vice versa. Now we know the reasons to name the book.

3. Part I: Systems engineering practice

The first article of the book, '*Methodology for an Integrated Definition of a System and its Subsystems: The case-study of an Airplane and its Subsystems*' by *Sergio Chiesa, Marco Fioriti & Nicole Viola*, demonstrates application of the current 'theory' of Systems Engineering on the example of an aircraft, one of the most complex computer-intensive modern systems. Reliability and safety requirements to any aircraft are extremely high that demands a very sophisticated process of development of all aircraft's sub-systems. From another point of view, the development is very expensive and it sometimes needs in compromises (trade-offs). These conditions need specific methodologies and well-grounded requirements to the product. The article presents such a methodology and in addition to its research value, it provides wonderful material for teaching Systems Engineering and Aviation students.

Aircrafts and missiles are extremely 'complex' objects to develop. However, space systems are usually even more 'complex' because in addition to crafts themselves they include a lot of specific ground services to launch and operate the crafts and to receive, collect, and process information sent by spacecrafts. All these demand some additional methods or even methodologies to develop a system that works and meets other requirements. The second article of this Part, '*Complex-systems Design Methodology for Systems-Engineering Collaborative environment*' by *Guido Ridolfi, Erwin Mooij & Sabrina Corpino*, presents a methodology that is designed for implementation in collaborative environments to support the engineering team and the decision-makers in the activity of exploring the design space of complex-system, typically long-running, models. The term 'complexity' is used in the real life without too much thinking what it means, 'complex'. However, for developers of a system, the complexity has to be measured (or estimated) in particular measurement units to understand what methods and solutions could better suit the requirements (trade-offs are needed as usual). The authors show that contribution of the human factor is fundamental for obtaining a final product with a high cost-effectiveness value. This means that any human activity in Systems Engineering processes needs specific methodological and tool support as much as possible. As a case study, an Earth-observation satellite mission is introduced in the beginning of the article and this satellite mission is used throughout the chapter to show step by step implementation of the suggested methods. This article is a good source for teaching material, as well as the first one.

Considering the system requirements, first of all, any system performs functions. As the system is viewed as being 'composed' of a few lower layers, the *Functional Analysis* is done on each layer for each sub-system. A particular case how it could be done can be a valuable source of material for systems' engineers and for students. The third article of Part I, *'Functional Analysis in Systems Engineering: Methodology and Applications'* by *Nicole Viola, Sabrina Corpino, Marco Fioriti & Fabrizio Stesina,* gives the opportunity to see practical applications of the Functional Analysis. Functional Analysis applies in every phase of the design process; it turns out to be particularly useful during conceptual design, when there is still a wide range of potentially feasible solutions for the future product. The precious role of Functional Analysis consists in individuating as many available options as possible, but not missing any ideas that may offer significant advantages. The article gives very vivid examples of application of Functional Analysis in development of various systems. Three of them deserve special mentioning: (1) Functional Analysis at sub-system level to define the avionic sub-system of an aircraft; (2) Functional Analysis at system level to define a satellite in Low Earth Orbit; (3) Functional Analysis at system of systems level to define a permanent human Moon base. The paper is a wonderful illustrative material for a range of engineering university courses.

As it has been already mentioned, nowadays *safety* is one of the most important properties of any complex system. As it is shown in the next articles of the book, *'A Safety Engineering Perspective'* by *Derek Fowler & Ronald Pierce, safety* is actually a set of attributes that have to be considered and measured separately. Authors show that the concept of *'reliability'* should not be mixed up or even considered together with the concept of 'safety'. Reliable system elements may contribute to non-reliability of a system just because they do not suit the requirements to this particular system. In other words, functionality of the system, of its components and their elements has to be carefully analysed and expressed on all layers of system hierarchy: requirements to a higher layer architecture component have to be carefully allocated to 'sub-systems' of the next layer, including ones to reliability and safety. The article introduces the principles of safety assurance and safety cases and showed how they should drive all the processes of a safety assessment, throughout the project life cycle.

Development of complex systems is extremely expensive. If it is a completely new kind of systems, there are no historical data to base a new project on. More or less, managers understand how to cost hardware and to a lesser extent, software. However, it is not the case for integration and interfaces of complex systems that needs new methods and tools for estimations. When the cost of the process of development of larger and more complex systems, a system of systems, and enterprises is estimated, managers' ability to make accurate (or at least adequate) estimates becomes less relevant and reliable. The following, fifth, article of the book, *'Life Cycle Cost Considerations for Complex Systems'* by *John V. Farr,* presents some of the methods, processes, tools and other considerations for conducting analysis, estimation and managing the life cycle costs of complex systems. It considers some estimation models and tools for hardware, software, integration at the system level, and project management. It briefly describes *Cost Management* as a separate task of a Systems Engineering project. The article emphasises that systems engineers are usually not trained for doing accurate system development cost estimation, and proper methods, processes, and tools could significantly help them in the task.

According to the definition of a system, the 'system' may have different forms, in particular, to be a service ('***Service***: *Useful work performed that does not produce a tangible product or result, such as performing any of the business functions supporting production or distribution*' [PMBOK]). Any service has first to be created, then it operates, it needs maintenance, repair, upgrade, take-back, and consultation. Any product during its life somehow influences the environment. When service is developed, the environmental problems have to be carefully analysed: what is the influence. The sixth article, '*Integrated Product Service Engineering - Factors Influencing Environmental Performance*' by *Sofia Lingegård, Tomohiko Sakao & Mattias Lindahl*, analyses widely-used strategies of *Service Engineering* and suggest improvements of the strategies. Some products are used for 20-40 years and definitely knowledge about the products is increased during the time. Characteristics of the product may turn out deviated from supposed ones in the development; environmental requirements may change over the decades; 'better' products with the same mission may be developed and so on, and so on... Unfortunately, in real practice, rather often little care is taken in product development (and in its specification) for future services, maintenance, and end-of-life-treatment. Traditionally, the initial focus is on developing the 'physical' product; once that is done, a possible service (intangible product) is developed, but this is hindered by the limitations set up in and resulted from the physical product. When *Integrated Product Service Offering* proposed by the authors is used, the development is accomplished in an integrated and parallel approach.

During the use of a complex system, usually, an extremely big amount of data is collected. What and how to do with the data to extract 'useful' information for planning new projects and developing new systems? It has been a rather 'normal' situation when people did not know what to do with the information and its collection and keeping detailed records were just a waste of money. The problems to properly use the data are: absence of available methods for that amount of data to be analysed, lack of computational resources, impossibility to interpret results of the analysis etc. New approaches are needed to cope with the problems. The seventh article of the book's Part I, '*Leveraging Neural Engineering in the Post-Factum Analysis of Complex Systems*' by *Jason Sherwin & Dimitri Mavris*, presents such an approach. They suggest considering the breadth of results and techniques emerging from neural engineering to bolster systems analysis for engineering purposes. In particular, instead of relying on an inconsistent mapping made by human experts to design analysis, why not understand some cognitive elements to expertise and, in turn, apply that comprehension to both systems analysis and manipulation? As the case study, methods of neural engineering to the post-factum analysis of Iraq's stability during 2003-2008 were applied. Such an analysis was never performed in a real context; however authors frame the problem within the context of its utility to a decision-maker whose actions influence the outcome of such a system.

Usually, when the Systems (or other kind of) Engineering is discussed in the literature, a set of processes consisting of activities and tasks is presented. But it is only one 'dimension' of the project management; there are two other: (1) work products to use and generate, and (2) people and tools involved. Any Systems Engineering organisation has potential capabilities for creating systems (or other products). *Capability Portfolio Management* allows an organisation to coordinate capabilities needed to correspond to potential projects (investments). The most Capability Portfolio Management processes are too complex to be used by inexperienced managers. The eighth article of the book, '*Abstracted Effective*

Capabilities Portfolio Management Methodology Using Enterprise or System of Systems Level Architecture' by *Joongyoon Lee,* suggests a simpler and more practical methodology for developers and enterprises. The process consists of only 16 sequential tasks that corresponds to *ISO/IEC 24744 Software Engineering - Metamodel for Development Methodologies,* ISO, 2007.

Systems are developed by engineers who are taught and trained for that. Potentially, there could be different approaches for that teaching. One is that sub-system specialists are taught how to develop sub-systems and there are someones who know how to integrate sub-systems in a system. Another approach is to get all system project participants familiar with development of systems, not just system components and their integration. The second one allows better understanding and communication. The ninth paper of the Part, *'System Engineering Method for System Design'* by *Guillaume Auriol, Claude Baron, Vikas Shukla & Jean-Yves Fourniols,* presents some educational materials, the process and the outcomes to teach an engineering approach. The case to illustrate the approach is rather practical; it includes commonly used sensors, wireless network, and computational facilities. Various issues can be raised during teaching on wireless sensor networks: electronic design, risks to humans, energy management, telecommunication technologies, etc. The case demonstrates all implementation and some management processes (in terms of ISO/IEC 15288) for a liner project life cycle model. The paper may be very useful as reading (or even a set of educational ideas) for students of various engineering courses.

For each development project engineers with particular knowledge and skills are needed. When project's team is formed, the project manager team have to be sure that project participants correspond to project requirements. How to test competencies of the project teams? There are some traditional approaches and the last, tenth, article of this part, *'Assessing the Capacity for Engineering Systems Thinking (CEST) and other Competencies of Systems Engineers'* by *Moti Frank & Joseph Kasser,* suggests a new tool for that. As there is no known way for directly 'measuring' thinking skills of individuals, an indirect way is needed, for example, IQ tests are pen-and-paper indirect tests for 'measuring' the intelligence of individuals. The tool combines questionnaires for three main concepts: (1) Success in a systems engineering position, (2) An interest in systems engineering positions and (3) Capacity for engineering systems thinking (CEST); they are all interconnected and interrelated. The will and interest to be a systems engineer basically means the desire and interest to be involved with job positions that require CEST. In other words, the authors hypothesise that there is a high positive correlation between the engineering systems thinking extent of an individual and his/her interest in what is required from successful systems engineers.

4. Part II: New systems engineering theories

According to the *U.N. telecommunications agency,* there were 5 billion mobile communication devices all across the globe in to the end of 2010 [BBC] and the quantity of produced mobile phones and rate of diffusion are still increasing. The devices are used by all people regardless of race, age or nationality but their requirements to the devices differ. In other words, quality of the devices (as correspondence to requirements) should be treated differently. From another point of view, the level of quality has to be 'high enough' for all categories of the devices and the levels need to be compared. For an effective communication between parties a common 'quality language' is needed and unitless process capability indices are widely used for this purpose. However, according to the statement of the author of the first article of Part II, '*Usage of Process Capability Indices during Development Cycle of Mobile Radio Product*' by *Marko E. Leinonen*, the existing process capability indices do not suit the modern practice in full. The article analyses the current approaches to definition and calculation of indices and proposes new equations for one-dimensional process capability indices with statistical process models based on calculations and simulations. In addition, process capability indices have been defined for multidimensional parameters which are analogous to one-dimensional process capability indices. One of the main difference between one and two-dimensional process capability indices analysis is that a correlation of the data with two-dimensional data should be included into the analysis.

Systems engineers communicate each other during a system's development and users communicate to the system during its operation/use. Effectiveness of the communications has a significant effect on the result of the system's development and success in the system's use. *Human Engineering* may be considered (within the context under consideration) as a sub-area of the Systems Engineering knowledge area. The second article of Part II, '*Augmented Human Engineering: A Theoretical and Experimental Approach to Human Systems Integration*' by *Didier Fass*, focuses on one of the main issues for augmented human engineering: integrating the *biological user's needs* in its methodology for designing human-artefact systems integration requirements and specifications. To take into account biological, anatomical and physiological requirements the author validates theoretical framework. He explains how to ground augmented human engineering on the *Chauvet* mathematical theory of integrative physiology as a fundamental framework for human system integration and augmented human design. The author proposes to validate and assess augmented human domain engineering models and prototypes by experimental neurophysiology. He presents a synthesis of his fundamental and applied research on augmented human engineering, human system integration and human *in-the-loop* system design and engineering for enhancing human performance – especially for technical gestures, in safety critical systems operations such as surgery, astronauts' extra-vehicular activities and aeronautics.

Nowadays e-Infrastructures become more and more spread out in the world, mainly for research and development. '*The term e-Infrastructure refers to this new research environment in which all researchers - whether working in the context of their home institutions or in national or multinational scientific initiatives - have shared access to unique or distributed scientific facilities (including data, instruments, computing and communications), regardless of their type and location in the world*' [e-IRG]. It is obvious that being, in some sense, a 'super-system', an e-Infrastructure cannot take into account all technologies used in 'sub-parts' of the structure, peculiarities of different group of researchers, different cultures and so on. A harmonised approach (a meta-model) is needed for creation suitable e-Infrastructures. The third article of Part II, '*A System Engineering Approach to e-Infrastructure*' by *Marcel J. Simonette & Edison Spina*, presents such. It aims to deal with the interactions between e-Infrastructure technologies, humans and social institutions, ensuring that the emergent properties of the system may be synthesised, engaging the right system parts in the right way to create a unified whole that is greater than the sum of its parts.

Generally, no big/complex system can be developed by organisation on its own; tens and even hundreds of other Systems Engineering and other kinds of Engineering organisation may be involved in the project. Then a rather complicated management task of dealing with numerous sub-contractors emerges. The two *Agreement Processes* of *ISO/IEC 15288-2008: Acquisition and Supply ones,* cover the task: '*These processes define the activities necessary to establish an agreement between two organizations. If the Acquisition Process is invoked, it provides the means for conducting business with a supplier: of products that are supplied for use as an operational system, of services in support of operational activities, or of elements of a system being developed by a project. If the Supply Process is invoked, it provides the means for conducting a project in which the result is a product or service that is delivered to the acquirer.*' The fourth article of Part II, '*Systems Engineering and Subcontract Management Issues*' by *Alper Pahsa,* presents a possible interpretation of activities and tasks of ISO/IEC 15288 Agreement Processes in terms of the INCOSE materials.

Tactical wireless radio frequency communication systems are a kind of communication systems that allow the interoperability and integration of Command, Control, Computers, Communications, and Information and Intelligence, Surveillance and Reconnaissance Systems in the field of information management control in modern armed forces. According to the current practice, the systems are rather too vulnerable. So, when they are under development and use, they need additional methods of analysis to decrease their vulnerability. The fifth article of Part II, '*System Engineering Approach in Tactical Wireless RF Network Analysis*' by *Philip Chan, Hong Man, David Nowicki & Mo Mansouri,* presents an approach to use mathematical Bayesian network to model, calculate and analyse all potential vulnerability paths in wireless radio frequency networks.

Engineering of systems usually includes involvement of many disciplines and knowledge areas. The disciplines have their own terminology and standards. Often the same terms in different disciplines have different semantics. The same situation is for standards; for example, process standards may present similar processes in more or less different way and in different terms. Harmonising the standards is a slow and difficult process and ISO and IEEE Working Groups have been done the activities for decades. It does not put any restraint on independent researchers to try to create their own synergetic models. The last article of the book, '*Creating Synergies for Systems Engineering: Bridging Cross-disciplinary Standards*' by *Oroitz Elgezabal & Holger Schumann,* is an attempt to merge standards related to Systems Engineering even though they officially refer to different knowledge areas.

5. References

INCOSE Handbook: Systems Engineering Handbook - A Guide for System Life Cycle Processes and Activities, Version 3.2. INCOSE. 2010.

INCOSE: International Council on Systems Engineering. http://www.incose.org/

ICS: Institute of Control Science http://www.ics-ras.com/

MIPT: Moscow Institute of Physics and Technology (State University) http://phystech.edu/about/

IFAC: International Federation of Automatic Control http://www.ifac-control.org/

ISO 15288: ISO/IEC 15288:2008, Systems and software engineering – System life cycle processes, 2nd Edition, Geneva: International Organisation for Standardisation, 2008.

IACP: Institute for Automation and Control Processes http://www.iacp.dvo.ru/english/institute/institute.html

R. Thayer: Richard H Thayer http://www.richardthayer.com/bio.htm
IEEE 1362: IEEE Std 1362-1998, IEEE Guide for Information Technology-System Definition-Concept of Operations (ConOps) Document, IEEE, 1998.
LMU: London Metropolitan University www.londonmet.ac.uk
PMBOK: ANSI/PMI 99-001-2004, *A Guide to the Project Management Body of Knowledge Third Edition, 2004.*
CBC: CBC Money Watch http://www.cbsnews.com/stories/2010/02/15/business/main6209772.shtml
e-IRG: e-Infrastructure Reflection Group http://www.e-irg.eu/

Part 1

Systems Engineering Practice

1

Methodology for an Integrated Definition of a System and Its Subsystems: The Case-Study of an Airplane and Its Subsystems

Sergio Chiesa, Marco Fioriti and Nicole Viola
Politecnico di Torino
Italy

1. Introduction

A modern airplane is without any doubts one of the clearest and most convincing example of "complex system". A modern airplane consists in fact of various types of elements of different technologies (structures, mechanics, electric, electronics, fluids, etc.). Each element has specific tasks to perform and all elements are harmonically integrated to constitute the whole system. Moreover the airplane is a particularly critical system because of quite obvious safety reasons, because of the relevance of its mission, because of high costs and eventually because of its long Life Cycle. Figure 1 shows an example of a modern transport aircraft.

Fig. 1. Alenia C 27 J

Let us consider the case of such an airplane, whose mission statement sounds like: "To transport in flight a certain payload from point A to point B". At a first glance the airplane can be seen as a single entity able to perform a well defined function but, getting more into the details, the airplane appears as consisting of various parts, all harmonically integrated and concurrently working to accomplish the same mission. For instance, taking into account Figure 1, different items, like the wing, the fuselage, the horizontal and vertical tails, the engine nacelles with propellers and the wheels of the landing gear (when the aircraft is on ground), can be easily individuated. By looking at the whole aircraft more into the details, other items can be identified or at least imagined, like the structural elements, the engines and many mechanical, electronic and fluidic installations, referable to the numerous and various technologies present onboard the aircraft.

1.1 Terminology

Before proceeding any further, it is worth clarifying the terminology related to the so-called "system view" and used in the remainder of the chapter.

Taking into account the functional decomposition of the aircraft, it is quite obvious, being the aircraft a complex system, that at the first level of the physical tree there are not single items but group of items, harmonically integrated to perform certain determined functions. Considering a rigorous approach from the terminology point of view, these groups of items should be identified as "subsystems". However, practically speaking, all first level building blocks of the aircraft physical tree (indicated in Figure 2 as subsystems) are usually defined as "systems" (like, for instance, the avionic system, the fuel system, the landing gear system, etc.), as they gather together many different equipments. This ambiguity confirms the following typical characteristic of the system view of complex systems: the concept of system can be applied at different levels. The aircraft system is therefore formed by "n" "subsystems", which in their turn may be thought of as "systems", consisting of the integration of different equipments. A further level of subdivision may also be introduced, in order to split each subsystem into sub-subsystems, made up of various equipments, as Figure 3 shows.

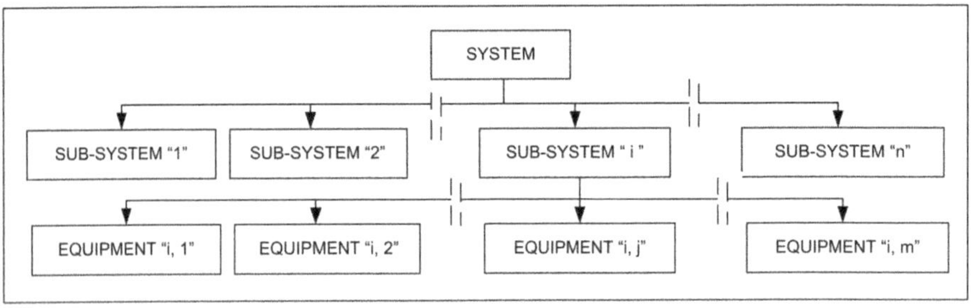

Fig. 2. System view terminology for the aircraft physical tree

Figure 3 illustrates the physical tree of the avionic system (more correctly "subsystem" from the terminology standpoint) of a modern transport aircraft. Because of its high complexity and of the great number of performed functions, the avionic system is in its turn decomposed into several systems (more correctly "sub-subsystems"), which have to accomplish different functions. In particular in the example presented in Figure 3 there are four systems to accomplish the navigation ("Navigation System"), flight controls ("Flight Control and Auto-Pilot System"), communications ("Communications System") and the detection ("Radar System") functions. For sake of brevity only the subdivision of the radar system into equipments is shown in Figure 3. There are two different types of radars: the weather radar and the altimeter radar. They both interface with the same integrated radar display and relative processor. Eventually it is worth noting that the equipments themselves, at least the complex ones, are not at all single entity but may be again further decomposed into modules, which quite often are Line Replaceable Units (LRU) modules, i.e. items that may be replaced quickly at an operating location, in order to minimize the aircraft down time for maintenance.

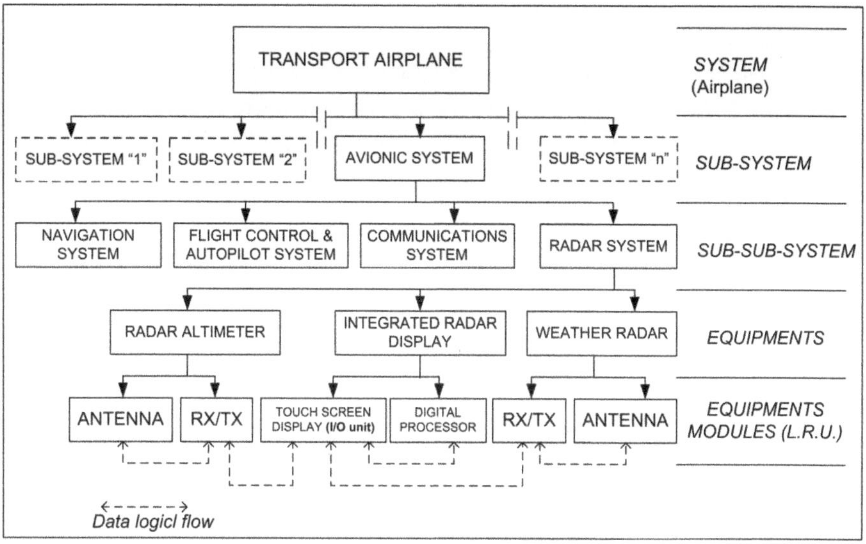

Fig. 3. Transport airplane avionic system physical tree

Please note that in the remainder of the chapter the subsystems are referred to as systems for the above mentioned reasons.

1.2 State of the art and trends of the aeronautical systems

Before presenting the methodology, it is worth describing the state of the art and the general trends of the aeronautic systems. Only the main systems onboard medium/large airplanes are here considered.

Figure 4 illustrates the main systems of a transport aircraft and all their interactions, in particular in terms of power exchange. Please note that the building blocks with dotted line have not been dealt with specifically in the present text. By looking at Figure 4 it is possible to note that:

a. structures and engines have been included into the decomposition of aircraft systems, even though they are not dealt with in the present work, as usually considered in the traditional approach to aircraft conceptual design.
b. In particular both the engines, which are in charge of aircraft propulsion, and, if present, the Auxiliary Power Unit-APU, which is a source of energy alternative to the engines, have been included into the decomposition of aircraft systems because, apart from being systems on their own, they have strong relationships with all other aircraft systems, both because of physical interfaces and because their size is strictly connected to the aircraft weights and aerodynamic characteristics, which are in their turn largely affected by all other onboard systems.
c. The Fuel System interfaces directly with the engines and the APU, as it lets them work properly. Same talks apply to the engine starting system.
d. Taking into account that electrical, hydraulic and pneumatic systems basically perform the same function, onboard systems may be more rationally designed to envisage only

the electrical system (as onboard the aircraft there are users that can be supplied only with electric power, like lights, electronics, etc.). This solution is represented by the actual successful trend of the so-called "All Electric Aircraft", which shows quite a few advantages, if compared to the traditional aircraft, in terms of simplicity and rationality. Other intermediate solutions do also exist as the so-called "More Electric Aircraft" testifies, where the engines power is initially transformed only into electric power and then partially into hydraulic and/or pneumatic power.

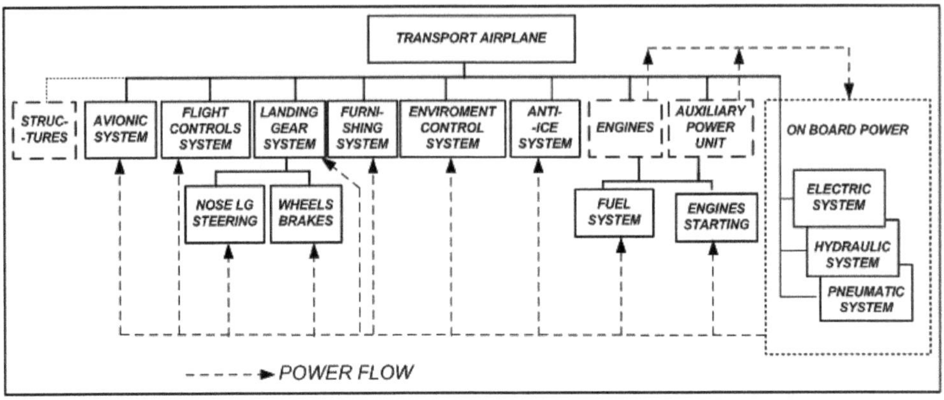

Fig. 4. Transport airplane system and its (sub-)systems

Table 1 summarizes functions, types and features of the main onboard systems.

SYS	FUNCTIONS PERFORMED	SYSTEMS CONFIGURATIONS	NOTES
AVIONIC SYSTEM	To acquire information (from airplane, from external environment, from other Entities, from Telemetry). To elaborate them. To give them to the Crew, to the airplane, to other Entities by means of Telemetry.	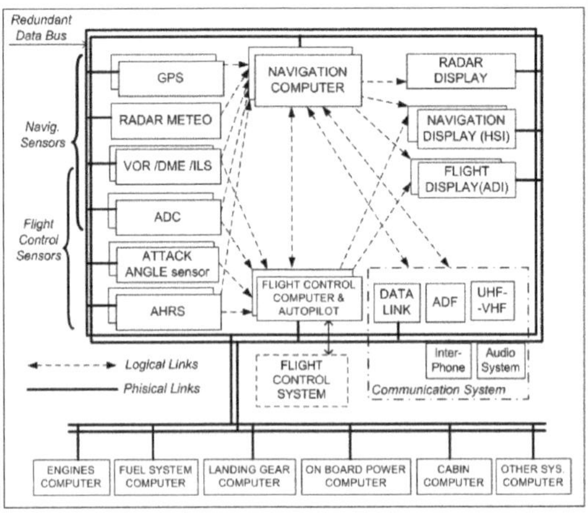	The avionics will be considered at the today state-of-the-art, taking into account usual kinds of equipments. The new trend to "Integrated Modular Avionics" is not considered in a very preliminary approach. The data exchange between Equipments is based on DATA BUS.

SYS	FUNCTIONS PERFORMED	SYSTEMS CONFIGURATIONS	NOTES
FLIGHT CONTROL SYSTEM	To modify aerodynamic actions on airplane (by changing its shape), in order to guide and control flight trajectories and to navigate (primary flight controls). To modify aerodynamic characteristics when necessary (secondary flight controls).	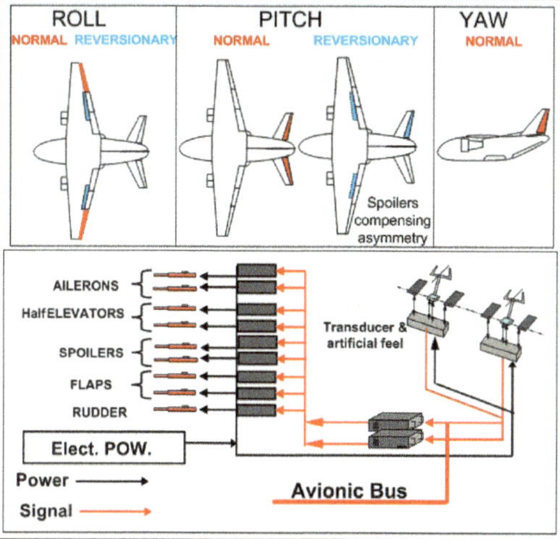	The modern flight control system is normally based on digital signal transmission (Fly-By-Wire) and on hydraulic or on electric power (following new trends of All Electric Aircraft). The system design is mainly based on actuator design, considering the possible solutions of hydraulic, electro-hydraulic and electric actuators.
SYS	FUNCTIONS PERFORMED	SYSTEMS CONFIGURATIONS	NOTES
LANDING GEAR (NLG Steering MLDG Wheels Brakes)	To allow the airplane to move on ground. To support the impact at touch-down. To allow extension and retraction of the system. To steer the nose landing gear. To apply the brake on main landing gear wheels.	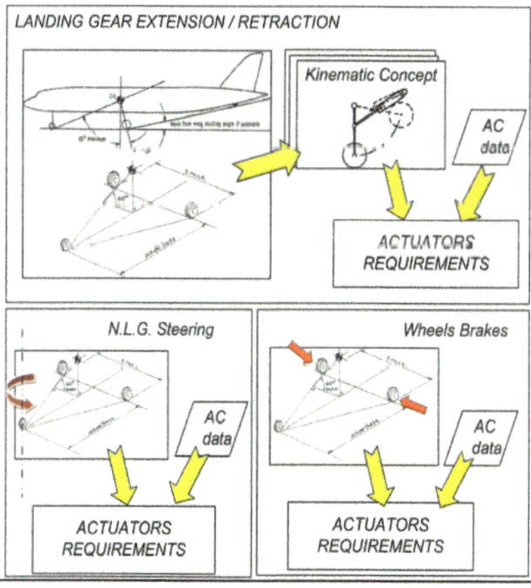	Also for this system, after accomplishing the architectural design (to be carefully integrated with the whole aircraft configuration), the typical design activity consists in sizing the actuators. Also in this case the actuators can be hydraulic (that is the state-of-the-art), electro-hydraulic and electric.
SYS	FUNCTIONS PERFORMED	SYSTEMS CONFIGURATIONS	NOTES
FURNISHING SYSTEM	To guarantee a pleasant and comfortable journey to the passengers, providing them with all services required.	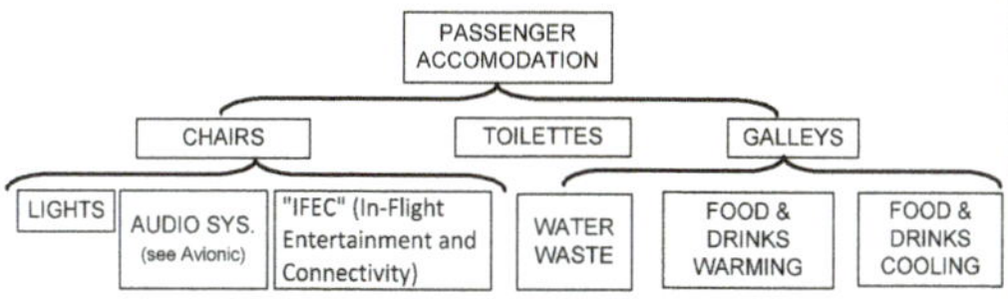	In a very preliminary approach, all several systems connected to the furnishing system can be simply considered by the point of view of weight and as power users.

SYS	FUNCTIONS PERFORMED	SYSTEMS CONFIGURATIONS	NOTES
ENVIRONMENT CONTROL SYSTEM	To provide all people onboard the aircraft with correct values of air total pressure, partial O_2 pressure and temperature.	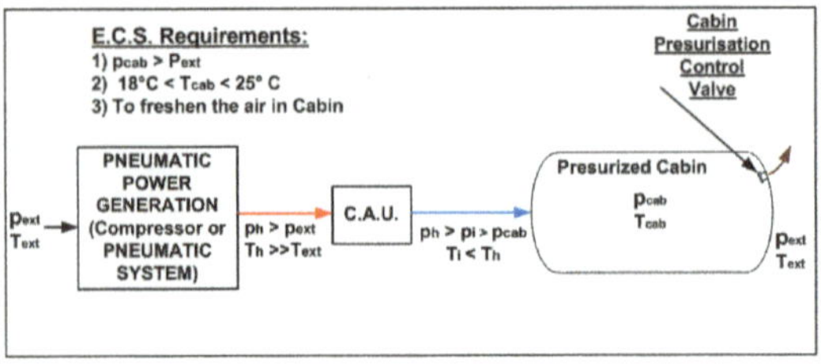	Two kinds of CAU can be envisaged: "vapor cycle" and "air cycle". If the CAU output temperature of the air is < 0°C, it is mandatory to introduce it in the cabin, after mixing with re-circulated cabin air.

SYS	FUNCTIONS PERFORMED	SYSTEMS CONFIGURATIONS	NOTES
ANTI-ICE SYSTEM	To avoid problems due to ice formation of the airplane external surfaces.	 The ice problem → increase C_{D0} → decrease C_{LMAX} → mobile devices jamming → to perturbe air intake flow → propellers dynamic unbalance 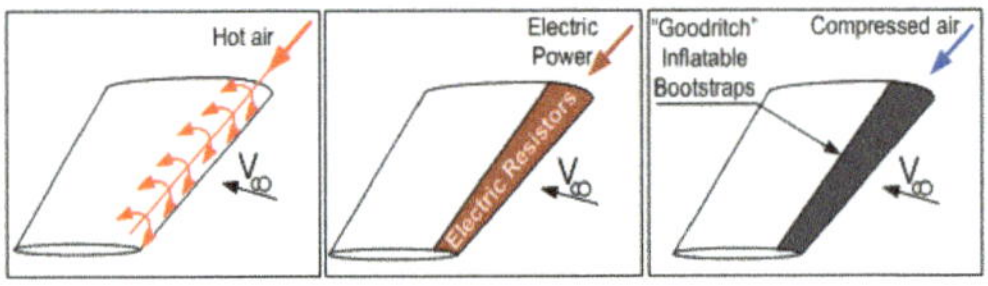 A new kind of anti-ice system on wing leading edge, characterised by very low electric power required, is the "Impulse System".	Apart from the anti-ice or de-ice actions illustrated in the figure beside, please consider the electric ice protection of hinges, compensation horn, small sensors, windshields and propellers.

SYS	FUNCTIONS PERFORMED	SYSTEMS CONFIGURATIONS	NOTES
FUEL SYSTEM	To perform pressure refuelling, allowing tanks venting. To store onboard all fuel necessary to engines and APU and to feed them when requested.	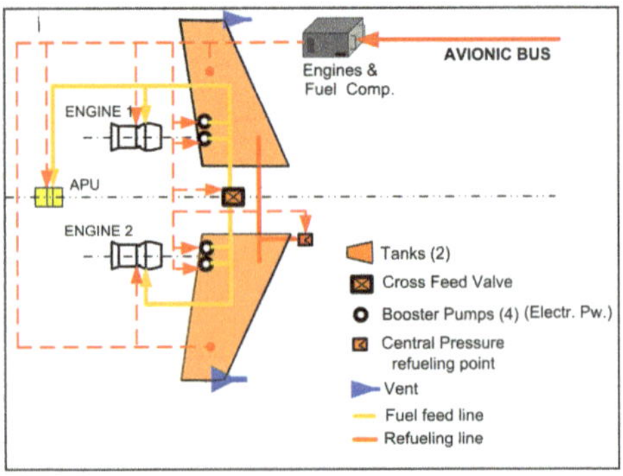	This system greatly affects aircraft configuration because of the extension and great volumes of its tanks. The booster pumps are usually electrically driven.

SYS	FUNCTIONS PERFORMED	SYSTEMS CONFIGURATIONS	NOTES
ELECTRIC SYSTEM	To generate electric power necessary onboard the aircraft. To transform part of it in different forms of electrical current as requested. To feed correctly the users.	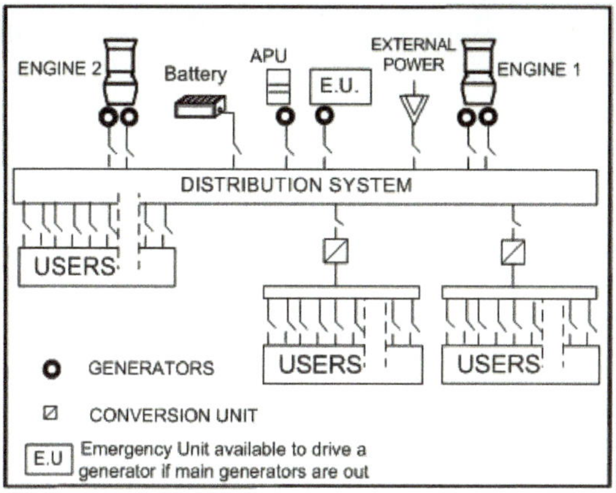	The amount of electric power generated onboard the aircraft is more and more increasing. This is particularly true if the electrical system will substitute the hydraulic and the pneumatic system. New forms of electric power (and generators) are now considered. Due to the reversibility characteristic of electric machines, engine starting is also considered.
SYS	FUNCTIONS PERFORMED	SYSTEMS CONFIGURATIONS	NOTES
PNEUMATIC SYSTEM	To generate pneumatic power necessary onboard the aircraft. To feed correctly the users.	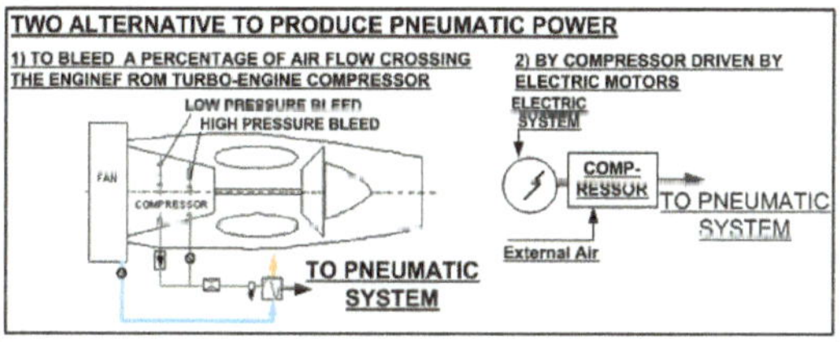	The bleed air from engines and APU is the state of the art of pneumatic power and it is particularly useful, if the air has to be introduced in pressurized cabins. To avoid engine's penalties, electric driven compressors can also be adopted.
SYS	FUNCTIONS PERFORMED	SYSTEMS CONFIGURATIONS	NOTES
HYDRAULIC SYSTEM	To generate hydraulic power necessary onboard the aircraft. To feed correctly the users (actuators).	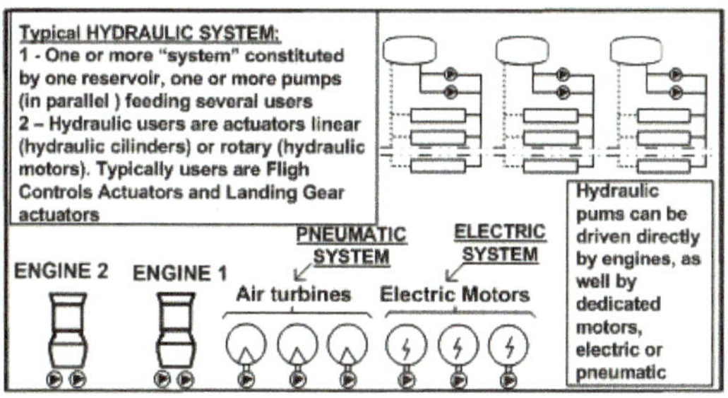	Hydraulic power is the state of the art form of power used to feed actuators. Electric actuators as well as hydraulic system supplied by electric motor driven pumps can be considered a valuable alternative to the conventional hydraulic system with engine driven pumps.

Table 1. Functions, types and features of the main onboard systems

2. Airplane system design

As already said, an airplane is a complex system, consisting of many different elements all harmonically integrated to form a unique entity, designed to perform a well defined mission.

Let us now examine the complex process, which, starting from the customer needs and moving on to the definition of requirements, proceeds with the development and then the manufacturing of the new airplane. Figure 5 schematically illustrates this complex process. Considering a reference frame with time on the x-axis and the level of details on the y-axis, it can be noted that, starting from the customer needs, the new product is first defined at system level, then at subsystem level and eventually, getting more into the details of the design flow, at equipment level. Every successive step, which corresponds to a new design phase, is an iterative process (see Figure 5) and the results of each phase are seriously affected by those of the previous phase. If we look at Figure 5, we can therefore understand that, starting from the customer needs and then the requirements definition, the process gets through all design phases (from the conceptual to the preliminary and eventually to the detailed design) following a typical top-down approach with an increased level of details from the system to the equipments. Then, once equipments have been defined and thus bought and/or manufactured, they are tested and integrated to form first the subsystems and eventually the whole system through the final assembly, according to a typical bottom-up approach. Once the final assembly has been completed, new activities at system level can be performed. After successfully accomplishing these activities, i.e. the system functional testing, the operative life of the new product can begin.

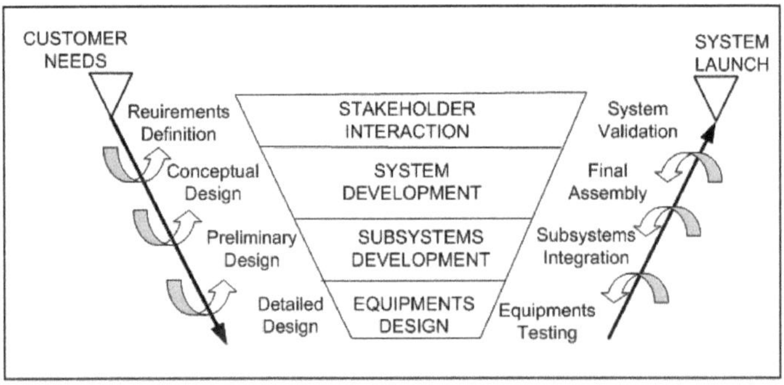

Fig. 5. The system design process

2.1 Airplane conceptual design

Taking into account the whole design process presented in Figure 5, it is quite clear that the main criticality of the conceptual design phase lies in the capability of generating (Antona et al., 2009) a first idea of the new product. A first idea of the future product implies:

a. architectural choices, i.e. definition of the global product's architecture in terms of shape and type of main elements and mutual location of the elements themselves. It is worth noting that the various alternatives can generate quite a few combinations, which are all potentially feasible and which shall then be traded to pick up the best ones. For sake of clarity, let us consider a modern medium passenger transport airplane. The possible alternatives for its architectural layout may be expressed in terms of:

- engines type: for instance state-of-the-art turbo-fan with high by-pass ratio and innovative turbo-fan with very high by-pass ratio;
- engines number: for instance two engines with high thrust or four engines with lower thrust;
- engines position: for instance located in nacelles directly attached to the underside of the wing or aft mounted;
- definition of all envisaged systems, without getting into the details of any of them.

b. quantitative choices, i.e. preliminary (please note that in aerospace field approximation even at this stage shall not exceed 10%-15% of the final value) definition of the most relevant characteristics of the future product, like, for instance, size, weight and performances. At this level of the design process the future product is thus addressed as a unique system. As far as its subsystems are concerned, they are just envisaged but not yet sized at this stage, even though their weight may be already estimated as percentage of the system empty weight, according to a typical top-down approach.

Once the concept of the future product has been generated, the level of details is still so poor that the manufacturing process could never begin. In order to enter production, the design of the future product has to proceed from the conceptual to the preliminary and eventually to the detailed design phase but this evolution requires a great deal of resources in terms of time, people and obviously money and cannot be pursued, unless the first idea of the future product has been declared feasible and competitive at the end of the conceptual design phase. It is worth remembering here that the conceptual design phase may sometimes also be called "feasibility study" or "feasibility phase".

At the end of the conceptual design phase we thus have a first idea of the future product that cannot yet be manufactured but can without any doubts be evaluated and compared with other similar potentially competing products, which may already exist or be still under development.

The conceptual design phase is therefore extremely relevant because:

- on the basis of the results of the conceptual design it is possible to decide whether or not to start the following expensive design activities;
- the choices that have the greatest impact upon the future product (i.e. architecture and main system characteristics, like size, weight, performance, cost, etc.) are taken during the conceptual design phase (see Figure 6).

At the same time the conceptual design phase is extremely critical because:

- it is the most fundamental phase of the whole design process;
- it is a particularly difficult and complex phase of the design process as decisions, that are crucial for the future product, have to be taken in a context which is generally poorly defined. Criticalities lie for instance in the capability of developing reliable mathematical models able to predict the behaviour of the future product, when the future product itself is still largely unknown.

Taking all these considerations into account, it seems absolutely valuable and interesting, both for pure methodological and more applicative aspects, to improve the conceptual design activities, specifically the aerospace systems conceptual design activities, in terms of accuracy and thoroughness of the results achieved.

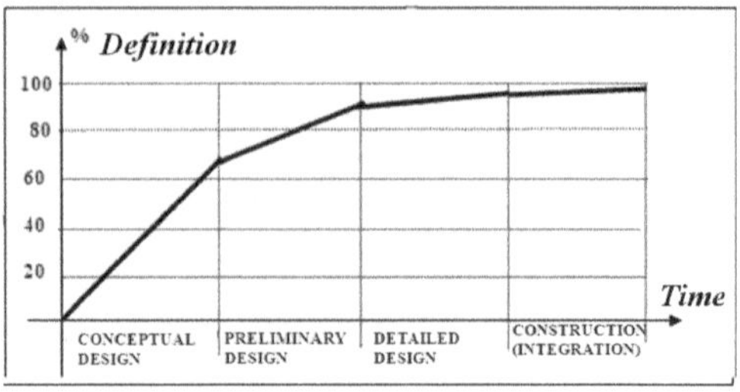

Fig. 6. Conceptual design relevance

2.2 Airplane systems conceptual design

Unlike the past, when the system view of the airplane was not at all evident and the results of the conceptual design were almost exclusively turned to the preliminary definition of the airplane main characteristics, today the systems engineering approach is widely accepted and appreciated and the results of the conceptual design include also initial basic choices for onboard systems. Please note that these initial choices for the onboard systems lay the groundwork for the next activities of systems development during the successive design phases, as shown in Figure 5. It is quite obvious that the capability of preliminary defining onboard systems already during the conceptual design phase implies more accurate and detailed results of the conceptual design itself. The initial definition of the onboard systems allows in fact achieving a more precise and reliable estimation of the whole system characteristics (like, for instance, the system empty weight, given by the sum of the onboard systems weights) and make the start of the successive preliminary design activities easier. However it is clear that more accurate and detailed results require a more complex conceptual design phase, which can be successfully faced today thanks to computer programs automation and to new powerful software tools.

Figure 7 schematically illustrates the main steps of conceptual design according to the traditional approach (airplane conceptual design) and to the proposed innovative approach (airplane & systems conceptual design).

As Figure 7 shows, the traditional approach to conceptual design envisages both architectural and quantitative choices, mutually interrelated, to generate the first idea of the future product (see also sub-section 2.1). According to this approach in conceptual design there are just the individuation of the onboard systems of the future product and the estimation of their weights. Unlike the traditional approach, the innovative approach, besides the architectural and quantitative choices, envisages also the preliminary definition of onboard systems, once the systems themselves have been individuated. For every onboard system, the preliminary definition implies:

- choice of systems architecture through block diagrams at main equipments level;
- initial sizing of such blocks, in terms of weight, volume and power required, on the basis of their performance requirements, in order to be able to start selecting them;
- preliminary studies of equipments and systems installation onboard the airplane, on the basis of main equipments weight and volume considerations. These preliminary studies on systems installation allow making more accurate estimation on the airplane mass properties;

- evaluation of mass and power budgets on the basis of weight and power required of each system equipment.

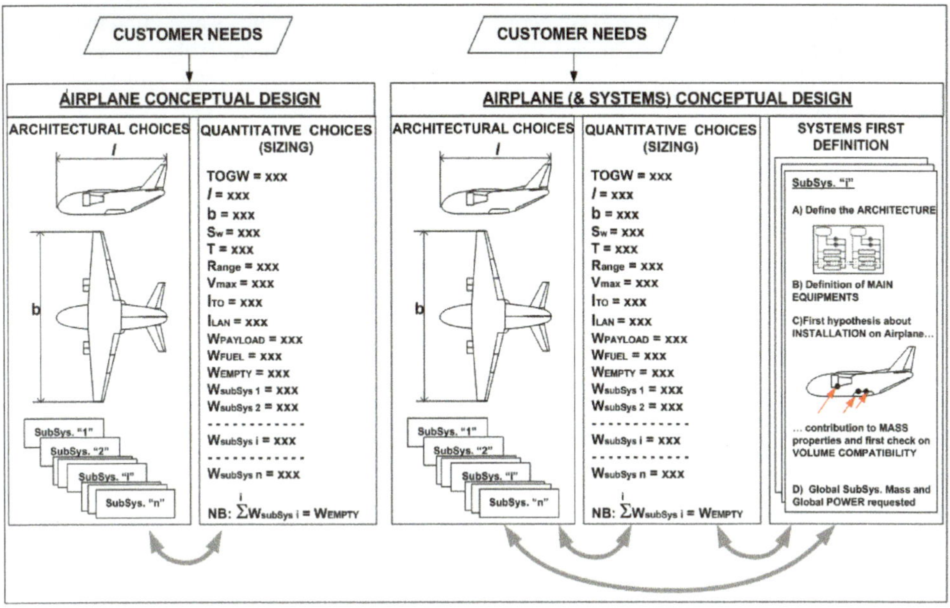

Fig. 7. Main steps of conceptual design according to the traditional and the innovative approach

Drawing some conclusions, we can say that, as the proposed new approach guarantees more accurate and detailed results and as the problem has not been extensively addressed so far (unlike what has happened in other disciplines, like, for instance, structures, aerodynamics and flight mechanics, whose mathematical algorithms have been integrated in a Multi Disciplinary Optimization, MDO, context in conceptual design), the development of a conceptual design methodology that pursues the new approach (see right hand side of Figure 7) appears extremely useful and valuable. ASTRID (Aircraft on board Systems Sizing And TRade-Off Analysis in Initial Design phase) is the acronyms of the innovative conceptual design methodology, proposed by the Authors. ASTRID is based on a dedicated software tool to easily perform iterations and successive refinements and to make the evaluation and comparison of various potentially feasible alternatives possible. ASTRID will be the main topic of the next section.

3. Airplane system innovative conceptual design methodology

Main features of the proposed new conceptual design methodology for the airplane system are the early investigation of avionics and onboard general systems and their integration with the traditional activities of conceptual design, i.e. the definition of system architecture and the accomplishment of system sizing, in terms of weight, volume, performances and system cost estimation. However, unlike the traditional approach to preliminary system sizing, avionics and onboard general systems, cannot be easily assessed through few and simple relationships. It is worth remembering here that, according to the traditional approach, the study of avionics and onboard general systems starts only after at least a preliminary concept of aircraft has been defined. The conventional sequence of design activities, characterized by aircraft conceptual design and then avionics and onboard general systems preliminary assessment, is still the current state-of-the-art, like a considerable number of valuable references, such as Daniel P. Raymer (Raymer, 1992) and Jan Roskam (Roskam, 1990), testifies. The same approach is pursued by two important software tools of aircraft design, RDS – "Integrated aircraft design and analysis" (by Conceptual Research Corporation, a company founded and lead by Daniel Raymer) and AAA – "Advanced Aircraft Analysis" (by DAR Corporation, founded by Jan Roskam), which have been developed on the basis of the works of, respectively, Daniel P. Raymer and Jan Roskam and have recently become widespread also at industrial level. The relevance of avionics and onboard general systems in aircraft conceptual design is witnessed by John Fielding from Cranfield College of Aeronautics (Fielding, 1999), who dedicates a great effort to the description of avionics and onboard general systems, but, as his work provides the reader with just an introduction to aircraft design issues, no specific methodology is reported in the text. On the basis of this preliminary assessment, the development of ASTRID seems to be highly desirable, in order to support the design process of new aircraft.

3.1 General context, goals and overview of ASTRID methodology

Before proceeding any further, let us briefly review the most common and widely used methodologies of aircraft conceptual design, focusing in particular on the way in which avionics and onboard general systems are taken into account. There are two main types of approaches:

- methodologies in which the aircraft Maximum Take-off Gross Weight (MTGW) is defined in such a way to match requirements (generally expressed in terms of performances) and it is broken down into pay-load, fuel and empty weight, being the empty weight often defined as a percentage of MTGW itself;
- methodologies in which the aircraft MTGW is estimated on the basis of requirements (for example the fuel weight depends on the range requirement) and the components of the empty weight are estimated on the basis of the Weight Estimation Relationships (WERs).

It can be noticed that in the first case every considerations about avionics and onboard general systems is postponed to a later stage of the design process, where, apart from all other requirements, avionics and onboard general systems shall be compliant with the previously defined constraint of global weight. Unlike the first case, in the second type of methodologies avionics and onboard general systems are taken into account since the very beginning of the design process at least from the point of view of weight, as their weight is established as part of the empty weight, either as percentage (in simplified methodologies) or as a result of WERs for the various systems (Staton, 1972) (Chiesa et al., 2000). It is interesting to observe that, on the basis of WERs for a single system, the same number of CERs (Cost Estimation Relationships) have been derived by several authors (Beltramo et al., 1979). Only in the second type of methodologies of aircraft conceptual design, some influences of avionics and onboard general systems on the overall aircraft design can therefore be expected since the very beginning of the design process, as the WERs of the various systems allow defining some crucial parameters of the systems themselves, like the number of fuel tanks of the fuel system, the number of actuators of the flight control system, the electric power that has to be supplied by the energy sources, etc.. Nevertheless other considerations on the following issues are still missing:

- performances of the systems and their capability of satisfying the requirements for the definition of the new aircraft;
- volume required by the systems and their installation onboard the aircraft, with the consequent influence on all other mass properties other than weight.

After reviewing the various existing methodologies of aircraft conceptual design, the main characteristics of the new methodology can be brought to evidence. Referring in particular to the influence of avionics and onboard general systems on the conceptual design of the whole aircraft, the new methodology envisages taking into account the design of avionics and onboard general systems since the very beginning of the design process through various successive refinements and iterations that affect also the main aircraft characteristics. The new tool shall not therefore be structured at level of the single systems (for example as ATA subdivision) but, for each system, at level of its main equipments (i.e., for instance, in the avionic system: weather radar, AHRS, ADC, VOR, radio UHF, etc.; in the electrical system: generators, TRUs, inverters, batteries, etc.). Thanks to this approach four main advantages that may lead to a better definition of the whole aircraft very early during the design process can be envisaged:

1. possibility of defining the various systems architectures, even if simplified, very early during the design process;
2. possibility of achieving a reasonable confidence of the capability of the systems to perform their assigned functions;
3. capability of carrying out installation study very early during the design process, thus being able to estimate the influences on the centre of gravity position and moments of inertia;
4. capability of preliminarily estimating safety and reliability and performing an initial assessment of maintainability/accessibility with optimization of the turn-around operations (Chiesa, 2007).

Focusing the attention on main equipments, by estimating their weights and costs, might lead to neglect the contribution to the overall weight and cost of the remaining parts of the systems, such as small components, like lines, pipes, wires, installation devices, etc. However the problem can be solved by means of a further estimate of these small components and/or by matching weight and cost estimations at main equipment/components level with results obtained by WERs and CERs at system level.

Before getting into the details of the logical steps that have to be taken to apply ASTRID methodology, a synthetic overview of the complete methodology is reported hereafter.

After preliminary estimating the aircraft global parameters, the main equipments of each system can be identified through, for example, the functional analysis, keeping in mind the various possible alternatives of architectures and taking into account the new emerging technologies. After the identification of the main equipments of each system and their interfaces, the inputs and outputs of each building block (i.e. main equipment) can be individuated. Specifically per each building block the number of inputs/outputs as well as which parameters have to be considered as inputs/outputs have to be established. It is quite obvious that the aircraft data, the design constraints (including the constraints of regulations) and the performance requirements, that characterize the equipments, are among the considered parameters, which on a statistical basis allow then to estimate the weight, volume, cost and any other possible feature of the equipment itself. Starting from the inputs/outputs of every main equipment, the relationships that allow calculating the value of the outputs on the basis of the inputs can then be defined through statistical relationships.

Notwithstanding the integration between the various systems, each system has to be considered at least initially separately for the identification of its main equipment. It appears therefore obvious that, in this phase, a logical sequence with which the tool addresses the various systems has to be established. In order to avoid or minimize iterations, for instance, the systems requiring power supply have to be considered first and later on those generating power.

Once the complete set of relationships between inputs and outputs of each main equipment and their sequence, which constitute a mathematical model, has been established, the design process proceeds with the application of the iterative loops for the refinement of the aircraft sizing and performance estimation.

The output of the convergence of this iterative loop is an optimized aircraft with optimized avionics and on-board general systems architecture.

3.2 ASTRID methodology

Purpose of the section is to describe in an easy and straightforward way the various steps that have to be taken to apply ASTRID methodology and the logical path that has to be followed to move from one step to the next one.

Figure 8 shows the flowchart of the complete methodology.

Main goal of the methodology is to identify the best global configuration, in terms of architecture and system sizing, of avionics and onboard general systems for a defined airplane concept, which may be either already frozen or still under development. It is worth noting that the former case implies more constraints with respect to the latter case for the avionics and onboard systems design. Moreover in the latter case the global aircraft design can still benefit from the data coming from the avionics and onboard systems design, in order to achieve a more accurate global conceptual design.

ASTRID is therefore a separate module that can however be integrated with the global aircraft concept definition thanks to specific building blocks dedicated to data exchange.

The methodology is characterized by the possibility of carrying out more designs of avionics and onboard general systems for the same aircraft concept, in order to trade then off the various designs and pick up the best ones. The methodology also allows addressing only some systems, in case others have still been designed.

Main expected result of every system module is the definition of the system architecture and the accomplishment of the system sizing at equipments level, with obvious advantages in terms of estimation of aircraft mass and power budgets. Per each system, it is also possible, if requested, to study the system installation onboard the aircraft at equipments level, with clear advantages in terms global aircraft mass properties and evaluation of the feasibility of the aircraft configuration itself.

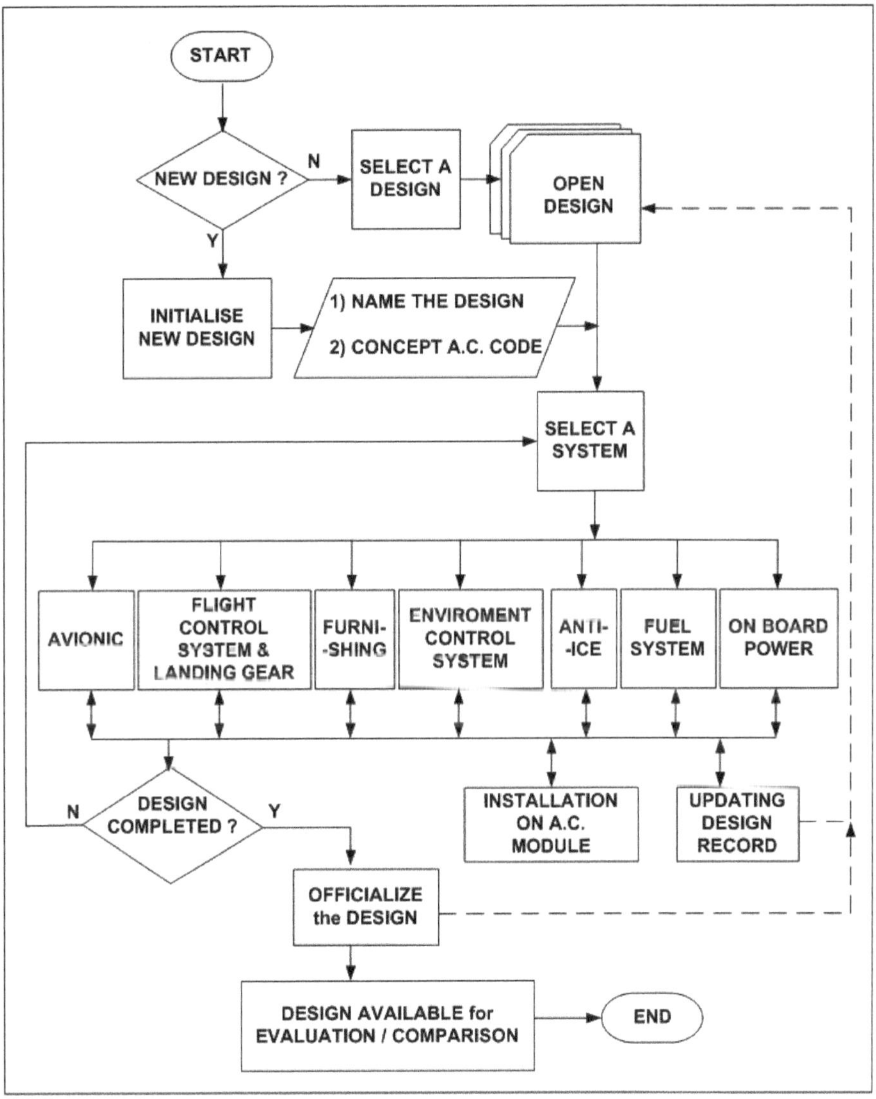

Fig. 8. ASTRID methodology flow-chart

Taking now into account avionics and onboard general systems, Table 2 sums up the activities to perform and the tools/algorithms to apply, in order to accomplish the design of each system.

(SUB) SYSTEM	ACTIVITIES	TOOLS
AVIONIC SYSTEM	a) definition of the Subsystems constituting the Avionic System b) for every Subsystem definition of list of Main Equipments c) Integration of the Equipments d) for every Main Equipment: -defintion of Driving Characteristic - Individuation of an existing "Off the Shelf" -If any, estimate Weight, Volumes, Power requested on statistical basis	Functional ANALYSIS (Fig. 9) Function Equipment Matrix (Fig. 10) Connection Matrix (Fig. 11) Avionic System scheme (Fig. 12) Statistical Data Base
FLIGHT CONTROLS & LANDING GEAR SYSTEM	a) Definition of the architecture of actuators of every Flight Control Surfaces and Landing Gear (and their related Systems: NLG steering and MLG wheels brakes) b) Establish requirements for the Actuators	Considering the linear hydraulic actuator, the graph that depicts force vs velocity allows us to search the optimal sizing of the linear actuator by defining "Stall Force" and "No Load Actuation Rate" and than sizing the actuator and Power requested. The same graph can be easily translated into torque vs angular speed, if the choice is a mechanical screw-jack driven by an Hydraulic Motor, and it is applicable also for **Electric Actuators** (Fig. 13)
FURNI- -SHING SYSTEM	Definition of kind, number and location of: a)chairs, b)Galleys, c)Toilets, d) **IFEC**	Estimate Power requested for: a) Lights, b) Audio Syst, c) **IFEC**, d) Food & Drinks heating, e) Food & Drinks cooling, f) Water waste
ENVIRO- -MENT CONTROL SYSTEM	Cabin Thermal Load "**q**" evaluated as function of Tcab in situation of max request of heating.... $q_A = q_A(T_{cab})$ ...and max request of cooling $q_P = q_P(T_{cab})$	Matching $\begin{cases} q_A = q_A(T_{cab}) \\ q = \dot{m}_{condA}\, C_p\, (T_{i\,HOT} - T_{cab}) \end{cases}$ $\begin{cases} q_P = q_P(T_{cab}) \\ q = \dot{m}_{condP}\, C_p\, (T_{i\,COLD} - T_{cab}) \end{cases}$ ...and define air mass flow requested (Fig. 14)
ANTI-ICE SYSTEM	1) Estimate the Surface to be protected **Sp** 2) Estimate POWER **PSZ** needed for "Small Zones" protection	(in case of "Hot Air Sys.") → $\dot{m}_{AI} = K_1\, S_p$ (in case of "GOODRITCH Sys.") → $\dot{m}_{AI}$ defined by the one of greater boot (in case of "Resistors Sys.")→ $P_{AI} = K_2\, S_p$ (in case of "Impulse Sys.")→ $P_{AI} = K_3\, S_p$ P_{sz} from estimate
FUEL SYSTEM	1) Pipes section definition for: a) Engine's and APU fuel feed b) Pressure refueling 2) Booster-pumps definition, choice of the types or estimation of weight and power required	- Continuity equation - Gravity and Pressure losses -Pump Characteristic curve **p=p(Q)** comparison with boundary coditions of running (Fig. 15)
ONBOARD POWER SYSTEM	1)MAIN ALTENATIVE — **BLEED** →	Verify compatibility of: 1) $\dot{m}_{condA} + \dot{m}_{AI}$ with $\dot{m}_{BLEED\,A}$ (n-1) 2) $\dot{m}_{condP}$ with $\dot{m}_{BLEED\,P}$ 2) $\dot{m}_{condP}$ with $\dot{m}_{BLEED\,APU}$
	1)MAIN ALTENATIVE — **NO BLEED** →	Define COMPRESSORS able to give the equivalent of BLEED, define the relevant electric motors and the requested Power P_A
	2) **HYDRAULIC SYSTEM** (if any hydraulic user are present) a) definition of Q nominal of Pumps →	**LOADS DIAGRAM TOOL:** (Fig. 16) Comparison between diagrams: $Q_{requested} = \left[\sum_i Q_i\right]_{t_f}^{t_0}$ & $Q_{available}$ = f (**tmiss**) → engine driven pump / Costant → electric motor driven pump
	b) Pipes section definition →	- Continuity equation - Gravity and Pressure losses
	c) Reservoir Capacity →	$C_{RESER} = K_{RESER}\, Q$
	3) **ELECTRIC SYSTEM** Choice of the kind of Electric Current to be generated and of the other (to be obtained from the one by Conversion) needed on board	**The choice of what and how much Power of every kind is needed is a choice at Users level** **28VDC** Generator/ Converter **115VAC - 400Hz** Generator/ Converter **115VAC - wide freq.** Generator **270VDC** Generator/ Converter **230VAC - wide freq.** Generator (Figg. 17, 18, 19)
	a) definition of Power "Tipe i" (Converter) b) definition of Power "Tipe j" (Converter) c) definition of Power "Tipe k" (Converter) d) definition of Power "Tipe l" (Converter) e) definition of Power "Tipe m"(Generators) → The GENERATOR capability to STARTING ENGINE will be verified	**LOADS DIAGRAM TOOL:** similar to the one seen for Hydraulic System PLEASE NOTE: the Power elaborated by CONVERTERS will be considered before as it will be a "USER" for GENERATORS (Figg. 17, 18, 19) BATTERY definition (Fig. 20)

Table 2. ASTRID methodology: systems design

Considering each system separately, the following considerations need to be highlighted:

- avionic system. Main activities of the conceptual design of the avionic system consist in identifying which and how many types of equipments will form the whole system. The design purses the typical functional approach. The functional tree, which is one of the main tasks of the functional analysis, allows defining the basic functions that the avionics system shall be able to perform. Figure 9 illustrates an example of functional tree (for sake of simplicity this example refers to the block diagram of the avionic system shown in Table 1), where the top level function "avionic system" has been decomposed into first level functions, which identify the various subsystems of the avionic system. For sake of simplicity only one of the first level functions, "to control flight behaviours", has been further subdivided into lower level functions. The so called basic functions, i.e. those functions that cannot be split any further, are in this case functions that can be performed by equipments. Through the functions/equipments matrix (see example in Figure 10) the basic functions are associated to equipments. Once the functions/equipments matrix is completed, all equipments of the avionic system are known. Figure 10 illustrates the functions/equipments matrix related to first level function "to control flight behaviours" of Figure 9. On the basis of performance requirements, either already available equipments can be individuated or new (not yet existing) equipments can be preliminary sized by statistically estimating their characteristics, like weight, volume, requested power per each flight phase. Once the basic equipments are identified, the links between each equipment can be established through the connection matrix (see example in Figure 11). Eventually the avionic system architecture is presented in the functional/physical block diagram (see example in Figure 12).

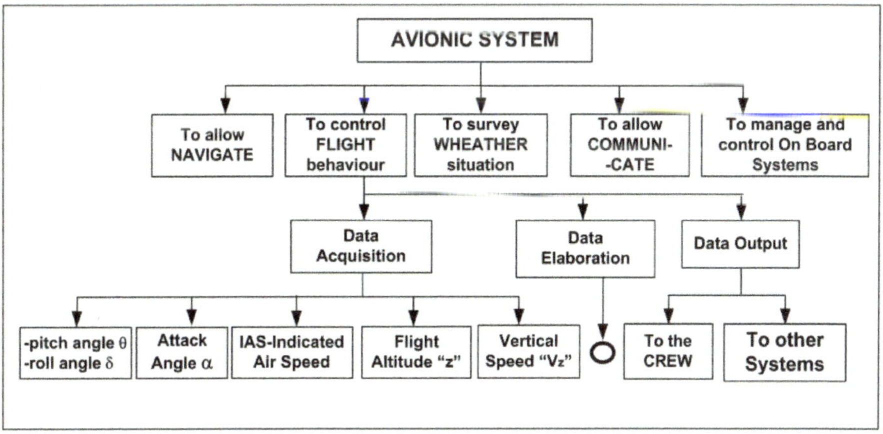

Fig. 9. Avionic system design: functional tree

Flight Controls & Landing Gear System. Even though flight controls and landing gear are separate systems, here we address them together as the main issue of their conceptual design is in both cases the definition and sizing of the various actuators (i.e. of their main (basic) equipment), thus leading to the system mass and power budgets. Main activities of the conceptual design of the flight controls & landing gear system consist in defining the architecture of flight control surfaces and landing gear actuators and then sizing the actuators themselves. Figure 13 illustrates schematically the applied algorithm for optimal actuator sizing. In particular Figure 13 focuses first on hydraulic cylinders (linear actuators) and hydraulic motors (rotary actuators). Considering the linear hydraulic actuator, the graph that depicts force vs. velocity allows us to achieve optimal sizing. The same graph can be easily translated into torque vs. angular speed, which is valid both for the hydraulic rotating actuator and for the electric rotary actuator (making the hypothesis of the presence of a current limiter), as Figure 13 shows. After completing the actuators sizing activity, it is fundamental to understand when the various actuators will work, i.e. during which flight phases (it is worth remembering, for instance, that generally primary flight controls actuators work continuously throughout all flight phases, while secondary flight controls and landing gear actuators work only during certain flight phases). Eventually, considering the power consumption of each actuator and the flight phases during which each actuator is supposed to work, the electric loads diagrams and thus the electric power budget can be generated

Basic Equipment	Basic Functions: Attitude Angle -pitch angle θ -roll angle δ	Angle of Attack α	IAS-Indicated Air Speed	Flight Altitude "z"	Vertical Speed "Vz"	Data Elabo-ration	Output DATA to the CREW	Output DATA to other Systems
AHRS (3 rate Gyro Unit)	✓							
α angle sensor		✓						
ADC-Air Data Computer			✓	✓	✓			
FCC-Flight Control Computer						✓	✓	✓
Flight Control Display							✓	
Interface Unit								✓

Fig. 10. Avionic system design: functions/equipment matrix

- Furnishing system. This system is made up of various equipments that may strongly affect the whole aircraft, in terms of mass and power required, especially in case a civil transport aircraft is considered. Main activities of the conceptual design of the furnishing system consist in identifying which and how many types of equipments will form the whole system, individuating their location and estimating their mass and power consumption. The estimates of mass and power consumption, based on the state-of-the-art technology available for the envisaged equipments, may have, respectively, a serious impact on the global aircraft concept and on the onboard power system sizing.
- Environment control system. Main activities of the conceptual design of the environment control system consist in preliminary estimating the thermal load, q,

between the cabin and the external environment, and then the required air mass flow, m_{cond}, to keep the temperature of the cabin within an acceptable range of values (usually between 18°C and 25°C). After estimating the thermal load, the required air mass flow can be computed, depending on the desired temperature inside the cabin, T_{CAB}, and on different operative scenarios, which range between the so-called cold case (case A in Table 2), when maximum heating is required, and the so-called hot case (case P in Table 2), when maximum cooling is required. The air mass flow, that can be provided at different temperatures (for instance, at high temperature, $T_{I\ HOT}$, in cold cases or at low temperature, $T_{I\ COLD}$, in hot cases), can in fact be computed by matching the two equations, which express the thermal load q_A or q_P in Table 2 and in Figure 14 and the heat load provided by the system (q in Figure 14) to maintain the desired temperature inside the cabin.

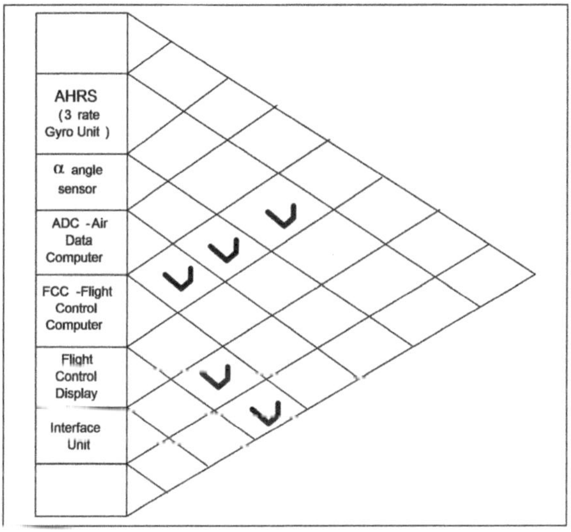

Fig. 11. Avionic system design: connection matrix

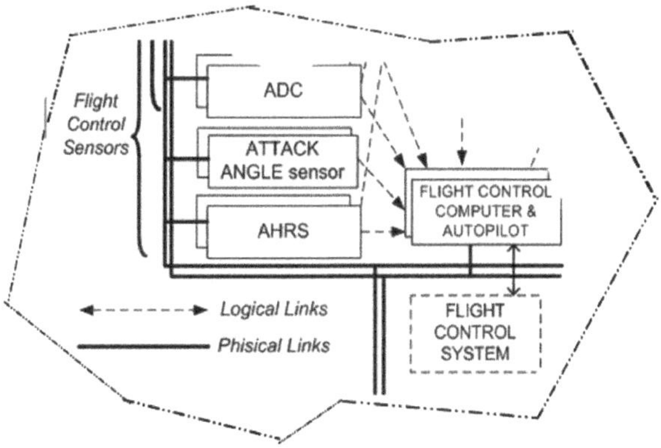

Fig. 12. Avionic system design: functional/physical block diagram

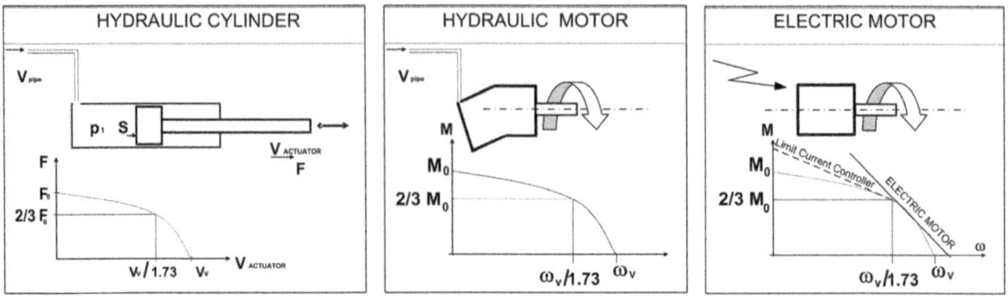

Fig. 13. Flight controls & Landing Gear system design: actuator sizing

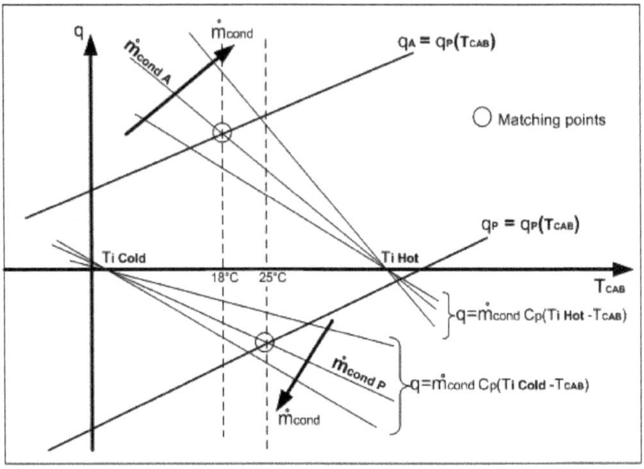

Fig. 14. Environmental Control System design: m_{cond} estimation

- Anti-ice system. Main results of the conceptual design of the environment control system consist in evaluating the surface of the aircraft that has to be protected to avoid ice formations (see SP, protected surface, in Table 2) and in estimating the power required to protect that surface. It is worth noting that the power required may be either pneumatic power (i.e. air mass flow, m_{AI} in Table 2) or electric power (P_{AI} in Table 2). Apart from wide aircraft surfaces, also small zones have to be taken into account in terms of electric power required for anti-icing (see P_{SZ}, power required for small zones, in Table 2).
- Fuel System. Once the aircraft architecture has been defined, the equipments of the fuel system that have the largest impact on the whole aircraft, i.e. the fuel tanks, have usually already been determined in terms of number, capacity and location onboard the aircraft. However fuel feed, pressure refuelling pipes and the fuel booster pumps, used to boost the fuel flow from the aircraft fuel system to the engine, have still to be identified as main results of the conceptual design of the fuel system (see Table 2). As fuel booster pumps are usually electrically driven, it is clear that their power consumption represents another important input to the whole aircraft power budget. The definition of the fuel booster pumps is based on their graphs (p=p(Q)) that depict pressure, p, as a function of fuel flow, Q (Figure 15), taking into account the requirements of maximum and minimum engine interface pressure (respectively pmax_engine and pmin_engine in Figure 15), as well as

the pressure losses along fuel pipes, and the requirements of maximum and minimum fuel flow (respectively Qmax and Qmin in Figure 15).

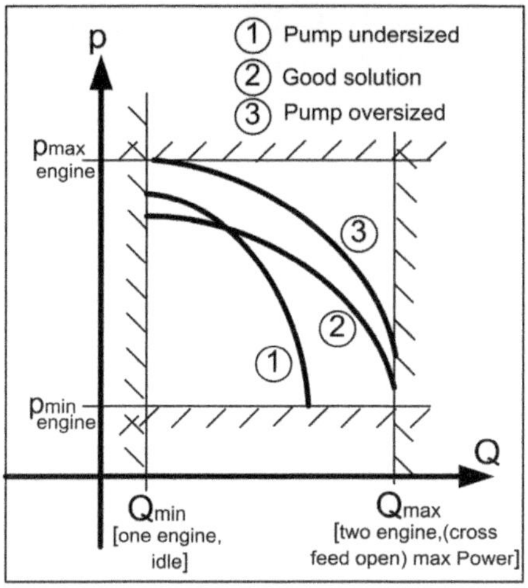

Fig. 15. Fuel system design: fuel booster pump definition

- Onboard Power System. The onboard power system may be consisting of the pneumatic (generally using bleed air from the engines compressors), the hydraulic and the electrical system, as in case of not brand new state-of-the-art aircraft. Conversely, as in case of new and future generation aircraft, the onboard power system may be either consisting of only the electrical system ("All Electric Airplane") or of mainly the electrical system ("More Electric Airplane"). Per each type of power (pneumatic, hydraulic and electric), different alternatives may be envisaged and defined as a result of the conceptual design of the onboard power system. The available solutions for the different types of power, reported in Table 2, are here listed:

a. pneumatic system. Pneumatic power may be supplied either as bleed air from the engines or as air coming from dedicated electrically driven compressors. In the latter case the disadvantages due to the increase of weight and power required, because of the installation of dedicated components, have to be traded with the advantages due to the use of the engines without any penalties of consumption related to the bleed air.

b. Hydraulic system (in case there are hydraulic users onboard the aircraft). The hydraulic system can be defined by estimating the reservoirs capacity and by sizing the pumps that can be either powered by the engines of by dedicated electric motors. Figure 16 shows an example of load diagram for hydraulic pumps: the hydraulic flow required, $Q_{required}$, throughout all flight phases is compared with the hydraulic flow available, $Q_{available}$, either in case of engine driven pump or in case of electric motor driven pump. As it can be noted both types of pumps may satisfy the hydraulic flow required. Apart from the pumps and the reservoirs, also pipes can be defined through the continuity equation, taking into account gravity and pressure losses.

c. Electrical system. The electrical system can be defined by sizing generators and electric power conversion units, which are in charge of converting part of the electric power from one form to another, according to the feeding requirements of various users. It is worth underling the importance of developing different solutions in terms of users and therefore generators, in order to be able to compare then these alternatives and pick up the best one. It has to be remembered that different forms of electric power imply different weight and cost. Among the various available solutions, the most common trend is to select the option that generates the type of power to feed the greatest number of users, in order to minimize the request of power conversion. As reported in Table 2, various solutions are available. The most common ones are listed hereafter:

i. 115 VAC 400 Hz generation and conversion of part of the electric power to 28 VDC;
ii. 115 VAC wide frequency generation and conversion of part of the electric power to 115 VAC 400 Hz, 270 VDC and 28 VDC;
iii. 270 VDC generation and conversion of part of the electric power to 115 VAC 400 Hz and 28 VDC;
iv. 230 VAC wide frequency generation and conversion of part of the electric power to 115 VAC 400 Hz, 270 VDC and 28 VDC.

While the first case is the today most common solution, the other three options represent the future trends, characterized by new forms of electric power to satisfy the ever increasing request of electric power onboard the aircraft. Figure 17, Figure 18 and Figure 19 show possible solutions of the electrical system architecture and the relative load diagrams of these new trends. It is worth noting that in the load diagrams the different forms of electric power, which have been converted from the main generation, are considered as users that contribute to the global electric power required. The generator is sized on the basis of the global electric power required during the various mission phases, while all power conversion units are sized on the basis of the amount of electric power they are requested to supply. As reported in Table 2, especially in case of "All Electric" philosophy, the capability of every engine driven generator of performing engine starting (thanks to the electric machines reversibility) shall be verified. When accomplishing the task of engine starting, the generator is powered by another generator driven by the APU, which, being in its turn a gas turbine engine of relative small dimensions, can be easily set working by a traditional 28 VDC electric starter, fed by the battery. Eventually the most appropriate battery has to be selected (see Figure 20), in order to be able to perform APU starting and to be able to feed vital users (according to regulations) even though for a limited time.

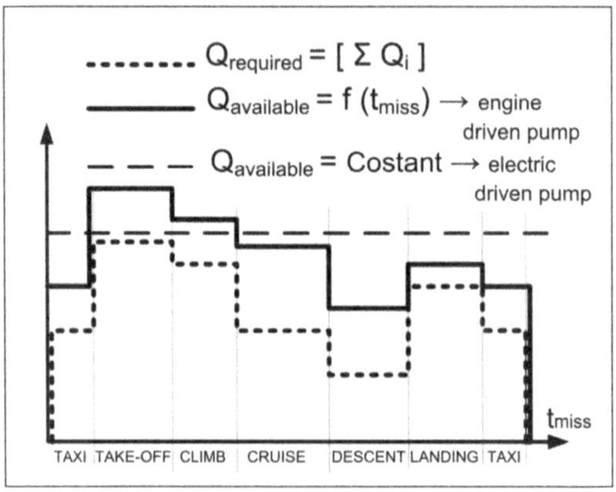

Fig. 16. Hydraulic system design: hydraulic pumps load diagram

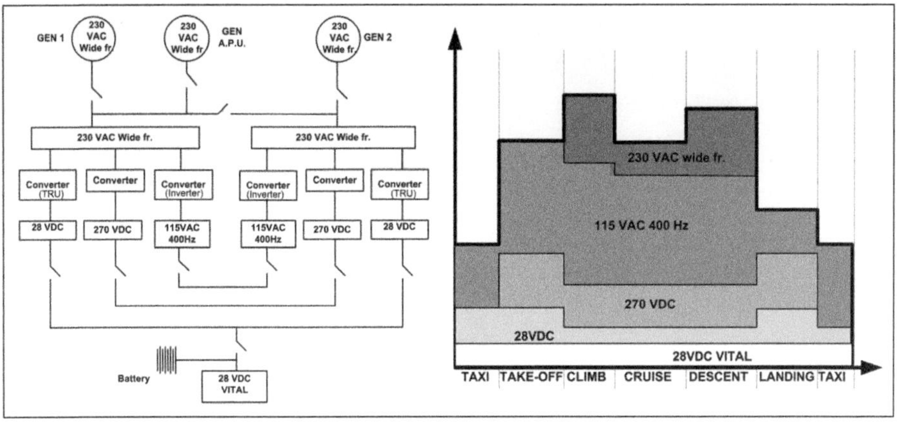

Fig. 17. Electrical System: 230 VAC wide frequency generation and example of load diagram

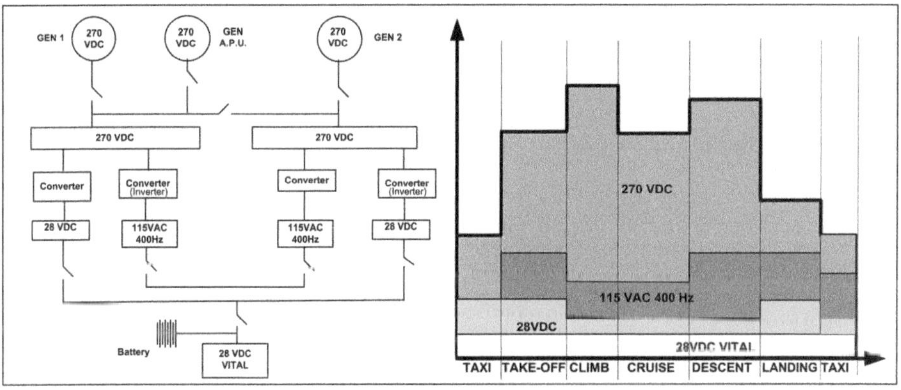

Fig. 18. Electric System: 270 VDC generation and example of load diagram

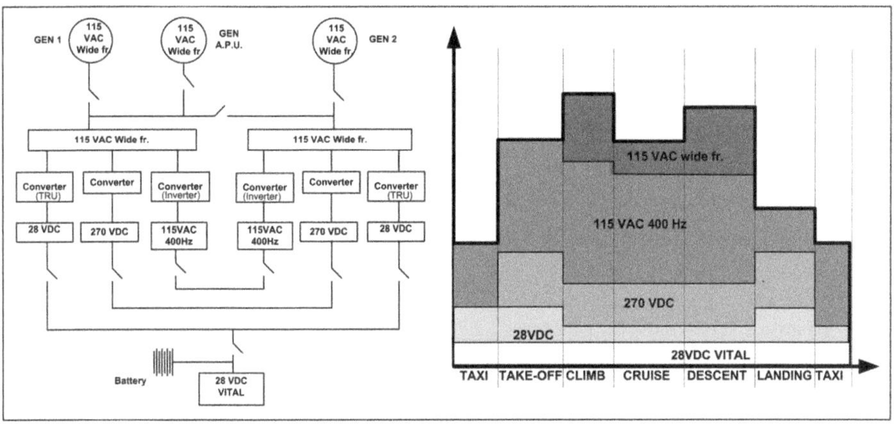

Fig. 19. Electric System: 115 VAC wide frequency generation and example of load diagram

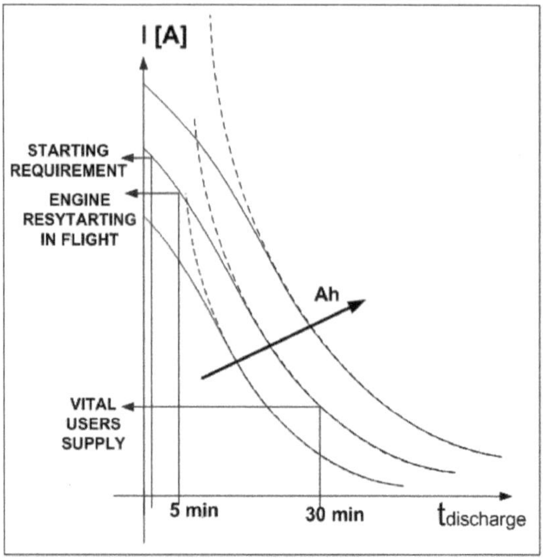

Fig. 20. Battery selection

4. Conclusions

After an overview of the airplane system conceptual design, the chapter focuses on an innovative conceptual design methodology, ASTRID, which allows assessing and preliminary sizing avionics and onboard general systems very early during the design process. The advantage is a better definition of the whole aircraft, in terms of weight, mass properties, power budget and consequently cost already during the conceptual design phase. A better quality of the design of new aircraft is likely to widely improve the development of future aircraft. The proposed innovative methodology can contribute to the achievement of this goal with limited cost.

5. Acronyms

AC = Aircraft
ADC = Air Data Computer
ADF = Automatic Direction Finder
ADI = Attitude Director Indicator
AHRS = Attitude Heading Reference System
APU = Auxiliary Power Unit
ASTRID = Aircraft on board Systems Sizing And Trade-Off Analysis in Initial Design phase
CAU = Cold Air Unit
CERS = Cost Estimation Relationships
DME = Distance Measuring Equipment
ECS = Environment Control System
GPS = Global Position System
HSI = Horizontal Situation Indicator
IFEC = In-Flight Entertainment and Connectivity

ILS = Instrumented Landing System
LG = Landing Gear
LRU = Line Replaceable Unit
MDO = Multi Disciplinary Optimization
MLG = Main Landing Gear
MTGW = Maximum Takeoff Gross Weight
NLG = Nose Landing Gear
SYS =System
TOGW = TakeOff Gross Weight
UHF = Ultra High Frequency
VAC = Voltage Alternate Current
VDC = Voltage Direct Current
VHF = Very High Frequency
VOR = VHF Omnidirectional Range
WERs = Weight Estimation Relationships

6. Nomenclature

b = wingspan
C_{D0} = parasite drag coefficient
C_{LMAX} = maximum lift coefficient
C_p = specific heat (costant pressure)
C_{RESER} = reservoir capacity
F = force
F_0 = maximum static force (stall force)
K = constant
l = fuselage length
l_{LAN} = landing distance
l_{TO} = takeoff distance
M = momentum
M_0 = maximum static momentum (stall momentum)
m_{AI} = anti-ice system air mass flow rate
$m_{BLEED\ A}$ = air mass flow rate bleed from engine or APU compressor or from dedicated compressor in case of max request of heating
$m_{BLEED\ P}$ = air mass flow rate bleed from engine or APU or dedicated compressor in case of max request of cooling
m_{condA} = air mass flow rate supplied by ECS in case of max request of heating
m_{condP} = air mass flow rate supplied by ECS in case of max request of cooling
P_{AI} = electrical power required by anti-ice system
p_{cab} = cabin air pressure
p_{ext} = external air pressure
p_h = pressure of pneumatic power generation output air
p_i = pressure of CAU output air
p_{max_engine} = maximum engine interface pressure
p_{min_engine} = minimum engine interface pressure
P_{SZ} = electric power required for small zones to avoid ice formation
q = cabin thermal load

Q = fluid volumetric flow rate
q_A = q in case of maximum request of heating
$Q_{AVAILABLE}$ = available hydraulic mass flow rate
q_P = q in case of maximum request of cooling
$Q_{REQUESTED}$ = required hydraulic mass flow
Sp = protected surface (by ice formation)
Sw = wing surface
T = thrust
T_{cab} = cabin air temperature
T_{ext} = external air temperature
T_h = air temperature of pneumatic power generation
T_i = output air temperature of CAU
$T_{I\,COLD}$= temperature of air supplied by ECS in case of max request of cooling
$T_{I\,HOT}$ = temperature of air supplied by ECS in case of max request of heating
t_{miss} = mission time
V_∞ = unperturbed air velocity
V_{max} = airplane maximum speed
V_V = no load rate

7. References

Antona, E., Chiesa, S., Corpino, S., Viola, N., (2009). L'avamprogetto dei velivoli. Memorie della Accademia delle Scienze di Torino. Classe di Scienze Fisiche Matematiche e Naturali. 65- 115. 32

Beltramo, M.N., Morris, M.A. & Anderson, J.L. (1979) Application of parametric weight and cost estimating relationship to future transport aircraft, Proceeding of 38th Annual Conference of the SAWE, New York, U.S.A., May 7 10, 1979

Chiesa, S. (2007) Affidabilità, sicurezza e manutenzione nel progetto dei sistemi, CLUT, ISBN 9788879922647, Turin, Italy

Chiesa, S., Maggiore, P., Corpino, S. & Pasquino, M. (2000) The Weight Estimation in Aerospace Engineering at the Polythecnic of Turin, Proceedings of SAWE Europe Conference, Bremen, Germany, November 8 9, 2000

Raymer, D. P. (1999) Aircraft Design: a conceptual approach (third edition), AIAA (American Institute of Aeronautics and Astronautics) Education Series, ISBN 1-56347-281-0, Reston, Virginia, USA

Roskam, J. (1990) Airplane Design, Vol.1 to 7, DARcorporation, ISBN 1-884885-24-1, Lawrence, Kansas, USA

Staton, R.N. (1972) Weight estimation methods, SAWE JOURNAL, Vol. 31, (April-May 1972)

2

Functional Analysis in Systems Engineering: Methodology and Applications

Nicole Viola, Sabrina Corpino, Marco Fioriti and Fabrizio Stesina
Politecnico di Torino
Italy

1. Introduction

Functional Analysis is a fundamental tool of the design process to explore new concepts and define their architectures. When systems engineers design new products, they perform Functional Analysis to refine the new product's functional requirements, to map its functions to physical components, to guarantee that all necessary components are listed and that no unnecessary components are requested and to understand the relationships between the new product's components. The chapter begins with the definition of the role of Functional Analysis in conceptual design (section 2) and then proceeds with the description of a proposed methodology (section 3 and sub-sections 3.1, 3.2 and 3.3) and with the presentation of its applications (section 4 and sub-sections 4.1, 4.2 and 4.3) at subsystem, system and system of systems levels. Eventually some conclusions are drawn.

The design process, in particular the design process of complex systems, can be split into three phases (Raymer, 1999):

1. Conceptual design.
2. Preliminary design.
3. Detail design.

Even though Functional Analysis applies to every phase of the design process, it turns out to be particularly useful during conceptual design, when there is still a wide range of potentially feasible solutions for the future product. The precious role of Functional Analysis consists in individuating as many available options as possible, without forgetting any ideas that may offer significant advantages. In the remainder of the chapter we refer specifically to the application of Functional Analysis during the conceptual design phase to explore complex systems.

2. Functional analysis in conceptual design

The conceptual design process is schematically illustrated in Figure 1, where the role of Functional Analysis is highlighted, as well as its interactions with all other building blocks of the conceptual design methodology. Starting from the mission statement, the mission objectives can be derived. Once the broad goals of the system, represented by the mission objectives, have been established, the system requirements can be defined. On the basis of the system requirements, the conceptual design process evolves through the system architecture and the mission definition. The system architecture definition consists of the two main tasks: Functional Analysis and System Sizing.

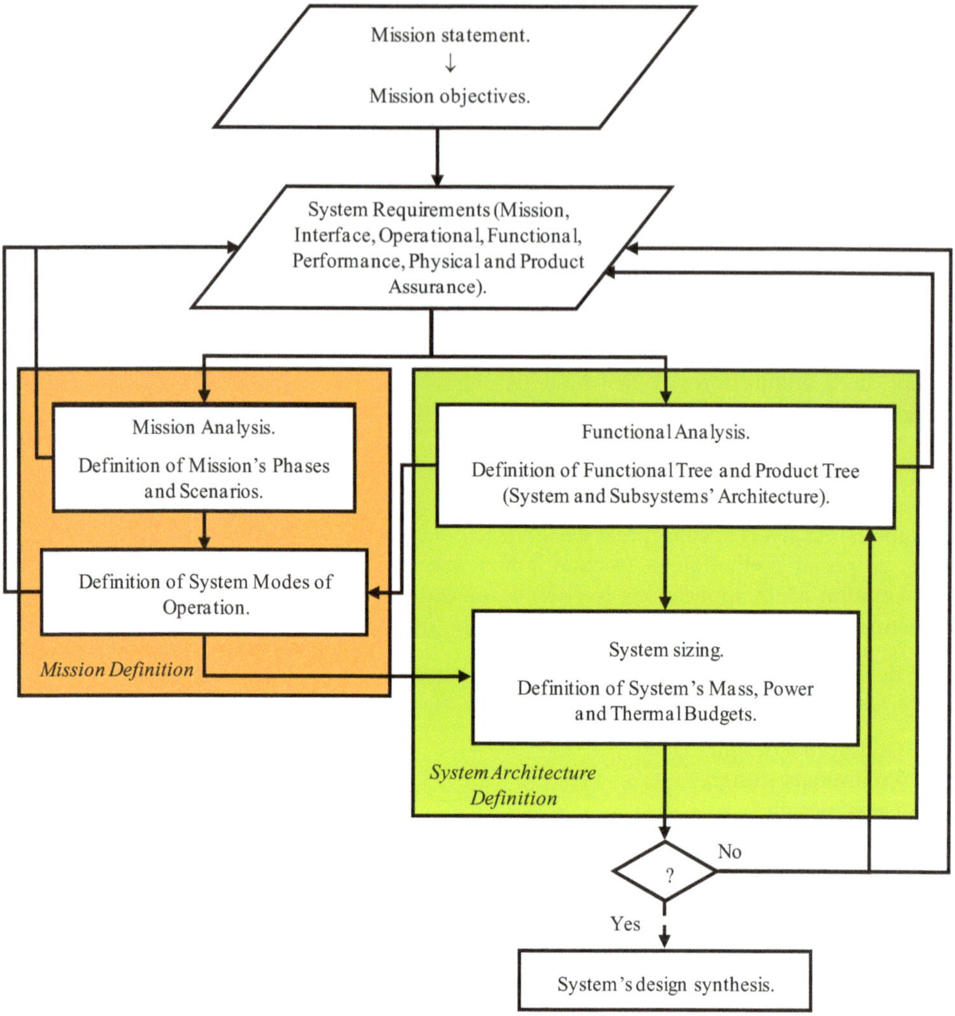

Fig. 1. Conceptual design process flow-chart

Primary results of Functional Analysis are the functional tree and the product tree: the former identifies the basic functions, which the system has to be able to perform, while the latter individuates all system physical components, which are able to carry out the basic functions. In other words, these components may be the equipments or the subsystems, which constitute the whole system. Once the components of the product tree have been identified, it is possible to investigate how they are connected to form the system. It is thus possible to develop both the functional block diagram (secondary or additional result of

Functional Analysis) and the physical block diagram of each subsystem and of the whole system. In order to complete the system architecture, the definition of the system budgets (mass, electric power, thermal power budgets, etc.) has to be carried out. However this task can be fulfilled only after the system modes of operation have been established. The modes of operation are part of the mission definition and can in their turn been set up only after the subsystems and their equipments have been identified. Once both the mission and the system architecture have been preliminary defined, before proceeding any further with the system design synthesis, it is important to verify whether or not all system requirements have been satisfied. Being the design activity typically a process of successive refinements, several iterations may be necessary before achieving the system design synthesis, thus freezing the system design.

Iterations may occur at every stage of the conceptual design process, thus resulting in a continuous trade or refinement of system requirements. In particular, as far as functional requirements (which are part of system requirements) are concerned, their refinement is mainly due to the feedback of Functional Analysis outputs and specifically of functional tree outputs. The basic functions, i.e. the bottom level functions of the tree, are in fact used to completely define or just refine the functional requirements. Unlike system requirements, which are detailed descriptions or quantitative expressions of the system itself, taking into account what we would like to achieve and what the budget allows us to achieve, mission objectives are the broad goals that the system shall achieve to be productive. Thus, whereas system requirements are traded throughout the design process, mission objectives may be slightly or not at all modified during conceptual design. For these reasons the top level function of the functional tree, i.e. the very first step of the Functional Analysis, can either be one mission objective or one top level functional requirement.

Functional Analysis as a fundamental tool of the design process is discussed by a number of references. Wertz and Larson (Wertz & Larson, 2005) present the Functional Analysis to decompose the functional requirements and focus only on one single task of Functional Analysis, i.e. the functional tree. NASA (NASA, 2007) and ESA (ESA, 2009) consider Functional Analysis as the systematic process of identifying, describing, and relating the functions a system has to be able to perform, in order to be successful, but does not consider it as a design tool to address how functions will be performed, i.e. to map functions to physical components. Particular emphasis is given to the possibility of capturing the technical requirements by performing Functional Analysis (ESA, 2009). In contrast we present Functional Analysis both to define the system functional architecture, through the development first of the product tree and then of the functional block diagram, and to define or refine the functional requirements, through the accomplishment of the functional tree. The following section describes into the details the tasks of the proposed Functional Analysis methodology.

3. Functional Analysis: Methodology

Starting from the mission objectives/top level system requirements or directly from the mission statement, the Functional Analysis allows identifying the physical components, the so-called building blocks, which constitute the future product, and how they are interrelated to build up the functional architecture of the future product. Moreover through Functional Analysis the functional requirements can be defined or anyway refined.

In conceptual design Functional Analysis can be applied at different levels: subsystem level (like the avionic subsystem of an aircraft, consisting of various pieces of equipment; see sub-section 3.1), system (like a satellite consisting of various subsystems; see sub-section 3.2) and system of systems level (like a Moon base, consisting of various systems; see sub-section 3.3). According to the considered level, the physical components or building blocks, which make up the future product, are therefore equipments, subsystems or systems.

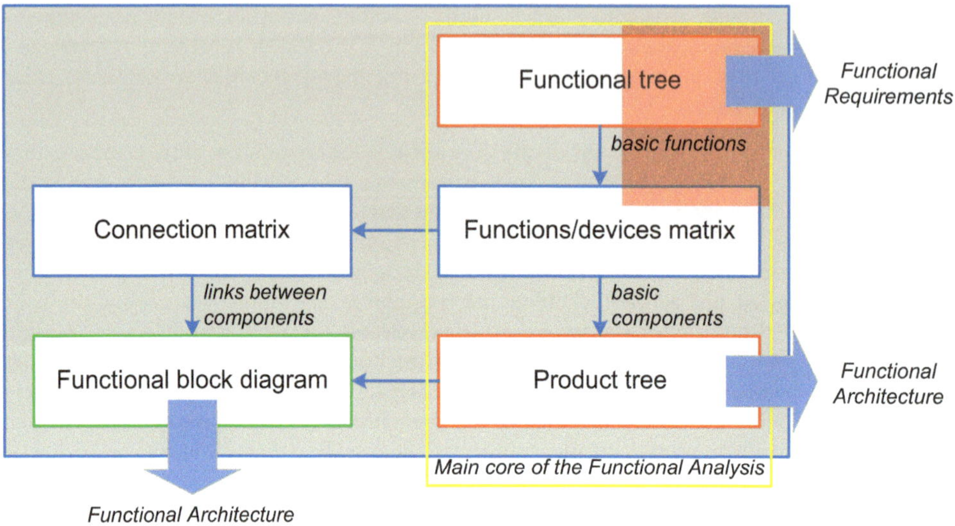

Fig. 2. The Functional Analysis

Figure 2 shows the flow-chart of the proposed Functional Analysis methodology, illustrating all its tasks, how the various tasks are related one another and the inputs/outputs of each task.

The tasks, which have to be accomplished in order to carry out Functional Analysis, are listed hereafter:

- functional tree;
- functions/components (or functions/devices) matrix;
- product (or physical) tree;
- connection matrix;
- functional block diagram.

On the basis of the mission objectives/top level system requirements the functional tree has to be developed as first step of Functional Analysis. Once the basic functions have been identified and the functional tree has therefore been completed, the functions/components matrix can be built and the basic components of the product tree can be individuated. Once the basic components have been determined, both the product tree and the connection matrix can be completed. Eventually, knowing all components (thanks to the product tree) and their relationships (thanks to the connection matrix), the functional block diagram can be fulfilled.

As highlighted in Figure 2, the main core of Functional Analysis is made up of the functional tree, the functions/devices matrix and the product tree. In fact through the functional tree and particularly through the identification of the basic functions, the functional requirements of the future product can be defined or refined, and through the product tree the building blocks of the future product can be determined, thus laying the major groundwork for the definition of the functional architecture of the future product. The functional architecture can then be completed once the relationships between the various components are clearly identified, i.e. after developing the connection matrix and the functional block diagram.

Primary outputs or objectives of Functional Analysis are therefore (see red boxes in Figure 2):

- functional tree;
- product tree.

Secondary output or objective of the Functional Analysis is (see green boxes in Figure 2):

- functional block diagram.

In the next sub-sections all tasks of Functional Analysis are considered separately and described into the details. In particular the most important rules, that have to be known to fulfil each task, are given and the procedure, that has to be followed, to move from one task to the next one, is explained.

3.1 Functional tree

The functional tree gives the possibility of representing a product by means of the functional view, which is alternative to the more common physical view. The functional and physical views are complementary not opposite views. In fact through the functional view we look at a new product asking ourselves "what does it do?", while through the physical view we look at a new product asking ourselves "what is it?", which is without any doubts the most immediate question that arises in our mind, when looking at something that is unknown. Both views are valid as they are fundamental approaches to analyze complex systems by subdividing them into parts, characterized by a poor or high level of details, depending on the need of thoroughness and/or on the level of the analysis itself.

The functional tree allows splitting the higher level functions, which stem from the mission objectives/top level system requirements, into lower level functions and eventually it allows identifying the basic functions that have to be performed by the future product. Higher level functions are complex functions that have to be decomposed into simpler functions, i.e. lower level functions, in order to accomplish the analysis. Therefore, starting from the so-called top level function, the functional tree generates various branches, which move from the most complex function to the basic functions, i.e. those functions at the bottom of the tree that cannot be split any further. Main output of the functional tree is therefore the identification of the basic functions through the decomposition of the higher level functions. The basic functions help defining or refining the functional requirements of the future product, as each basic function can be rewritten as a functional requirement. As an example, the basic function of Figure 3 "To detect infra-red (IR) threads" can be rewritten as "The system shall be able to detect infra-red (IR) threads". Figure 3 shows an example of functional tree. The top level function is "To perform defence", particularly the defence of a military aircraft. The blue box represents the top level function, while the red boxes represent the basic functions. Starting from the top level function and getting down to the basic functions, two successive levels of functions decomposition can be noted: the first (green boxes in Figure 3) and the second level functions (yellow boxes in Figure 3).

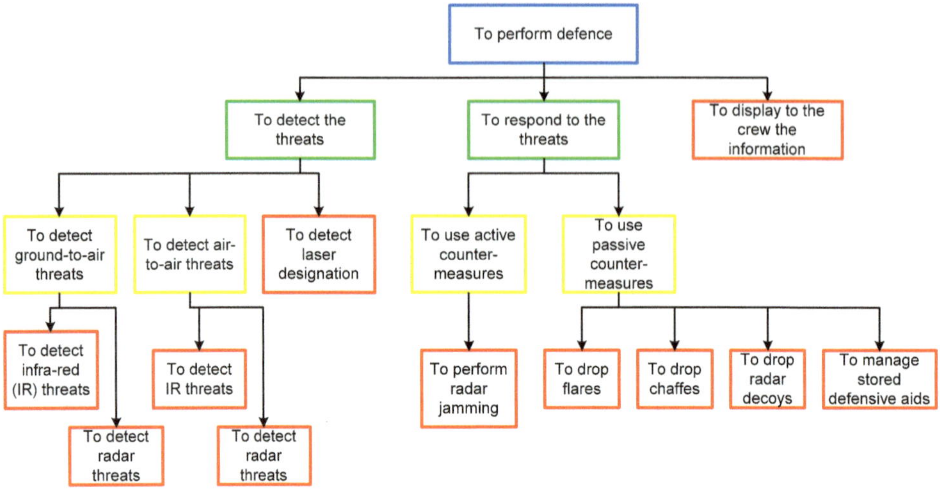

Fig. 3. Example of functional tree

In order to carry out the functional tree, the next rules have to be followed:

1. each function shall be expressed by means of verb and noun.
2. The definition of each function shall be as general as possible. Pursuing maximum generality, when describing functions, allows fostering the search of alternative solutions, in order not to forget any valuable options. This fundamental rule can be satisfactorily applied at the highest levels of the functional tree. However the lower are the levels of the tree, the less general are the functions' definitions. It is true in fact that the more you go down the three branches, the simpler become the functions and the more you get into the details of your analysis, thus making choices between available solutions. For example, if we look at the functional tree depicted in Figure 3, we note that the definitions of the first level functions are still very general, as they represent the logical decomposition of the top level function into the following sequence: "to get information" ("to detect the threats" in Figure 3), "to process information" (this function is included into "to respond to the threats" in Figure 3) , "to provide something with that information" ("to respond to the threats" in Figure 3) and/or "to provide somebody with that information" ("to display to the crew the information" in Figure 3). Then dropping to lower levels of the tree, we see that the basic functions refer to specific threats or counter-measures.
3. Lower level functions shall be either part of higher level functions or additional functions.
4. Lower level functions shall derive from higher level functions by asking "how" that higher level function can be performed. Therefore we move from the top to the bottom of the tree, through its various branches, asking ourselves "how". Viceversa we move

from the bottom to the top of the tree by asking ourselves "why". Looking again at the example reported in Figure 3, we may decompose the top level function "to perform defence" by asking ourselves: "how can the defence (of a military aircraft) be performed?". The answer to this question, that will thus represent the first subdivision of the top level function, may be (as shown in Figure 3): "to detect the threats" (i.e. "to get information"), "to respond to the threats" (i.e. "to process information" and "to provide something with that information") and "to display to the crew the information" (i.e. "to provide somebody with that information").

5. In case functions cannot be decomposed any further, they shall be reported at the bottom of the tree as basic functions. As an example, in Figure 3 the three functions, that are located in the row immediately beneath the top level function, represent the first decomposition of the top level function itself. Two out of these three functions are first level functions and are further subdivided, while the remaining function ("to display to the crew the information" in Figure 3) is already a basic function, as it cannot be split into other sub-functions. It is worth noting that the choice of carrying on decomposing a certain function or of stopping decomposing it depends on the level of details of the whole analysis (see next rule).
6. The individuation of basic functions shall depend on the level of details of the whole analysis. This implies that, if the target of Functional Analysis is, for instance, the definition of the functional architecture of a system at main equipments level, the basic functions of the functional tree shall be those functions related to specific equipments, like the example shown in Figure 3.
7. If we conceive different (physical) solutions at a certain level of the functional tree, the tree shall change from that point downwards but not upwards. This implies that, starting from the same mission objective, different functional trees can be generated not only because different people are working at it but because at a certain level of the tree (typically at lower levels) you may be obliged to make choices between alternative solutions. In this case, depending on the number of available options, a few functional trees shall be developed: they will be exactly the same from the top level function up to a certain level and will be different from that level to the bottom.

Eventually, getting back to the comparison between the function and physical views, the main advantages/criticalities of the functional view (i.e. typical approach of the functional tree) are reported hereafter.

The most significant advantages can be summarised as follows:

- the development of the functional tree, starting from mission objectives/top level system requirements, implies a thorough analysis of the mission objectives/top level system requirements themselves. This guarantees that the product, defined on the basis of the functional tree, meets all customer's needs and this is particularly important, if we remember that the functional tree is a design tool, useful to develop a new product. It is worth remembering here that, when we carry out the functional tree, we know very little about the new product. We just know the mission objectives and typically we have a preliminary draft of the system requirements but we ignore all elements that will constitute the new product. Thanks to the functions/devices matrix and then to the product tree, we will be able to say what elements will constitute the new product.

- The abstract view, typical of the functional tree, fosters the search of alternative solutions, thus avoiding biased choices.
- The functional view is absolutely coherent with the systems engineering view, which looks at the system as the integration of various elements.

The most significant criticalities can be summarised as follows:

- starting from the same mission objective/top level system requirement, different functional trees can be developed, depending on the people working at it and on the envisaged solutions. It is clear therefore that carrying out a functional tree is a typical design activity, which requires the widespread knowledge of the systems engineering designer, whose mind is not confined to any specific discipline but can embrace the whole multi-disciplinary system as integration of various parts.
- As typically the available options may be many, the main risk resides in the possibility of forgetting some concepts that may offer significant advantages for the future product.

3.2 Functions/devices matrix and product tree

Once the basic functions have been identified, it is possible to choose the components that will perform those functions by means of the functions/components (or functions/devices) matrix. The functions components matrix is therefore used to map functions to physical components.

The functions/components matrix can be built simply by matching the bottom of the functional tree, consisting of all basic functions, with one column of components able to perform those functions. Starting from the column containing the first basic function under consideration, the component able to perform that function can be defined by simply answering the question: "which component is able to perform this function?". This component can then be written down in the first row of the column of devices. The same process applies to all basic functions. Starting from the analysis of the first basic function, new components progressively fill in the column of devices. Eventually all basic components are determined. Table 1 shows a possible solution for the functions/devices matrix related to the functional tree illustrated in Figure 3. Following the procedure reported above, we take into account the first basic function on the left hand side of the functions/devices matrix, "to detect infra-red (IR) threats". If we ask ourselves which component or better which equipment is able to perform this function, we may answer that both the missile warning receiver and the infra-red (IR) warning receiver are able to fulfil the task. Then we write down both equipments in two separate rows of the functions/devices matrix and tick the intersections between these rows and the column of the basic function under consideration. Applying the same procedure, we gradually complete the functions/devices matrix, thus identifying all basic equipments.

Looking at Table 1 and remembering the logical decomposition of the top level function reported in Figure 3 ("to get information": "to detect the threats", "to process information": "to respond to the threats", "to provide something with that information": "to respond to the threats" and "to provide somebody with that information": "to display to the crew the information"), we note that:

- sensors are equipments that detect threats (sensors are highlighted in red colour in Table 1);
- processors are equipments that process information (processors are highlighted in orange colour in Table 1);
- passive or active counter-measures are equipments that respond to threats (passive and active counter-measures are highlighted in blue colour in Table 1);
- displays are equipments that display to the crew the information (displays are highlighted in green colour in Table 1).

		Basic functions								
		To detect infra-red (IR) threats	To detect radar threats	To detect laser aiming	To perform radar jamming	To drop flares	To drop chaff	To drop radar decoys	To manage stored defensive aids	To display to the crew the information
Basic devices	Missile warning receiver	X								
	Infra-red (IR) warning system	X								
	Radar warning receiver		X							
	Laser warning system			X						
	Jammer decoy dispenser				X			X	X	
	Jamming system				X					
	Chaff/flares dispenser					X	X		X	
	Store management system								X	
	Multi Function Display (MFD)									X

Table 1. Example of functions/devices matrix

Thanks to the functions/devices matrix we now know the basic components or building blocks, which constitute the future product. By simply grouping together the basic components, the product or physical tree of the new product can be generated. Unlike the functional tree, which has a typical top-down approach, the development of the product tree follows a straightforward bottom-up process. As we do know, according to the considered level, i.e. subsystem, system or system of systems level, the building blocks are respectively equipments, subsystems or systems. In case, for instance, the building blocks are equipments, they may be grouped into subsystems to form the whole system or, better, the product tree of the whole system, like illustrated in Figure 4.

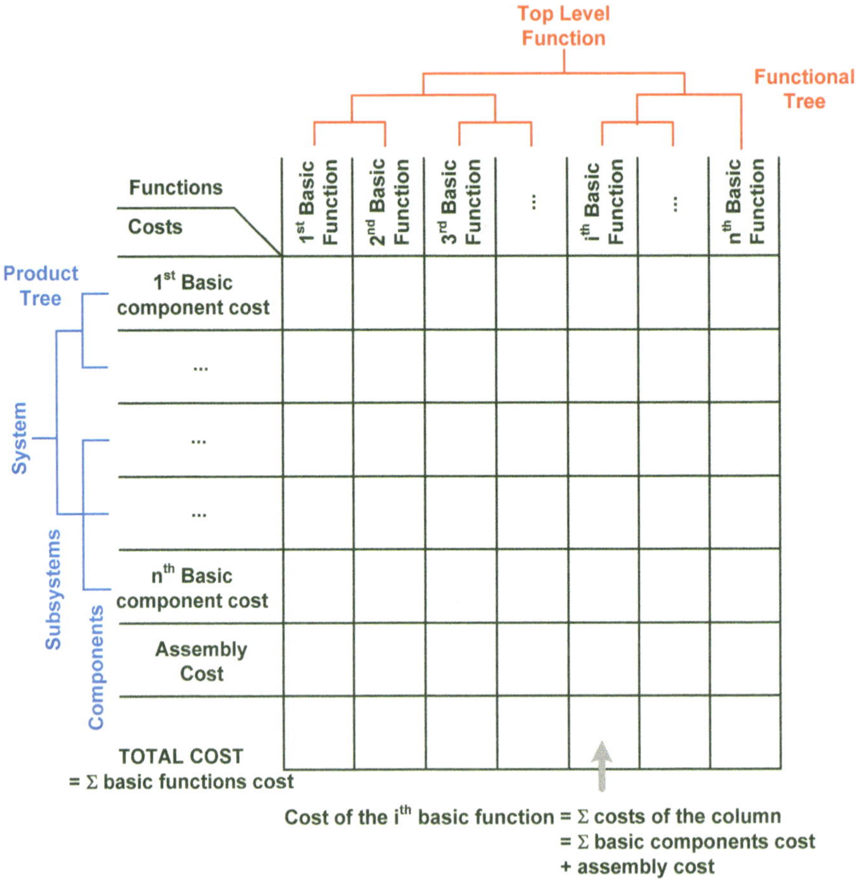

Fig. 4. Product tree and costs/functions matrix

In particular Figure 4 also shows the so-called functions/costs matrix, which is exactly the same as the functions/devices matrix except for the fact that here there is a column of costs instead of a column of devices. Quite obviously the functions/costs matrix can be built only after the functions/devices matrix, i.e. once the basic components have been identified. Main difference of the functions/costs matrix with respect to the functions/devices matrix lies in the consideration of the assembly cost. In fact, apart from the cost of each single basic component, the cost due to the assembly has to be taken into account, in order to estimate the cost of each single function and consequently the cost of the whole product.

3.3 Connection matrix and functional block diagram

Once the basic components have been identified, the links between the various components within the system can be determined. This goal is achieved by means of the connection matrix, which, as the name implies, highlights the connections between all building blocks.

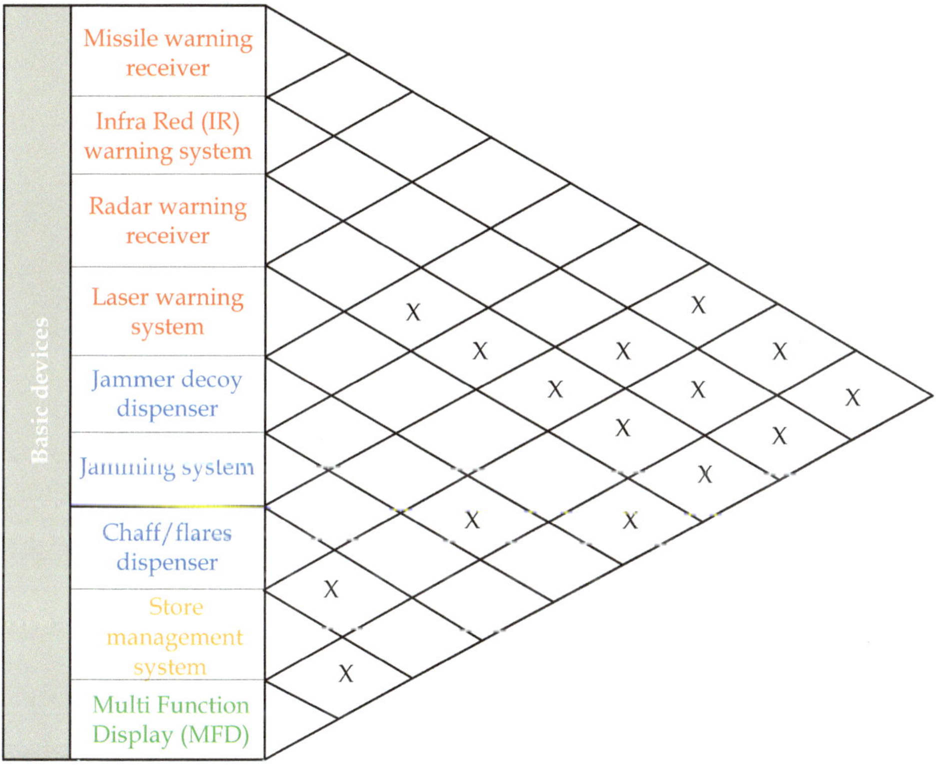

Fig. 5. Example of connection matrix

The connection matrix can either be a triangular (see Figure 5) or a square matrix, where both rows and columns have the same basic components. Starting from the first row and then proceeding down the column of basic devices, all components have to be analyzed, in order to understand whether or not there are connections between them. In case two components have a connection because, for instance, they are requested to exchange information, then the box where the two components intersects has to be ticked. As we can see, for example, all boxes where sensors (highlighted in red colour in Figure 5) and displays (highlighted in green colour in Figure 5) intersect have been ticked to show that sensors and displays exchange information.

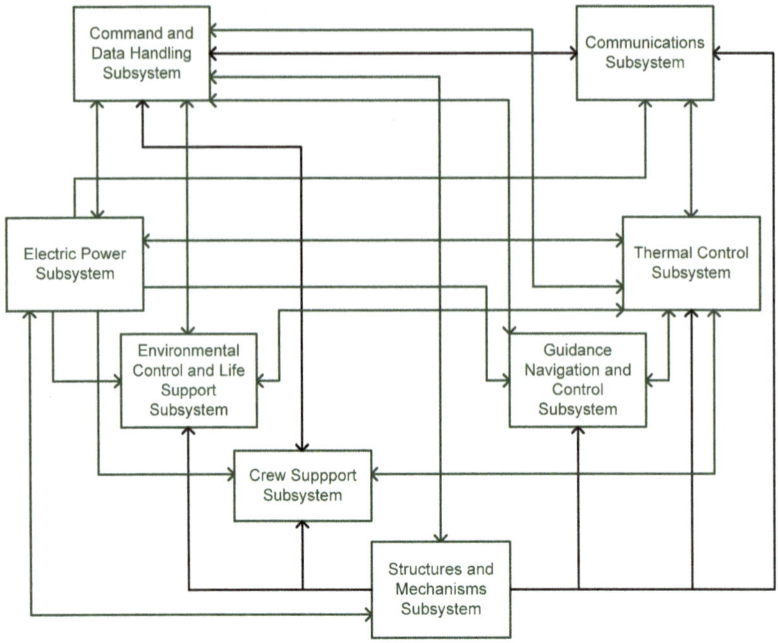

Fig. 6. The functional block diagram

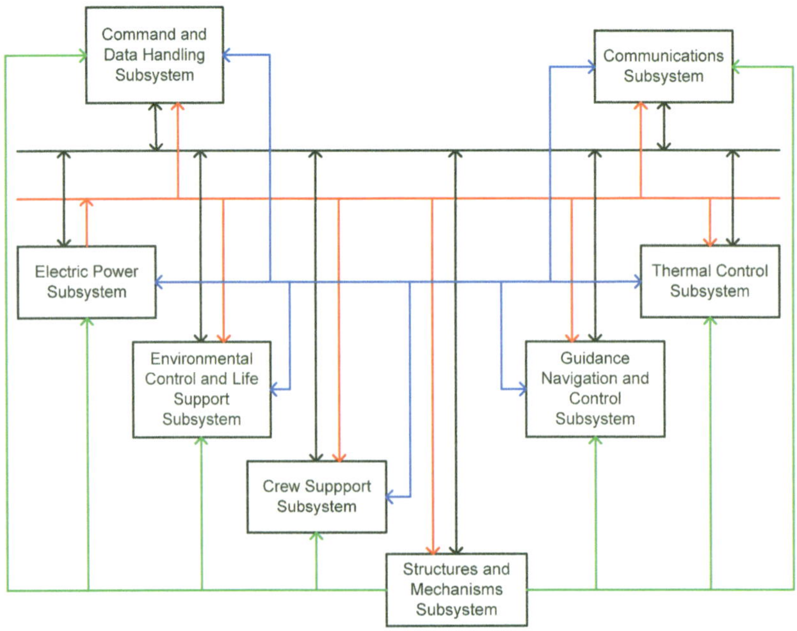

Fig. 7. The physical block diagram

It is worth underlying that nothing is said explicitly about the nature of the connections. For instance, in Figure 5, which shows a possible solution for the connection matrix related to the functional tree of Figure 3 and to the functions/devices matrix of Table 1, the type of connection between all equipments is a pure signal of information between sensors, processors, passive or active counter-measures and displays.

A different representation of the same concept, expressed by the connection matrix, is obtained through the functional block diagram, where building blocks, that need to be connected, are linked through point-to-point connections. In case these connections are arrows and not just simple lines, the functional block diagram provides the reader with additional information, if compared to the connection matrix, as it highlights not merely connections but where these connections are pointing to, i.e. if they are half duplex or full duplex connections. Just by looking at a functional block diagram, it is therefore possible to understand that, for instance, sensors are transmitting information to displays and not viceversa. Like in the connection matrix, also in the functional block diagram nothing is said about the nature of the connections, which may be, for instance, power, signal or fluid lines. This information is instead provided by the physical block diagram, which may be considered as complementary to the functional block diagram.

Figure 6 shows an example of functional block diagram for a complex system, consisting of various subsystems. This system, named Permanent Habitable Module (PHM) (Viola et al., 2007) is the first module of a permanent future human settlement on the Moon, designed to sustain the presence of three astronauts on the lunar surface. All main subsystems are highlighted in different boxes and the connections between them are shown. For sake of clarity, Figure 7 illustrates the physical block diagram of the same system presented in Figure 6. Four different types of connections between the building blocks have been envisaged: structures (green lines in Figure 7), power (red lines in Figure 7), signal (black lines in Figure 7) and fluid lines (blue lines in Figure 7).

Structures guarantee, specifically by means of secondary and tertiary structures, the anchorage of all subsystems and particularly of all their equipments to the primary structure. A power line supplies the various building blocks with the necessary power. As far as the signal lines are concerned, it is worth noting that, unlike the functional block diagram where there are point-to-point connections, in the physical block diagram there is a main bus to transmit commands and receive feedbacks to/from the various subsystems. Eventually the building blocks that need an active cooling interface to dissipate heat are connected by a ducting fluid line with the Thermal Control Subsystem.

In the next section three different applications of the Functional Analysis methodology are presented and discussed.

4. Functional Analysis: Applications

As the Functional Analysis can be applied at different levels, three different examples of applications of the methodology are presented in the following sub-sections:

- Functional Analysis at subsystem level to define the avionic subsystem of an aircraft;
- Functional Analysis at system level to define a satellite in Low Earth Orbit (LEO);
- Functional Analysis at system of systems level to define a permanent human Moon base.

4.1 Functional Analysis at subsystem level: The avionic system of a Very Light Business Jet aircraft

This sub-section deals with the application of the proposed Functional Analysis methodology at subsystem level to define the avionic system of a Very Light Business Jet (VLBJ). The VLBJ segment is constituted by civil transport jet-powered aircraft with maximum takeoff weight ranging from 2 to 4,2 tons, cruise speed of about 600 - 700 Km/h and payload capability varying from 4 to 8 passengers. The VLBJ avionics has been chosen as useful example because of its new functionalities and characteristics, which are not implemented in the avionic system of other civil aircraft. In fact VLBJs are designed to be certified as single pilot operations. This is made possible by advanced avionics automation, functional integration and easy-to-use capability.

Considering the aircraft mission profile, the environment where the aircraft will have to operate (air traffic control, landing and takeoff aids, navigation aids) and passengers and pilot requirements, the following macro-functions can be identified:

- to allow navigation.
- To perform flight controls.
- To allow communications.

For sake of simplicity only the macro-function "to allow navigation" will be dealt with here, in terms of functional tree and functions/devices matrix.

The complete functions decomposition of the top level function "to allow navigation" is reported hereafter.

1. To allow navigation

1.1 To acquire data

1.1.1 To identify weather situation
1.1.2 To detect magnetic field
1.1.3 To acquire surrounding terrain altitude
1.1.4 To acquire airplane data
1.1.5 To acquire airport data
1.1.6 To acquire flight plan data

1.1.6.1.1 To acquire data about navigation aids (VOR-DME) ground station

1.1.6.1.1.1 To memorize waypoints (VOR-DME stations)
1.1.6.1.1.2 To acquire radial and distance
1.1.6.1.1.3 To calculate flight coordinates

1.1.6.1.2 To acquire data for autonomous navigation

1.1.6.1.2.1 To memorize waypoints coordinates
1.1.6.1.2.2 To calculate flight coordinates

1.1.6.2 To acquire climb, descent and approach trajectory
1.1.6.3 To acquire landing path
1.1.6.4 To acquire different approach trajectory
1.1.6.5 To acquire missed approach procedure
1.1.6.6 To acquire holding procedure

- **1.1.7** To acquire waypoints altitude
- **1.1.8** To memorize flight plan

1.2 Data processing

- **1.2.1** To calculate optimal trajectory for all flight segment (climb, cruise, descent)
- **1.2.2** To calculate trajectory in case of critical failure
- **1.2.3** To calculate flight speed for each segment of mission profile
- **1.2.4** To calculate heading, attitude, distance and time to reach each waypoints when they are VOR-DME station
- **1.2.5** To calculate heading, attitude, distance and time to reach each waypoints when they are not VOR-DME station
- **1.2.6** To determine lateral and vertical deviation of actual trajectory in comparison to the proper one for climb, cruise and descent segments
- **1.2.7** To determine lateral and vertical deviation of actual trajectory in comparison to the proper one for approach and landing segments
- **1.2.8** To provide surrounding terrain altitude

1.3 Data management

- **1.3.1** To manage waypoints database
- **1.3.2** To manage airports database
- **1.3.3** To store and update navigation data
- **1.3.4** To store and update weather data
- **1.3.5** To verify acquired and calculated data accuracy

1.4 To display information

- **1.4.1** To display flight route and waypoints
- **1.4.2** To provide visual and acoustic warning in case of traffic collision
- **1.4.3** To display heading
- **1.4.4** To display true heading
- **1.4.5** To display environment data
- **1.4.6** To provide visual and acoustic warning in case of potential ground collision
- **1.4.7** To display surrounding terrain altitude
- **1.4.8** To display weather situation
- **1.4.9** To display approach and landing correct trajectory
- **1.4.10** To display trajectory in case of missed approach

On the basis of the basic functions listed above, the functions/devices matrix can be created, as shown in Table 2, which, for sake of simplicity, illustrates only part of the complete functions/devices matrix related to the top level function "to allow navigation". It is worth remembering that, as in this case the Functional Analysis is applied at subsystem level, the basic components are the main subsystem equipments. The functions/devices matrix has thus been called functions/equipments matrix.

Eventually Figure 8 illustrates the functional block diagram of the complete avionic system, where both half duplex and full duplex connections between equipments are highlighted.

		Basic functions							
		To identify weather situation	To detect magnetic field	To acquire surrounding terrain altitude	To acquire airplane data	To acquire airport data	To memorize waypoints (VOR-DME stations)	To acquire radial and distance	To calculate flight coordinates
Basic equipment	Weather Radar System (WX)	X							
	ADAHRS (ADS+AHRS + Magnetometer)		X						
	Syntetic Vision System			X					
	Flight Management System (FMS)			X	X	X	X		X
	Flight Computer			X					
	Navigation Computer			X					X
	Automatic Direction Finder (ADF)							X	
	VHF Omni Range (VOR)							X	
	Distance Measurement Equipment (DME)							X	

Table 2. Part of the complete functions/equipments matrix

4.2 Functional Analysis at system level: The cubesat e-st@r

In this sub-section an example of the methodology is given by the application of the Functional Analysis to a Cubesat project. The e-st@r (Educational SaTellite @ politecnico di toRino) program is taken as case-study. The project is an educational initiative carried out by students and researchers of Politecnico di Torino within an ESA program aiming at the launch and orbit operations of nine cubesats, developed by as many European Universities, to promote space activities among young generations. E-st@r program guidelines are illustrated in Figure 9.

The mission statement sounds as follows: "*Educate aerospace-engineering students on complex systems design and management, team work, and standards implementation. Achieve insight in the development of enabling technologies for low-cost access to space*".

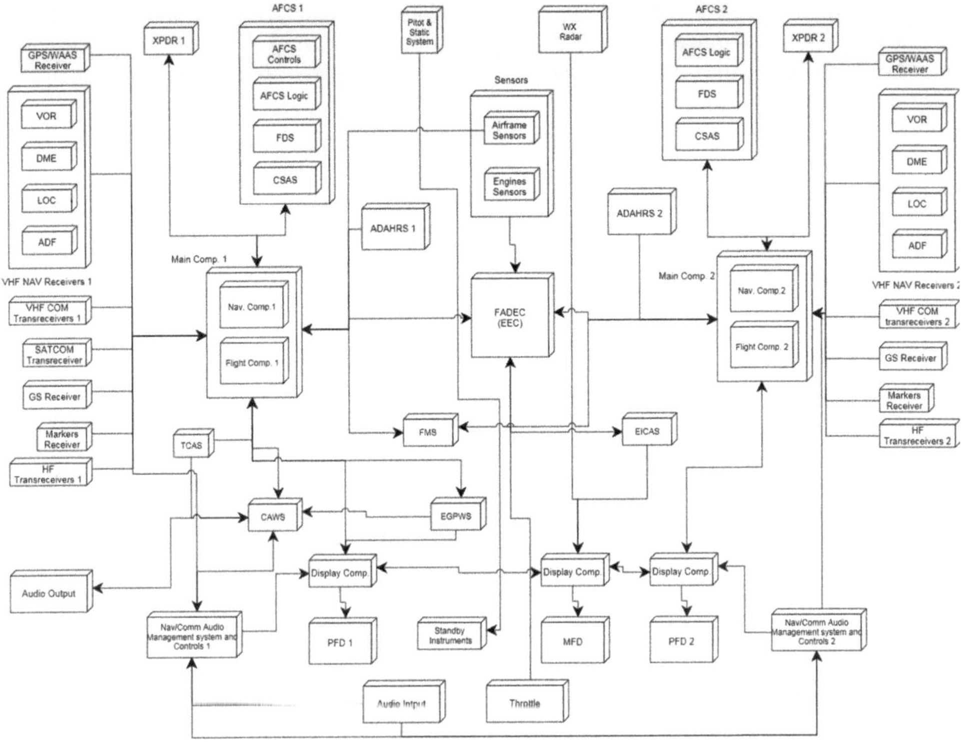

Fig. 8. VLBJ avionics functional block diagram

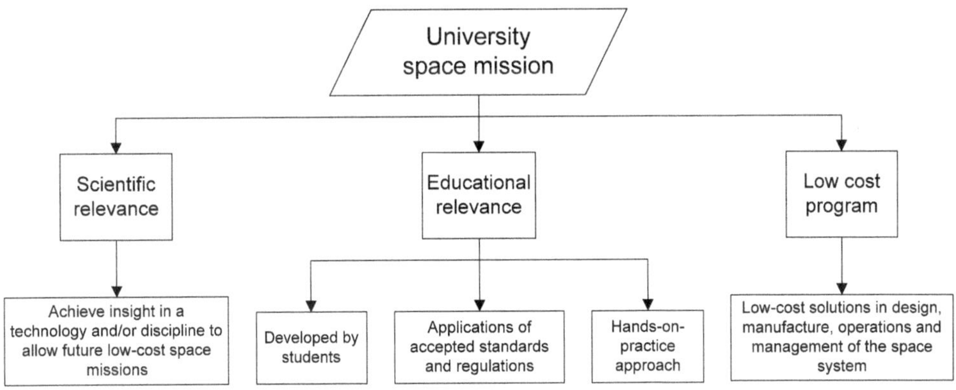

Fig. 9. e-st@r program guidelines

The following assumptions can be derived from the mission statement:

- the program shall be carried out by students. They shall design, manufacture, verify and test, and operate a space system.
- The program shall be carried out in compliance with current regulations and applicable standards.
- The program shall have educational relevance, which means that students must learn by practice.
- An experiment shall be included in the space system. The experiment shall be simple and cheap, but at the same time it must permit to achieve insight in a discipline and/or technology to be used in the future to allow low-cost space mission.
- The program driver shall be the research for low-cost solutions in design, manufacture, operations and management of space systems.

Notwithstanding the necessity of keeping cost down and taking into account the educational spirit of the e-st@r program, which implies the will of enhancing the interests and competences of the students, e-st@r has also scientific objectives, which reflect real interests of the scientific and industrial communities. Taking into account all high-level requirements and constraints, as a result of a trade-off analysis it has been decided that the system would accomplish a mission aimed at testing an active Attitude Determination and Control System (ADCS).

In conclusion, the mission scenario can be summed up as follows: a cubesat shall be inserted into a LEO by the end of 2012. The cubesat shall be piggybacked by the Vega LV during its Maiden Flight. Mission duration for this kind of project shall be in the range of 3-12 months. The cubesat shall be operated from ground in a simply and cheap way. High grade of operations autonomy is desirable. Students shall be designers, developers, manufacturers, operators and managers of the entire mission. The mission shall demonstrate some kind of non-space technologies and try to space-qualify them. The primary payload shall be a simple active ADCS. As secondary payload, the test of commercial items is considered. The mission data shall be available to the cubesat community and to radio-amateurs union. No commercial purposes shall be pursued.

Functional Analysis methodology has been used to derive the requirements for the system and to determine which subsystems are needed to carry out the mission. The second iteration of the Functional Analysis allows deriving next level requirements for equipments and components. A part of the complete functional tree for the e-st@r mission is shown in Figure 10. The mission segments are identified by the first level functions (i.e. "To connect ground and space segments", "To do on ground operations", "To reach the orbit", "To do in orbit operations" and "To comply with space debris mitigation regulations") and they reflect the mission architecture's elements.

The elements of the e-st@r mission architecture are reported hereafter:

- Space segment: Cubesat = payload and bus.
- Ground segment: one main ground control station + one backup ground control station (mobile and transportable). Radio amateur network. Cubesat laboratory at Polito.
- Launch segment: Vega LV and CSG (French Guyana)
- Subject: data measurement.

- Orbit: direct injection into LEO (approx 300x1450km, 70°).
- Operations: students at main and backup GCSs. Data processing at Cubesat lab for deeper investigation, or in case of emergency
- Communications.

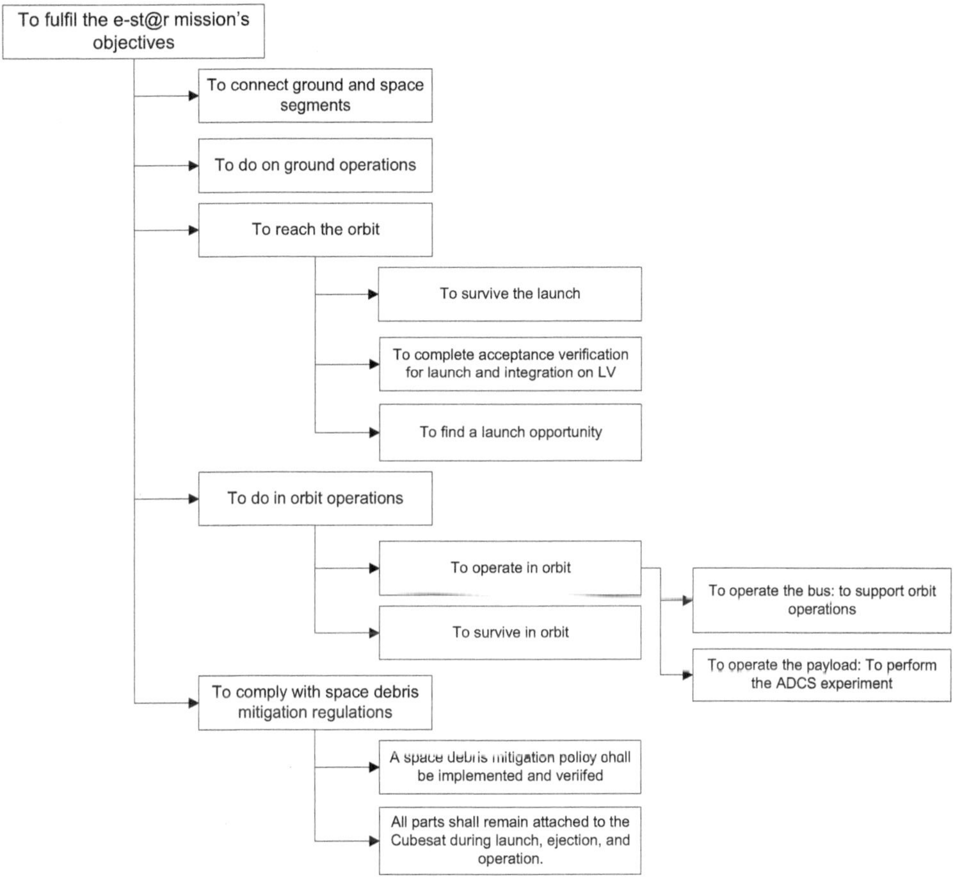

Fig. 10. Part of the complete functional tree for e-st@r mission

As an example, the product tree of two elements of the e-st@r mission architecture, the space (i.e. the cubesat, made up of payload and bus) and the ground segment, is shown in Figure 11. It is worth noting that, while the space segment can be directly linked through a functions/devices matrix to the first level function "To do in orbit operations" (see Figure 10), the ground segment can be directly linked to the first level function "To do ground operation"(see Figure 10). Eventually Figure 12 illustrates the physical block diagram of the cubesat. The block diagram shows all subsystems (apart from structures) and their connections. The design and sizing of the subsystems in phase A have been carried out using common available methods (Wertz & Larson, 2005), (Fortescue et al., 2003).

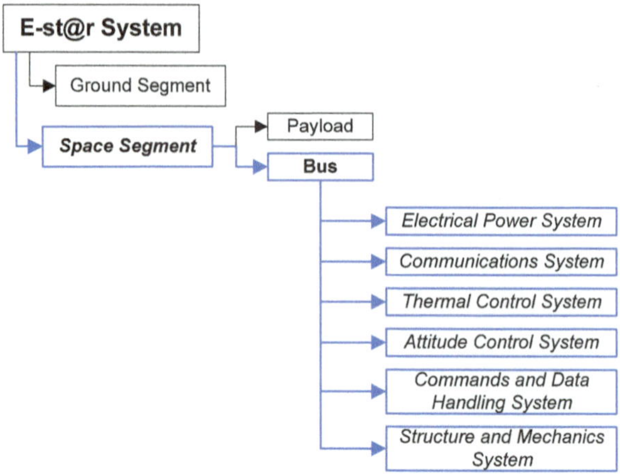

Fig. 11. Product tree of the e-st@r system: the ground and the space segment

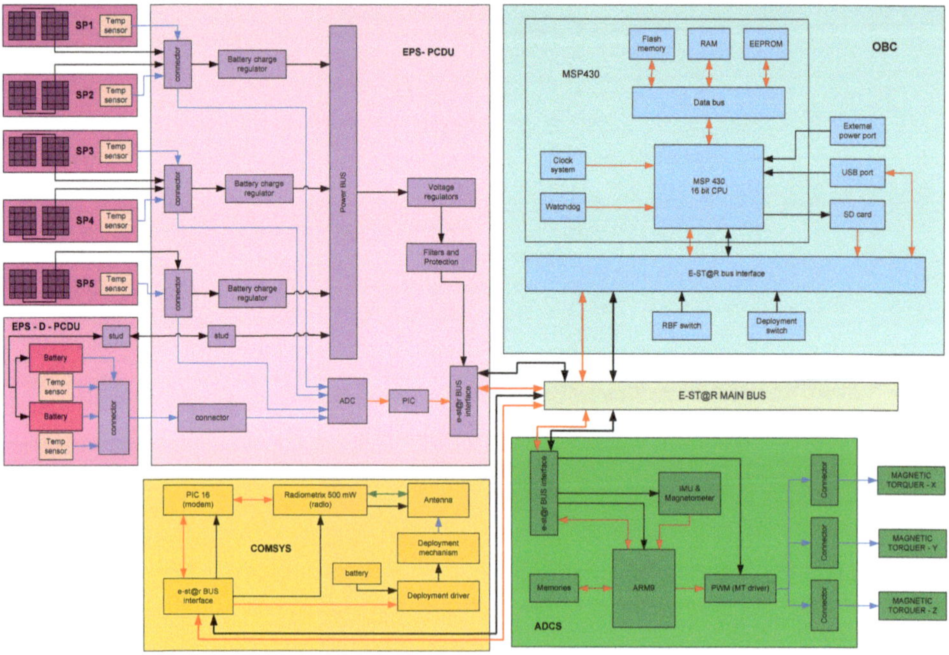

Fig. 12. Physical block diagram of the e-st@r cubesat

4.3 Functional Analysis at system of systems level: the permanent human Moon base PHOEBE

The system of systems here considered is a permanent human Moon base. The Functional Analysis methodology has been applied, in order to accomplish the primary objectives, i.e. in order to develop the functional tree and the product tree of the Moon base.

The Moon base has been given the name PHOEBE, which stays for: Permanent Human mOon Exploration BasE (Viola et al., 2008).

The mission statement is reported hereafter: *"To establish a permanent lunar base for a nominal crew of 18 astronauts (maximum 24 during crew rotation) with a turnover time of 6 months, to support scientific research, In-Situ Resources Utilization (ISRU) development, surface exploration and commercial exploitation; its evolution will provide an outpost for further space exploration".*

After the definition of the mission statement, nine mission objectives have been determined. The main top level system requirements are schematically represented in Figure 13, where they can be traced back to their correspondent mission objectives.

Once the top level system requirements, which stem from the mission statement and mission objectives, have been defined, the design process has proceeded with the accomplishment of the Functional Analysis, in order to determine all building blocks, i.e. the systems or modules, of the Permanent Human Moon Base that satisfy the top level system requirements. Main results of the Functional Analysis are presented hereafter. In particular Figure 14 illustrates the so-called "first level" functional tree, where the top level function "To carry out a Permanent Human Moon Base" has been split into 10 first level functions. Each first level function has then been divided into lower level functions to identify the basic level functions, i.e. those functions that can immediately be connected to one building block of the Moon base.

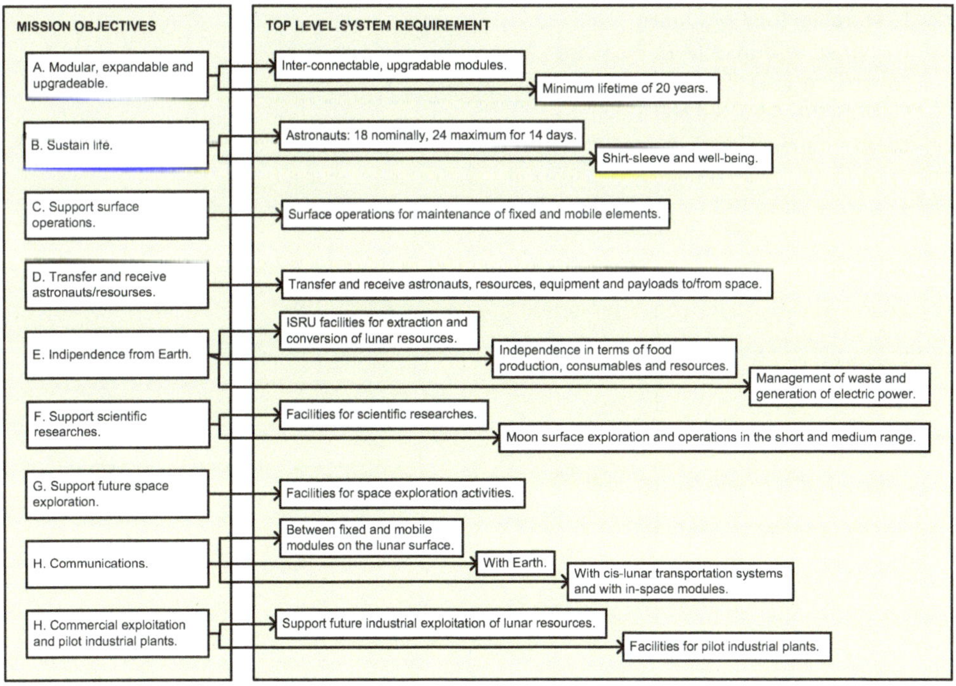

Fig. 13. Mission objectives and top level system requirements

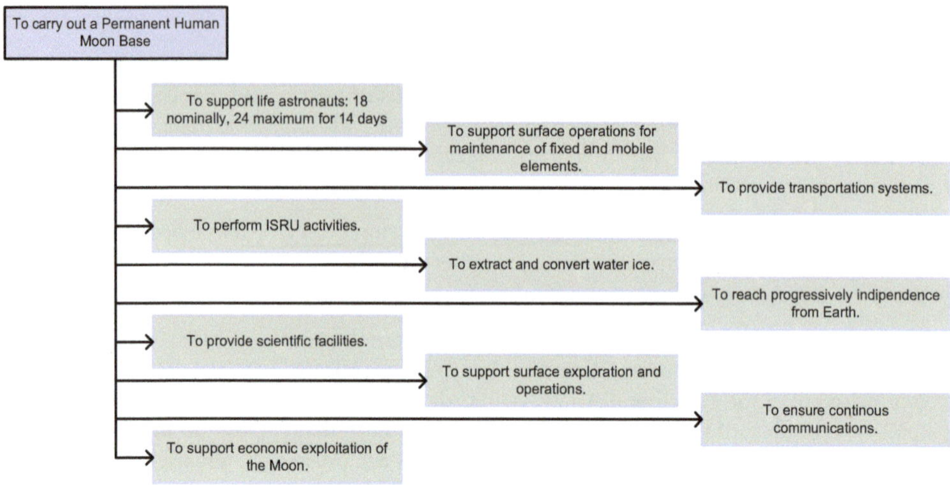

Fig. 14. PHOEBE First level functional tree

For sake of clarity, Figure 15 shows how the first level function "To reach progressively independence from Earth" has been decomposed into its basic functions:

- to provide plants growth facilities.
- To store the food produced.
- To extract resources from the waste.
- To retrieve TBD (To Be Defined) consumables.
- To store the retrieved consumable.
- To provide the electrical power.

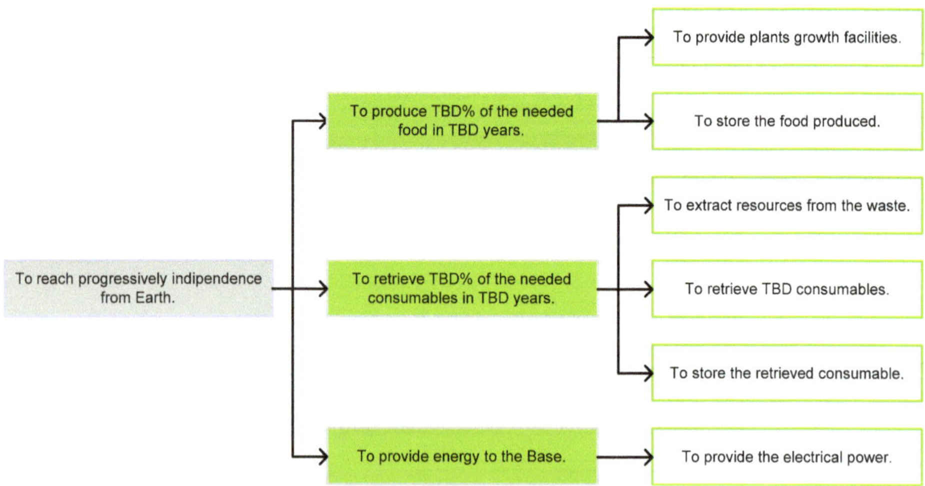

Fig. 15. Functional tree of the first level function: "To reach progressively independence from Earth"

Once the basic functions have been identified, it is possible to choose the building blocks of the Moon base that will perform those functions. Considering for instance the basic functions presented in Figure 15, the corresponding building blocks can be obtained through the functions/building blocks (or functions/devices) matrix (see Table 5).

		Functions					
		To provide plants growth facilities	To store the food produced	To extract resources from the waste	To retrieve TBD consumables	To store the retrieved consumable	To provide the electrical power
Building blocks	Green House	X		X	X		
	Stowage Module		X				
	Storage Module					X	
	Processing Plant				X		
	Material Science Laboratory				X		
	Power Plant						X

Table 3. PHOEBE: example of functions/building blocks matrix

As addressed in Table 3, six building blocks have been identified:

- the Green House, which is a source of food and consumables (oxygen recycling, inert gases) and produces TBD% of what is needed in the Moon base. It is made up by all the facilities necessary for plants growing, waste recycling and for retrieving TDB consumables;
- the Stowage Module, where the food can be stored;
- the Storage Module, where the consumables can be stored;
- the Processing Plant, which, apart from fulfilling the functions of processing resources and water ice, has also the capability of retrieving the consumables as well as the Green House;
- the Material Science Laboratory, which, apart from fulfilling the functions of processing resources and water ice and providing materials experiments and test facilities for space exploration technology, has also the capability of retrieving the consumables as well as the Green House;
- the Power Plant, which provides the Moon Base with electric power.

Applying the same methodology to all the other first level functions listed in Figure 14, the complete product tree of the Moon base, i.e. all PHOEBE systems, can be obtained, as Figure 16 illustrates, where the various building blocks have been grouped into four different categories or segment: transportation, in-space, mobile and fixed segment.

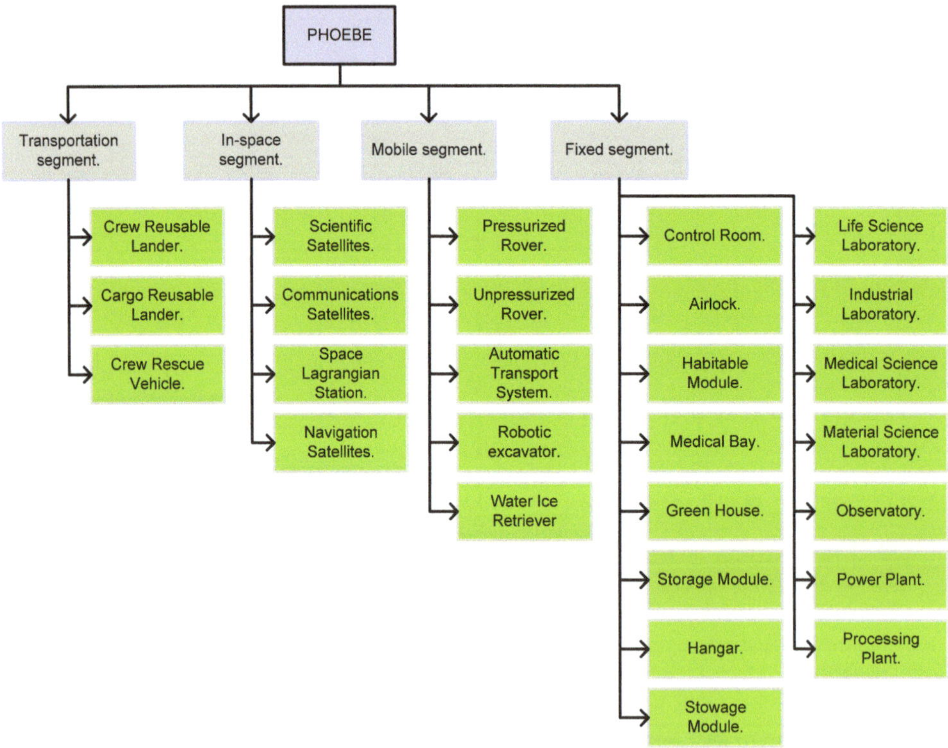

Fig. 16. PHOEBE product tree

The identification of all systems of the Moon base and the understanding of their related functions are the final results of the presented Functional Analysis at system of systems level.

5. Conclusion

The Functional Analysis is without any doubts one of the most fundamental tool of systems engineering design to develop a new product, as it guarantees a thorough analysis of the requirements, it fosters the search of alternative solutions, thus avoiding or at least limiting the risk of forgetting valuable options, and eventually it allows identifying the physical components of the future product and their relationships. It is therefore of paramount importance for every systems engineer to learn how to apply Functional Analysis to explore new concepts and then satisfactorily come out with innovative architectures.

After a brief introduction to underline the precious role of Functional Analysis within the conceptual design process, the chapter describes into the details all steps that have to be taken and all rules that have to be followed to accomplish Functional Analysis. Eventually the chapter presents three different applications of the methodology at subsystem, system and system of systems level.

6. Acknowledgment

The authors wish to thank all students that have worked at the e-st@r program and all students that have attended the SEEDS (SpacE Exploration and Development Systems) Master.

7. References

ESA-ESTEC (Requirements & Standards Division), (2009). *Space Engineering Technical Requirements Specification*. ESA Requirements & Standards Division Technical Report ECSS-E-ST-10-06C, European Space Agency for the members of ECSS, Noordwijk, The Netherlands.

Fortescue, P., Stark, J. & Swinerd, G. (2003). *Spacecraft Systems Engineering (Third Edition)*, John Wiley & Sons Ltd, ISBN 0-471-61951-5, Chichester, West Sussex, England.

Larson, W. J., Wertz, J. R. (2005). *Space Mission Analysis and Design (third edition)*, Microcosm Press and Kluwer Academic Publishers, ISBN 1-881883-10-8, E1 Segundo, California and Dordrecht/Boston/London.

NASA, (2007). *Systems Engineering Handbook*. NASA Technical Report NASA/SP-2007-6105 Rev1, ISBN 978-0-16-079747-7, Washington, DC, USA.

Raymer, D. P. (1999). *Aircraft Design: A Conceptual Approach (Third Edition)*, AIAA (American Institute of Aeronautics and Astronautics) Education Series, ISBN 1-56347-281-0, Reston, Virginia, USA.

Viola, N., Messidoro, P. & Vallerani, E. (2007). Overview of the first year activities of the SEEDS Project Work, *Proceedings of 58th International Astronautical Congress*, Hyderabad, India, 24-28 September 2007

Viola, N., Vallerani, E., Messidoro, P. & Ferro, C. (2008). Main results of a permanent human Moon base Project Work activity 2006-2007, *Proceedings of 59th International Astronautical Congress*, Glasgow, Scotland, 29 September - 3 October, 2008

3

Complex-Systems Design Methodology for Systems-Engineering Collaborative Environment

Guido Ridolfi[1,2], Erwin Mooij [2] and Sabrina Corpino[1]
[1]*Politecnico di Torino*
[2]*Delft University of Technology*
[1]*Italy*
[2]*The Netherlands*

1. Introduction

In the last decades man-made systems have gained in overall complexity and have become more articulated than before. From an engineering point of view, a complex system may be defined as one in which there are multiple interactions between many different elements of the system and many different disciplines concurring to its definition. However, the complexity seen from the system perspective is only partial. In more general terms complexity does not only regard the system *per se*, but it is also related to the whole life-cycle management of the system. This encompasses all the activities needed to support the program development from the requirements definition to the verification, validation and operation of the system in the presence of a large number of different stakeholders. These two interrelated views of complexity, being *bottom-up* in the first case and *top down* in the second, both converge to the system defined as an entity formed by a set of interdependent functions and elements that complete one or more functions defined by requirements and specifications.

Systems Engineering processes have been increasingly adopted and implemented by enterprise environments to face this increased complexity. The purpose is to pursue time and cost reduction by a parallelization of processes and activities, while at the same time maintaining high-quality standards. From the life-cycle management point of view the tendency has been to rely more and more on software tools to formally applying modeling techniques in support of all the activities involved in the system life-cycle from the beginning to the end. The transition from document-centric to model-centric systems engineering allows for an efficient management of the information flow across space and time by delivering the right information, in the right place, at the right time, and to the right people working in geographically-distributed multi-disciplinary teams. This standardized implementation of model-centric systems engineering, using virtual systems modeling standards, is usually called Model Based Systems Engineering, MBSE.

On the other side, looking at the problem from the perspective of the system as a product, the management of complexity is also experiencing a radical modification. The former adopted approach of sequentially designing with separate discipline activities is now being replaced by a more integrated approach. In the Aerospace-Engineering domain, for instance, designing with highly integrated mathematical models has become the norm. Already from the preliminary design of a new system all its elements and the disciplines involved over the entire life-cycle are taken into account, with the objective of reducing risks and costs, and possibly optimizing the performance.

When the *right people* all work as a team in a multi-disciplinary collaborative environment, the MBSE and the Concurrent Engineering finally converge to the definition of the system. The main concern of the engineering activities involved in system design is to predict the behavior of the physical phenomena typical of the system of interest. The development and utilization of mathematical models able to reproduce the future behavior of the system based on inputs, boundary conditions and constraints, is of paramount importance for these design activities. The basic idea is that before those decisions that are hard to undo are made, the alternatives should be carefully assessed and discussed. Despite the favorable environment created by MBSE and Concurrent Engineering for the discipline experts to work, discuss and share knowledge, a certain lack of engineering-tool interoperability and standardized design methodologies has been so far a significant inhibitor, (International Council on Systems Engineering [INCOSE], 2007). The systems mathematical models usually implemented in the collaborative environments provide exceptional engineering-data exchange between experts, but often lack in providing structured and common design approaches involving all the disciplines at the same time. In most of the cases the various stakeholders have full authority on design issues belonging to their inherent domain only. The interfaces are usually determined by the experts and manually fed to the integrated models. We believe that the enormous effort made to conceive, implement, and operate MBSE and Concurrent Engineering could be consolidated and brought to a more fundamental level, if also the more common design analytical methods and tools could be concurrently exploited. Design-space exploration and optimization, uncertainty and sensitivity analysis, and trade off analysis are certainly design activities that are common to all the disciplines, consistently implemented for design purposes at the discipline-domain level. Bringing fundamental analysis techniques from the discipline-domain level to the system-domain level, to exploit interactions and synergies and to enable an efficient trade-off management is the central topic discussed in this chapter. The methodologies presented in this chapter are designed for their implementation in collaborative environments to support the engineering team and the *decision-makers* in the activity of *exploring* the design space of complex-system, typically long-running, models. In Section 2 some basic definitions, terminology, and design settings of the class of problems of interest are discussed. In Section 3 a test case of an Earth-observation satellite mission is introduced. This satellite mission is used throughout the chapter to show the implementation of the methods step by step. Sampling the design space is the first design activity discussed in Section 4. Then in Section 5 and Section 6 a general approach to compute sensitivity and to support the engineering team and decision makers with standard visualization tools are discussed. In Section 7 we provide an overview on the utilization of a unified sampling method for uncertainty and robustness analysis. Finally, we conclude the chapter providing some recommendations and additional thoughts in Section 8.

2. Basic definitions

The discussion and the methodologies presented in this chapter are based on the assumption that the activity of designing a complex system is performed by a team of designers (the engineering team), using **mathematical models** to determine the physical and functional characteristics of the system itself. A mathematical model is a set of relationships, i.e., equations, providing figures-of-merit on the **performance(s)** of the system to the engineering team when certain **inputs** are provided. The inputs are represented by the **design variables**, i.e., factors that are responsible for influencing the performance(s) of the system. For this motivation, the design variables will also be called **design factors**, or more generally inputs, or simply variables. The domain of existence of the design variables forms the **design space**, where they can assume certain **values** between a minimum and a maximum. The **design-variable range** determined by the minimum and the maximum can of course only be as large as the domain of existence of the variable. Mimima and maxima for the design variables are usually set by the engineering team to limit the analysis to a specific region of the design space or to avoid infeasible conditions. For instance, the design range of the eccentricity, e, of a closed orbit about the Earth should not exceed the interval $0 \leq e < 1$. In the upper-left Cartesian diagram of Fig. 1 a hypothetical design space, formed by two variables, is shown. The limits of the variable ranges are represented by the dash-dotted lines. The subspace of the design space determined by all the design-variable ranges is addressed as the **design region** of interest, and it is represented by the rectangle formed by the dash-dotted lines and the vertical axis of the Cartesian diagram.

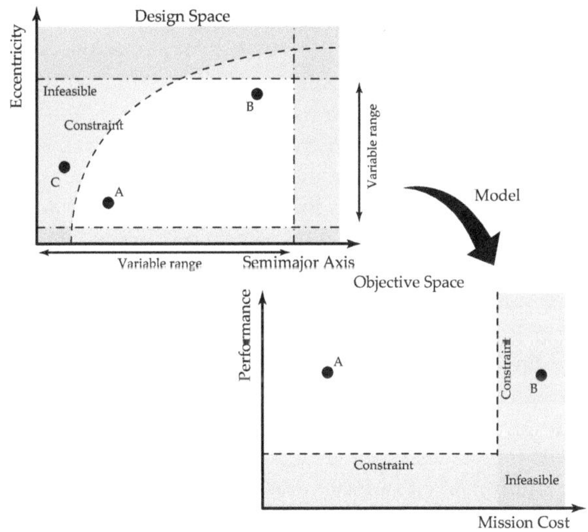

Fig. 1. Schematic representation of the design space and the objective space of the model.

Design variables can be **continuous** or **discrete**. A continuous variable can assume all the values between the minimum and the maximum. A discrete variable, instead, can assume only few specific values in the design-variable range. In this case the values are called **levels**. Discrete variables can be further distinguished into two classes, namely **ordinal** or **categorical**. The *length* of a solar array on a satellite system, for instance, is a continuous variable. It can assume, in principle, any value between a minimum and a maximum set to limit the weight or to provide a minimum performance under certain circumstances. The *number of cells* used to build the array is an ordinal variable. It can only assume the levels represented by the natural numbers, and certain characteristics increase (decrease) when the number of cells increases (decreases), e.g., the total mass. The *type of solar cell*, instead, is a categorical variable. This means that it can only assume certain levels (e.g. *type#1*, *type#2*, and so on), but in this case the order is not important. It is not always the case that, for instance, the *efficiency* of the solar cells increases going from the first type to the second type and so on. It depends on the order in which they appear in a database, for instance, that may be an arbitrary choice of the engineering team. The model of the system may be also subject to other sources of variability representing the non-deterministically known parameters typical of the operating environment of the system. The residual atmospheric density on orbit, the solar radiation, the orbit injection errors, just to mention a few, are factors that may not be directly controlled at a design stage therefore they must be taken into account in a statistical sense. These factors are called **uncontrollable**.

One of the main tasks of the engineering team during the **design process** of the system is to set the values and/or the levels of the design variables in such a way that the performance(s) of the system assume a certain optimal level under *certain circumstances* (**optimal design**), and/or such that the final system is insensitive to variations of the uncontrollable factors (**robust design**). The performance(s) of interest are called **objective(s)** of the analysis. The space in which the objectives can be represented, i.e., the domain of the images of the mathematical equations of the model, is called **objective space**. Thus, the model is responsible for relating points in the design space with points in the objective space. The term *certain circumstances* is used to indicate the constraints and boundary conditions of the analysis. As already mentioned, the **boundary conditions** are represented by the design-variable ranges, the dash-dotted lines of Fig. 1. The **constraints**, instead, are determined by an infeasible condition in the objective space, e.g., the mass of the satellite exceeding the mass that the launcher is able to deliver in a given orbit. Further, the constraints can also be determined by infeasible conditions on the design space, when certain combinations of the values or levels of the design variables are not allowed. This may happen, for instance, with the eccentricity and the semimajor-axis of an Earth-orbiting satellite. Their combined values must ensure that the perigee altitude of the orbit is at least larger than the radius of the Earth. Constraints may be linear or non-linear, continuous or discrete. The dashed lines in Fig. 1 represent the constraints in the design space (non linear in this case), and in the objective space (linear in this case). The thick dots in Fig. 1 represent the **design points**. In the design space they are a representation of the values of the design variables, while on the objective space they represent the corresponding set of output values. Considering a deterministic model, there is a one-to-one correspondence between one point in the design space and one point in the objective space. However, the engineering team must make sure to provide design points that do not violate constraints in the design space. For instance, an orbit with a semi-major axis of 7000 km and eccentricity of 0.7 would lead to a negative value of the satellite altitude at perigee (i.e., non existing orbit) thus with the impossibility of computing relevant parameters such as, for instance, *time-in-view* at perigee passage on a specific region on Earth. Therefore, in Fig. 1 the design point C does not have a corresponding image on the objective space. In this case, the semi-major axis and the eccentricity are classified as correlated inputs.

The problem of developing and implementing the mathematical model of a complex system is beyond the scope of this chapter. However, a brief discussion on the type of modelization approach is beneficial for a better understanding of the discussed design methodologies. The development of a mathematical model is tackled considering two main sub-problems, namely *problem decomposition*, (Sobieszczanski-Sobieski, 1989), and *problem formulation*, (Cramer et al., 1993; Tedford & Martins, 2006). In the literature, authors propose several model-decomposition techniques. However, two main classes may be identified, namely *Hierarchical Decomposition* and *Non-Hierarchical Decomposition* methods, (Sobieszczanski-Sobieski & Haftka, 1995). Non-Hierarchical Decomposition methods (NHD) are advised when there is no clear separation between two or more elements/disciplines, i.e. when the coupling between them is not negligible *a priori*. The formulation of the complex-system design problem is related to the allocation of the resources to the various elements of the architecture. Single- and multiple-level formulations are discussed in the literature, (Cramer et al., 1993; Tedford & Martins, 2006; Yi et al., 2008). The former must be executed on a single machine, the latter, instead, allows for more flexibility in allocating the computational resources. The mathematical models of a collaborative environment are most likely developed using a NHD approach, because it is the most general one, and with a multi-level architecture because resources are usually geographically distributed. An example of the multi-level architecture of a complex-system design problem is presented in Fig. 2. It represents the architecture most likely adopted in a collaborative environment with team-members responsible for element analysis and others responsible for system analysis. The thick lines represent input/output interfaces.

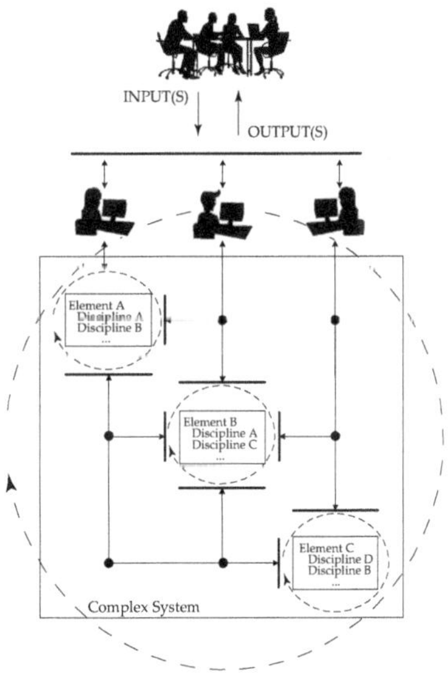

Fig. 2. Schematic of the Collaborative Bi-Level (COBiL) formulation for complex systems models.

3. Design case: Earth-observing satellite for natural disaster and land-use monitoring

Earth-observation satellites can observe areas over a wide range rather quickly. It is expected that their observation data combined with information obtained by aircraft and helicopters will be useful for a regular disaster condition assessment. This would make rescue operations more effective, would allow for extracting topographical information reflecting latest land-usage changes, and identifying disaster risks.

In this chapter, the preliminary design of an Earth-observation mission to support the world-wide disaster management process and land-usage monitoring is deployed and discussed to show a practical implementation of the proposed design approaches. The following mission statement is considered as driver for the design process:

Design an Earth observation mission to provide world-wide disaster-management capabilities, for over a period of 7 years

The limited available space and at the same time the willingness to effectively convey the message of this chapter led us to take several assumptions to determine boundaries of the analysis presented here. A satellite system with an optical payload (staring sensor) is considered. The main purpose is to achieve a compromise between the design variables in such a way to obtain the best possible image resolution, at minimum cost. The satellite shall revisit the same area on the Earth surface within 24 hours, and shall be able to send the acquired data back, in real time, to any equipped ground station (the reference ground station is considered with 1 m aperture antenna diameter) with a link margin of at least 4 dB. The selected launcher is of the class of the *Delta II 6920/25*, with a maximum payload on polar orbit of 2950 kg. A highly inclined, circular orbit has been selected, with $i = 98°$. The main mission geometry parameters and few of the equations implemented for computing the coverage and the resolution are presented in Fig. 3.

In Table 1 the design variables taken into account in the analysis, and their intervals or levels (in case of discrete variables) are summarized.

The mathematical model of the satellite system is composed of all its main subsystems (i.e., payload, Attitude Dynamics and Control System (ADCS), communication system,

Design Variables	Code	Intervals Min	Intervals Max	Levels	Design Variables	Code	Intervals Min	Intervals Max	Levels
Number of days (rep. ground track)	A	1	3	3	Number of slew maneuvers [-]	G	10k	30k	–
Number of orbits (rep. ground track)[a]	B	1	3	3	Transmitting output RF power [W]	H	5	30	–
Instrument aperture diameter [m]	C	0.3	1	–	Antenna diameter [m]	I	0.1	1	–
Min. ϵ [deg]	D	5	50	–	Type of solar array [-]	J	1	2	2
Max. slew angle [deg]	E	0	50	–	Type of thrusters [-]	K	1	2	2
Min. maneuver time [s]	F	60	180	–	Payload heritage [-]	L	1	2	2

Table 1. Settings of the design variables.[a] When $A = 1$, $B = 13, 14$ or 15. When $A = 2$, $B = 28, 29$ or 30. When $A = 3$, $B = 43, 44$ or 45.

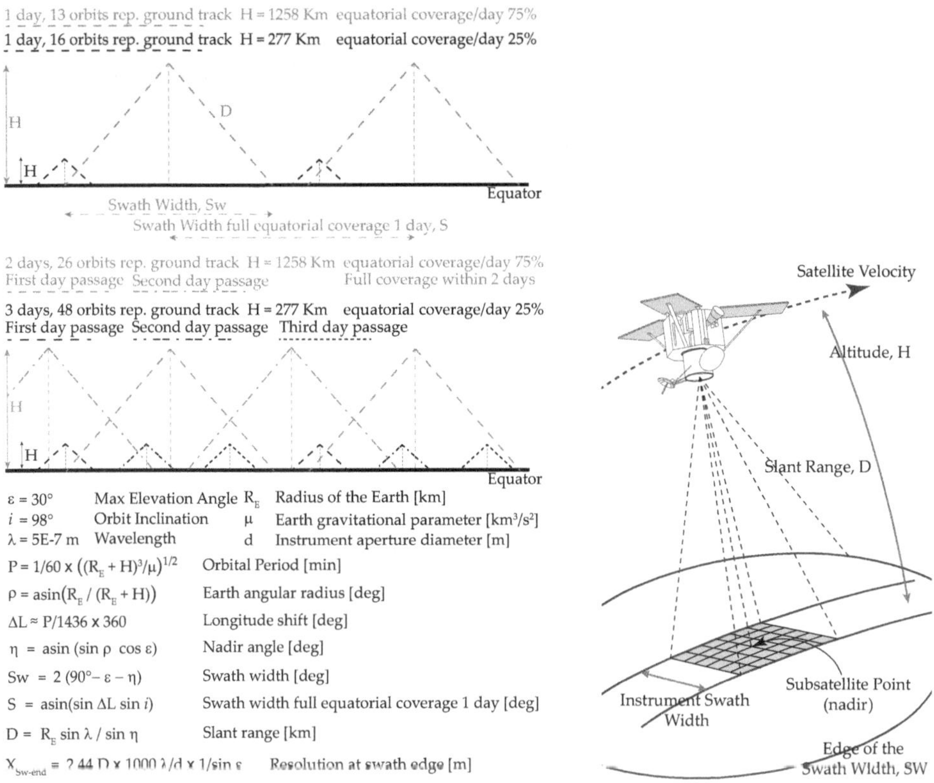

(a) Repeating ground tracks and optical-instrument resolution

(b) Satellite ground track representation and geometry on the Earth surface

Fig. 3. Satellite mission geometry. Equations adapted from (Wertz & Larson, 1999).

power and avionics system, propulsion system, structure and thermal-control system) and a ground control station model. The cost is computed using the *Unmanned Spacecraft Cost Model* presented by Wertz & Larson (1999). Cost is mostly related to the mass and power consumption of the satellite, the type of technology used (e.g., type of payload or type of attitude control), and on the technology heritage of its components (the higher the cheaper). From database, two types of solar arrays and two types of thrusters are taken into account. The two types of solar arrays present an efficiency, η, of 0.14 and 0.2, and a power density of 115 [W/kg] and 100 [W/kg] respectively. The two thrusters are the *STAR48A* and the *IUS-SRM2* with a specific impulse of 250 [s] and 300 [s], (Wertz & Larson, 1999), and a percentage of inert mass with respect to the propellant of 0.13 and 0.21, respectively. The two levels of payload heritage foresee an *adapted* design from an existing one and a *new* design, respectively. The *new* design is more expensive, but allows for a better management of the acquired data on board, i.e., reduced data rate. The results of the analysis are discussed in the following sections, for every design step and for every type of design methodology presented.

4. Sampling the design space

Sampling the design space is the first step necessary when the mathematical model of a system needs to be studied. A sample is a set of points in the design region (a $k-$ *dimensional* hyperspace) whose coordinates are the values of the design variables taken from their variability ranges, $(x_1, x_2, \cdots, x_k)$, in their marginal (for independent inputs) or joint (for correlated/coupled inputs) distribution, see the black dots in Fig. 1.

The simplest, and possibly most straightforward approach to sampling is to generate a sequence of random points in the design region, as shown in Figure 4(a). The Latin Hypercube Sampling (LHS),developed by McKay, McKay et al. (1979), is an alternative method seen as a subclass of the stratified-sampling class. The LHS provides full stratification of the design region, thus increased design-space coverage characteristics if compared to the generic stratified sampling and the random sampling, see Figure 4(b). However, good space-filling characteristics are not always guaranteed, in the sense that points in the design space may still form separate and disordered bunches. Viana et al. (2010) propose an algorithm for near-optimal Latin hypercube designs (i.e., maximizing the distance between the samples) without using formal optimization, see Figure 4(c). This method provides results with a negligible computational effort if the number of design variables k is not so large. According to our experience using this algorithm, it requires the generation of matrices with at least 2^k elements, irrespective of the number of samples actually required. The number of matrix entries to be stored to compute the near-optimal LHS can become cumbersome already for 20 variables. The Sobol LP_τ sequence,Sobol (1979), is a quasi-random sampling technique that provides *low-discrepancy* sample points. Here discrepancy indicates a measure of *non-uniformity* and proximity between the samples. In Bratley & Fox (1988) and Press et al. (2007) there are useful indications on how a Sobol LP_τ sequence, or its variant proposed by Antonov & Saleev (1979), can be computed. The (modified) Sobol LP_τ sequence has the particular characteristic of providing a sequence of points for which successive points at any stage *know* how to fill in the gaps in the previously generated distribution, Press et al. (2007), see Figure 4(d). This aspect is particularly useful for the re-utilization of previous sample points when additional points shall be sampled to improve the quality of the results, as will be demonstrated later in the case of regression analysis. The modified Sobol LP_τ sequence demonstrates that the additional sample points, the circles in Fig. 4, are placed in such a way to fill the gaps following a pre-defined pattern, allowing for a more efficient re-utilization of the samples previously generated.

4.1 Design of experiments

An experiment is a test performed to evaluate the outcome of a process given certain settings of the factors believed to influence it. The *experiments* considered in this chapter are all *computer experiments* performed on the mathematical model in correspondence of the sample points. However, the Design of Experiment (DoE) practice has older origins than the computer era, indeed it was first introduced by Fisher in 1935. The sampling methods belonging to the category of DoE can be distinguished in Factorial Designs (full or fractional), Orthogonal Arrays and other methods, amongst which, for instance, Central Composite Design (CCD). The common characteristic of these sampling methods is that they are all deterministic. The samples are placed on the design space according to a certain pre-defined geometry, so that

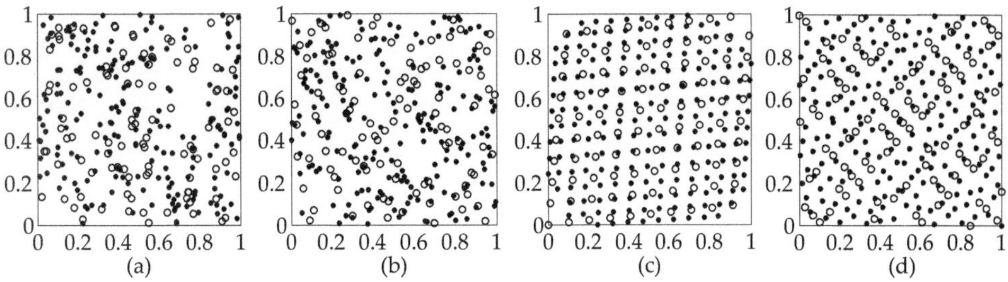

Fig. 4. Scatterplots of sampling points in a 2-dimensional design space based on (a) random sampling, (b) Latin Hypercube sampling, (c) sub-optimized Latin hypercube sampling, (Viana et al., 2010), (d) modified Sobol LP_τ sequence. • Initial sample, 100 points. ○ Additional sample, 100 points.

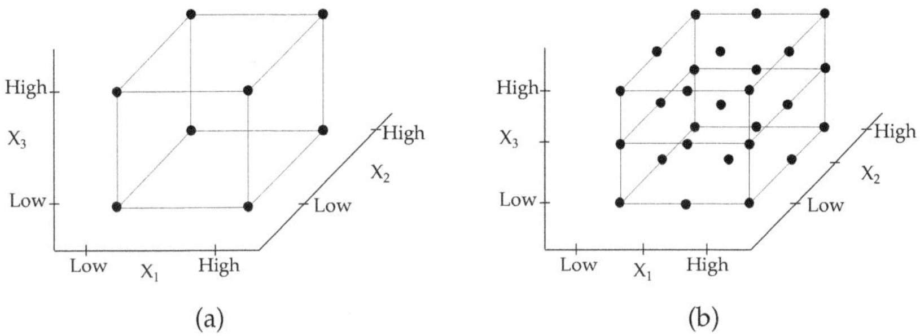

Fig. 5. Full factorial design with (a) 2 variable-levels and (b) 3 variable-levels in a 3-dimensional design space.

also *ordinal* and *categorical* variables can be used in the analysis, rather than only *cardinal* (i.e., continuous) variables as in the previously described sampling techniques. In this case the values of the variables are more properly called *levels*.

4.1.1 Factorial design

Factorial design, or *full* factorial design, is a sampling method that foresees one experiment for each possible combination of the levels of the factors. If factor A has a levels, factor B has b levels and factor C has c levels, the total number of experiments is $N = a \cdot b \cdot c$. There are special cases of factorial design where for all the factors only 2 or 3 levels are considered. They are usually called 2^k and 3^k factorial designs respectively, where k indicates the number of factors. The experimental structure obtained for 2^k and 3^k factorial designs is shown in Fig. 5 where the dots indicate the sample points.

Full-factorial design requires a number of experiments that increases with the power of the number of factors. Thus, already in the case of 2^k or 3^k factorial designs, the experimentation (i.e., the simulation of the model) can become cumbersome very soon. Therefore, fractional factorial designs were introduced as an attempt to reduce the computational effort for the analysis. As the name suggests, fractional-factorial designs only foresee a fraction of the

Experiment	Factors Assignment						
	A	B	C	D	E	F	G
1	1	1	1	1	1	1	1
2	1	1	1	2	2	2	2
3	1	2	2	1	1	2	2
4	1	2	2	2	2	1	1
5	2	1	2	1	2	1	2
6	2	1	2	2	1	2	1
7	2	2	1	1	2	2	1
8	2	2	1	2	1	1	2

Table 2. L_8 orthogonal array

number of experiments required by a full-factorial design with the same number of factors and the same number of levels. For instance a *one-half* fractional factorial design, or 2^{k-1} design, requires half of the experiments of the original 2^k design.

All the designs belonging to the category of DoE are also called matrix designs. Indeed their visualization, and their construction, is better understood if represented in the form of a matrix with the factors in the columns and the experiments to perform in the rows. A graphical structure for more than 3 variables becomes hard to visualize, see Table 2. A 2^{k-1} design is also called *Resolution 5* design (for $k > 4$). It is also possible to generate fractional-factorial designs that require less experiments than Resolution 5. However, the smaller the number of experiments, the lesser the information that can be obtained, as will be discussed in Section 5.2. Box et al. (1979) provide a thorough discussion on DoE, in general. Montgomery (2001), instead, present a complete overview of factorial designs, methods for obtaining several kinds of designs and their implications. For more detailed analysis we advise to refer to their original work.

4.1.2 Orthogonal arrays

Orthogonal Arrays, OAs, are special matrix designs originally developed by Taguchi (1987). OAs can be used as Resolution 3, Resolution 4, and Resolution 5 designs by properly arranging the columns of the design matrices, (Phadke, 1989). The term *orthogonal* is related to the balancing property, which means that for any pair of columns, all combinations of factor levels are present an equal number of times. In Table 2, the 1s indicate the *low* levels, while 2s indicate the *high* levels of the design factors.

The L_8 orthogonal array of Table 2 is only one amongst the many OAs discussed in (Taguchi, 1987). It is possible to build also three-, four-, and five-level OAs, and also mixed-levels OAs for factors having a heterogeneous number of levels, (Phadke, 1989). An efficient algorithm to generate three-level OAs is discussed by Mistree et al. (1994) while standard tables for other types of orthogonal arrays can be found in (Taguchi, 1987) and (Phadke, 1989).

4.1.3 Other experimental designs

The major distinction amongst the experimental designs is usually made between first- and second-order designs, as already hinted before. In the first case the design variables can assume only two levels, while in the second case at least three levels per design variable

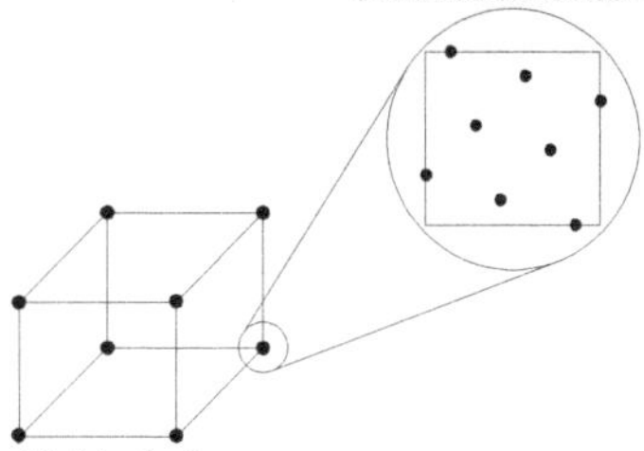

Fig. 6. Mixed-hypercube sampling with 3 discrete and 2 continuous variables.

are considered. The development of second-order designs is mainly related to the need of obtaining information on the curvature of the design space for fitting second-order response surfaces. Box et al. (1979) present a method to compute fractional 3^k factorial designs, the Box-Behnken designs, obtained by combining two-level factorial designs with balanced incomplete block designs. The Central Composite Design, CCD, introduced by Box & Wilson (1951), is build instead using a 2^k factorial design, plus a central point (in the geometric center of the design hyperspace), plus $2k$ points on the axis of every design variables at a distance α from the center. In a hyperspace normalized in the interval $[-1, 1]$, a CCD with $\alpha \neq 1$ will present 5 levels for each variables, while with $\alpha = 1$ it will only require the variables to assume 3 different levels. The interested readers may refer to Box et al. (1979) and Montgomery (2001) for a good overview and discussions on the many types of available experimental designs.

4.2 The mixed-hypercube approach

The purpose of the mixed-hypercube approach is to exploit both stratified sampling and DoE to efficiently sample the design space for obtaining information on the effect of both the continuous and the discrete design variables on the performance(s) of interest. The main idea is to separate the continuous variables and the discrete ones in two groups. A matrix design is then created for the discrete variables while for every row of the matrix design a Sobol sequence is generated for the remaining continuous variables. An example with three discrete and two continuous variables is presented in Fig. 6.

The advantage of using a matrix design instead of a space-filling technique for the discrete variables is that it allows to deterministically select the levels of the factors. When only few factor-levels can be selected (e.g., in a database there is a certain number of batteries, or only a limited number of thrusters is considered in the analysis of a satellite system) the maximum number of simulations is determined by a full factorial design. Therefore, its relative Resolution 5, 4, and 3 designs are the best way of obtaining samples by avoiding to disrupt the balance characteristics of the sampling matrix. The modification of a random or pseudo-random technique for sampling only at certain levels does not immediately provide such a balance, especially when the number of samples is kept low. On the other hand, in case of continuous variables matrix designs alone are less flexible in *filling* the design region and less suitable for the *re-sampling* process than the Sobol technique. The proposed mixed-hypercube sampling approach allows for covering the design region more uniformly when compared to all the other techniques mentioned in this section, already with a low number of samples. The sensitivity-analysis technique described in Section 5, will directly benefit from these characteristics, since convergence of the variance is obtained with a reduced computational effort, for instance. Further, response surfaces for the continuous variables, and linear and interaction graphs for the discrete ones can be directly computed from the outcome of the simulations, with no additional data manipulation, see Section 6. A more detailed description of the implications of using specific implementations of the mixed-hypercube sampling method in combination with the design approaches presented in this chapter is discussed in the following sections.

5. Sensitivity analysis

Sensitivity analysis can be defined as the study of the *effect* of a certain input x on a given output Y. This *effect* can be the result of a local measure, e.g., the measure of a derivative as for instance $(\partial Y/\partial x)_{x=x^*}$, which requires an infinitesimal variation of the input x around a specific value x^*. However, the measure of sensitivity can also be obtained when the input ranges in a specified finite interval. In this case sensitivity analysis is valid over the entire interval of variation spanned by the input factor rather than only a single point. Therefore this type of sensitivity analysis is often called *global*. The settings of the problem of designing a complex system by selecting the most appropriate combination of input-factor levels is particularly suitable for the implementation of global sensitivity analysis. Indeed, in this context sensitivity analysis is aimed at finding the set of relevant factors in the determination of the output, providing information that is valid over the entire design region, even if it represents only a (small) subset of the design space. The main design questions that can be answered by using the global sensitivity analysis technique are the following:

Amongst all the design factors of the system model, what are those actually influencing the performance of interest? To what extent do these factors influence the performance?

The answer to these questions, already at an early stage of the design, could bring several advantages to the engineering team. First, it allows to identify the *design drivers*, i.e., those factors or group of factors that shall be carefully assessed, because they will be the main responsible for determining the performance of the system. The extent of the influence identified may be useful for checking the adequacy of the model being used for the analysis and for corroborating the underlying analysis assumptions.

5.1 Sensitivity indices

The relative importance of the factors can be determined on the basis of the reduction of the (unconditional) variance of the output Y, $V(Y)$, due to fixing that factor to a certain (yet unknown) value. A global quantitative measure for the *importance* of the factors, based on their contribution to the variance of the response, was first introduced by Sobol (1993). In (Sobol, 1993) and in (Sobol, 2001), the author presents a formal demonstration of his approach and a method for computing the sensitivity indices (sometimes called Sobol indices). Consider $Y = f(X)$ as the model of interest. Y is the response vector while $X = (x_1, x_2, \cdots, x_k)$ is the vector with the k independent input factors. The method of Sobol discussed here and the regression-based sensitivity analysis described later in this section are in general valid for independent input factors. The case with correlated inputs implies that the correlation structure must be taken into account during the sampling of the design space, leading to higher computational cost, (Saltelli et al., 2004). An effective technique for imposing the correlation between input variables has been proposed by Iman & Conover (1982). However, in the case of systems design using mathematical models, dependencies between factors are very often accounted for within the model itself, leaving only the independent factors as design variables. Sometimes instead, input variables can still be considered independent if the design ranges are carefully selected. In the case of the semi-major axis and the eccentricity discussed in Section 2 one could limit the value of the eccentricity to the maximum possible with the minimum semi-major axis, for instance.

To compute the sensitivity, a sample of N points is taken from the model Y (performing N evaluations of the model Y). The unconditional variance $V(Y)$ can be decomposed as shown in Eq. (1), (Sobol, 1993). The expression in Eq. (1) is the ANOVA (Analysis Of Variance) representation of $V(Y)$, (Sobol, 2001).

$$V(Y) = \sum_i V_i + \sum_i \sum_{j>i} V_{ij} + \cdots + V_{12\cdots k} \tag{1}$$

All the terms of Eq. (1) are conditional variances of the factors indicated by the subscript indices. For instance V_i is the fraction of $V(Y)$ due to factor x_i only. V_{ij}, instead, represents the contribution of the interaction of x_i and x_j to $V(Y)$. The Sobol sensitivity indices are defined as in Eq. (2), (Sobol, 1993). S_i, S_{ij}, or $S_{i,j\cdots k}$ are sometimes called *first-order sensitivity indices*. They refer to the contribution to the variance of the single factors of Eq. (1). An additional measure of sensitivity is represented by the so-called total-effect sensitivity indices, S_{Ti}. A total-effect sensitivity index takes into account the unconditional variance of a certain variable x_i considering the first-order and all the higher-order effects in which it is involved. The total-effect sensitivity indices can be computed using Eq. (2) where V_{-i} indicates the contribution to the variance due to all factors but x_i and all the higher-order effects in which it is involved (Saltelli et al., 2004).

$$S_i = \frac{V_i}{V(Y)} \qquad\qquad S_{Ti} = 1 - \frac{V_{-i}}{V(Y)} \tag{2}$$

Global sensitivity indices can be estimated using qualitative or quantitative methods, it depends on the purpose of the analysis, on the complexity of the problem and on the available computational resources. A qualitative approach, like the method of Morris, (Morris, 1991), allows to determine the relative importance of the factors with a relatively limited computational effort. It is not possible to obtain a precise measure of the percent contribution of the factors to the unconditional variance, thus these methods are usually used as a preliminary analysis to detect and fix the unimportant factors. Therefore, qualitative methods are also called *screening methods*. Techniques like the method of Sobol, (Sobol, 1993), or the FAST (Fourier Amplitude Sensitivity Test), (Cukier et al., 1978), require a large number of model evaluations to provide quantitative sensitivity indices of the design factors, especially the terms like V_{ij}, or $V_{ij\cdots k}$. The regression-based sensitivity analysis method described in the following section provides a quantitative measure of the global sensitivity indices with a limited computational effort. The sensitivity indices computed with this method are based on the decomposition of the variance computed by a regression model, providing information on the first-order as well as on higher-order effects of the factors on the response.

5.2 Regression based sensitivity analysis

If the design region of interest is not stretched out so much, a polynomial regression model is often sufficient to accurately describe the behavior of the system. This is true for typical models of engineering systems, especially when the source of complexity is represented by the large number of elements and their interrelated behavior rather than the mathematical models of every single component. However, also when the complexity is related to the highly nonlinear and non-smooth behavior of the mathematical equations linking the design variables, in a relatively small portion of the design space a polynomial regression model is still able to describe the system and explain most (if not all) of the variability of the data.

The Regression-Based Sensitivity Analysis (RBSA) method proposed here, is general enough to be applicable to regression models of any order. However, the choice of the regression-order depends on several aspects that will be discussed throughout this section. For ease of the discussion the method will be explained using the second-order model presented in Eq. (3) as a reference.

$$Y = \beta_0 + \sum_{i=1}^{k} \beta_i x_i + \sum_{i=1}^{k} \beta_{ii} x_i^2 + \sum_{i=1}^{k-1} \sum_{j=i+1}^{k} \beta_{ij} x_i x_j \tag{3}$$

Here, β_i, β_{ii} and β_{ij} are the so-called regression coefficients that are calculated by fitting a response surface through the points sampled from the model, using the least-squares method. The estimate of the regression coefficients can be computed with the least-squares estimator, for instance:

$$\hat{\beta} = \left(\mathbf{X}\mathbf{X}'\right)^{-1} \mathbf{X}'\mathbf{Y} \tag{4}$$

The fitted model is therefore represented by the following equation:

$$\hat{\mathbf{Y}} = \mathbf{X}\hat{\beta} \tag{5}$$

Given a set of observations of a mathematical model, the variance of the data can be computed with the well-known equation:

$$\hat{V} = \frac{\sum_{i=1}^{N} \left(Y_i - E(\mathbf{Y})\right)^2}{N-1} \tag{6}$$

where $E(\mathbf{Y})$ is the expected value, or mean value, of the model output. The expression at the numerator of Eq. (6) is called sum of squares. Since in this case all the observations are taken into account we will refer to it as the total sum of squares, SS_T. The sum of squares of the regression only, instead, can be computed as follows:

$$SS_R = \sum_{i=1}^{N} \left(\hat{Y}_i - E(\mathbf{Y})\right)^2 \tag{7}$$

The SS_R, represents the portion of the total variability that can be explained by the regression model. In case the regression model perfectly fits the data then $SS_T = SS_R$. When residuals are present, in case of lack-of-fit, the portion of the total variability not explained by the regression model can be computed in the form of the error sum of squares, SS_E:

$$SS_E = \sum_{i=1}^{N} \left(Y_i - \hat{Y}_i\right)^2 \tag{8}$$

The regression sum of squares, as already mentioned, indicates how much of the observed variability is explained by the fitted model. To obtain the sensitivity indices of all factors that contribute to the total variability of the regression model, the regression sum of squares should be divided into its components, as done in Eq. (1). The main idea is to associate a sensitivity index to the additional variability calculated when a factor is added to the regression model. In Eq. (9) an alternative form of Eq. (5), combined with Eq. (4), is presented.

$$\hat{\mathbf{Y}} = \mathbf{X}\left(\mathbf{X}\mathbf{X}'\right)^{-1}\mathbf{X}'\mathbf{Y} = \mathbf{H}\mathbf{Y} \tag{9}$$

The matrix $\mathbf{X}\left(\mathbf{X}\mathbf{X}'\right)^{-1}\mathbf{X}'$ is called the *hat* matrix. It transforms the vector of the observed responses $\mathbf{Y}$ into the vector of the fitted values $\hat{\mathbf{Y}}$. Using the hat matrix, the total, regression, and error sums of squares can be expressed with the following relationships:

$$SS_T = \mathbf{Y}'\left[\mathbf{I} - \frac{1}{N}\mathbf{J}\right]\mathbf{Y} \qquad SS_R = \mathbf{Y}'\left[\mathbf{H} - \frac{1}{N}\mathbf{J}\right]\mathbf{Y} \qquad SS_E = \mathbf{Y}'\left[\mathbf{I} - \mathbf{H}\right]\mathbf{Y}$$

where $\mathbf{I}$ is a $N \times N$ identity matrix, and $\mathbf{J}$ is a $1 \times N$ vector of ones. Given these settings the RBSA is easy to compute. Let us consider a model in the form of Eq. (3) with three variables only. The compact notation Y_{full} denotes the model computed taking into account all the three factors, 2-factor interactions and quadratic terms, Eq. (10). In this notation $Y_{-x_1x_2}$ denotes the model computed excluding the factor x_1x_2., Eq. (11). The sensitivity index for the factor x_1x_2 can thus be computed as shown in Eq. (12).

$$Y_{full} = \beta_0 + \beta_1x_1 + \beta_2x_2 + \beta_3x_3 + \beta_{11}x_1^2 + \beta_{22}x_2^2 + \beta_{33}x_3^2 + \beta_{12}x_1x_2 + \beta_{13}x_1x_3 + \beta_{23}x_2x_3 \tag{10}$$

$$Y_{-x_1x_2} = \beta_0 + \beta_1x_1 + \beta_2x_2 + \beta_3x_3 + \beta_{11}x_1^2 + \beta_{22}x_2^2 + \beta_{33}x_3^2 + \beta_{13}x_1x_3 + \beta_{23}x_2x_3 \tag{11}$$

$$S_{x_1x_2} = \frac{V(Y) - V_{-x_1x_2}}{V(Y)} = \frac{SS_T - SS_R\left(Y_{-x_1x_2}\right)}{SS_T} \tag{12}$$

The conditional variance term $SS_R(\mathbf{X}_{-x_i})$ can also be computed and interpreted as the variance determined by excluding the ith design variable from the model. It is equivalent to the notation V_{-i} used before. In this case the sensitivity indices provide a measure of the total contribution of the variable x_i to the variance of the performance, considering all the interactions and higher order effects in which x_i is involved, see for instance Eq. (13) and Eq. (14). The sensitivity indices S_i are computed for all the terms of the model indicated in Eq. (3) while the total sensitivity indices S_{Ti} are computed for every design variable.

$$Y_{-x_1} = \beta_0 + \beta_2x_2 + \beta_3x_3 + \beta_{22}x_2^2 + \beta_{33}x_3^2 + \beta_{23}x_2x_3 \tag{13}$$

$$S_{Tx_1} = \frac{V(Y) - V_{-x_1}}{V(Y)} = \frac{SS_T - SS_R(Y_{-x_1})}{SS_T} \tag{14}$$

The validity of the sensitivity indices computed with RBSA depends on the lack-of-fit of the regression model with respect to the sample data. Indeed, particular attention must be paid to the ratio between the regression and the total sum of squares. If SS_R is close to SS_T, then the regression model is able to account for a large part of the output variance, and as a consequence the sensitivity indices are meaningful measures. If this is not the case, lack-of-fit is present meaning that important terms are missing from the initially assumed regression model. However, this information is still important to decide whether to proceed with sensitivity analysis anyway or to modify the initial assumption and increment the order of the regression model by adding extra terms, i.e., higher-order terms like cubic or higher order interactions. Regression models of higher order require a larger number of samples to estimate the effect of all the terms included in the model.

The minimum number of samples for building a regression model is equal to the number of factors present in the model plus one. However, we suggest to collect a set of additional samples that may vary from 4 to 6 times the number of variables to allow for the values of the SS_T and SS_R to stabilize. At first sight, this iterative approach may seem inefficient, due to the re-sampling of the design region. However, if the design space is sampled using the mixed hypercube approach presented in the previous section, the samples taken in one iteration can be efficiently re-used also for the subsequent one. For continuous variables this is demonstrated in Fig. 4. For discrete variables the possibility of reusing the previous samples to compute new results is due to the deterministic structure of a factorial design. Going from a Resolution 3, to Resolution 4, Resolution 5, or eventually to a full factorial design guarantees that the additional samples are different from the previous ones allowing to maintain the balancing structure of the matrix design.

When working with factorial design, the problem of *aliasing*, or *confounding*, is often experienced. The aliasing effect is the impossibility of discerning the effect of two different factors or interactions of factors. Observing Table 2, it is clear that the effect of factor C is equal to (is confounded with) the effect of interaction AB. In fact, column C is obtained by *xor* operation between columns A and B. In general, for a Resolution 3 design no main effects are confounded with any other main effect, but main effects are confounded with two-factors interactions (and higher order) that may also be confounded with each other. The design in Table 2 is a Resolution 3 design, for instance. For a Resolution 4 design no main effects confounded with any other main effect or with any two-factor interaction, but two-factor interactions can be confounded with each other and with higher-order interactions. Resolution 5 designs allows for experimentation with no main effect or two-factor interaction confounded with any other main effect or two-factor interaction, although two-factor interactions can be confounded with higher-order interactions, (Box et al., 1979) and (Montgomery, 2001). For this motivation, when selecting the type of matrix design for the discrete variables in the mixed-hypercube sampling approach it is necessary to match the *resolution* of the matrix design with the number of samples required to compute the desired effects. For instance, a Resolution 3 design is sufficient to compute linear effects only, more sample points are needed to take into account also the interactions (Resolution 4 and 5 for

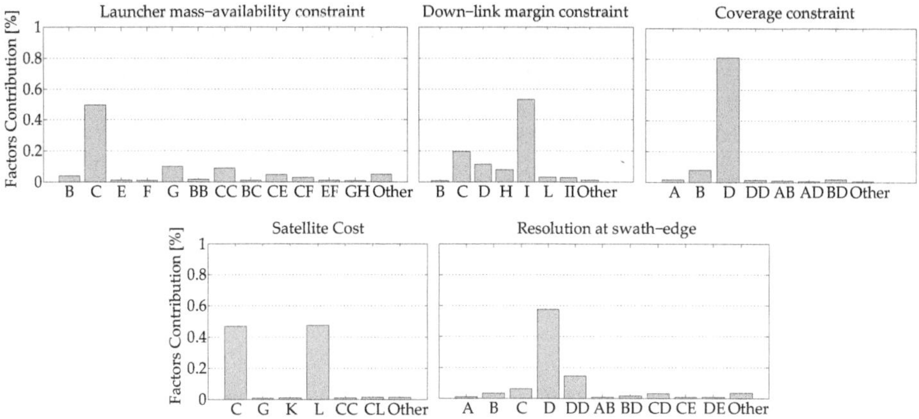

Fig. 7. Bar plots indicating the first-order sensitivity indices computed with the RBSA method.

2-factor interactions, full factorial for higher-order interactions) and, as mentioned already, even more than two levels per variable are required to estimate quadratic effects.

5.3 The Earth-observation mission, sensitivity analysis

In Fig. 7 the results from the sensitivity analysis on the model of the Earth-observation mission, computed using RBSA, are presented. The first-order sensitivity indices are visualized for the constraints (top three graphs) and the objectives (lower two graphs) discussed in Section 3. The results are obtained using a second-order model, see Eq. (3), re-sampled for additional cubic terms of the factors. Two full-factorial designs (3-level and 2-level) have been used for the discrete factors (A) and (B), and (J), (K), and (L), respectively (Table 1). For the continuous variables, instead, the Sobol sequence required 60 samples. The bars represent the percent (divided by 100) contribution of the factors indicated on the horizontal axis of the graphs, their interactions (when the product of two factors is indicated), and their quadratic effects (when the product of the factor by itself is indicated) to the variability of the constraints and the objectives. Cubic effects were limited. Their contribution and the contribution of all the other effects that are not explicitly shown in the bar plots, have been encapsulated in the bars named *Other*.

The first conclusion is that the factors (E), (F), (G), (J), and (K) have a limited effect on the objectives and constraints, probably less then one would expect since some of them are related to the propellant utilization on-board, which is usually a mass driver, thus with an effect on the cost. They can eventually be fixed to a certain level/value with a minor impact on the mission. The other design variables, instead, present contrasting behaviors. The instrument aperture diameter (factor C), for instance, affects the mass of the satellite and the satellite cost (the larger the diameter the larger the mass and the cost, reasonably) but also the down-link margin. The minimum elevation angle for the observation (factor D) has an effect on coverage (the smaller D is, the better) and on the resolution at the edge of the swath (the larger D is, the better). However, factor (D) also has some influence on the down-link margin constraint.

The effect of factors (C) and (D) on the down-link margin constraint, rather than the more obvious impact of the antenna diameter (factor I) and the transmitter RF power output (factor H), can be explained as follows. After these results were obtained, a close investigation on the model lead us to the relationship between the instrument aperture diameter and the *angular resolution*, that is related to the *pixel angular resolution*, thus to the *number of pixels* and finally to the *real-time data rate*, which causes the influence on the link margin. The elevation angle, instead, is related to the atmospheric attenuation that increases as the path to the receiver increase (so as the minimum elevation angle decrease). Many conservative assumptions were made for this applicative case. One of them is actually the fact that communication takes place with a ground station at the edge of the instrument swath width. The results of the sensitivity analysis will be used in the subsequent phase of the design methodology, as presented in the following section.

6. Graphical support to the engineering team

The information gathered during the sensitivity analysis is a roadmap for the engineering team to efficiently direct the design effort. The non-influential design factors can be fixed to a pre-determined level, because they will not affect the performance much, *de facto* reducing the dimensions of the design search-space. However, the influential design variables and the behavior of the system under the effects caused by their variation and their interactions shall be investigated in more detail. Indeed, the same samples used for sensitivity analysis can be used again to compute and present the response surfaces and the variable-trends linking the most influential design factors to the performance, in case of continuous variables. For discrete variables, linear and interaction graphs are computed and presented instead. The design questions that need an answer at this stage of the design process of a complex system are the following:

What is the shape of the design-region? What are the best parameter settings to optimize the objectives and meeting the constraints? What are the best system alternatives?

6.1 Response surfaces for continuous variables

The subject of Response Surface Methods, RSM, includes the procedures of sampling the design space, perform regression analysis, test for model adequacy and optimize the response, (Kuri & Cornell, 1996). The first three steps of the RSM are already in place, as previously discussed. The iterative approach of the RBSA, besides giving quantitative information on the sensitivity indices, also provides the regression coefficients, computed with Eq. (4), related to the best-found sample-fitting regression model. Thus, at this stage of the methodology, a surrogate model that links the design variables to the performance is available, see Eq. (5). Therefore, it is possible to visualize the trends of the objectives and the constraints as a function of the continuous design variables for each combination of discrete-variable levels. Response surfaces, and their bi-dimensional representation called contour plots, can effectively represent the shape of the subspace formed by two continuous variables. When only one continuous variable is of interest, single-variable trends are a valid alternative to contour plots.

Contour plots and single-variable trends could in principle also be computed for discrete variables, since the regression coefficients are available from the RBSA. However, the regression of a continuous function for intermediate discrete-variables levels would not be significant. To visualize the average effect of the discrete variables on the objectives and the constraints, linear and interaction graphs can be computed instead with the method shown in the following subsection.

6.2 Linear and interaction graphs for discrete variables

Consider the analysis of a system with M discrete factors $[A, B, \cdots, M]$, each with a different number of levels $[a, b, \cdots, m]$, and L continuous ones. Thus, there are $M + L = K$ design variables that form a $k-$ *dimensional* design space. Referring to Figure 6, the matrix-design for the discrete variables would be a $a \times b \times \cdots \times m$ hypercube (considering a full-factorial) while, concerning the Sobol sequence for the continuous factors, let us assume that l sample points are required for each combination of discrete design-variable levels. Once the design space has been sampled and the simulations executed, the responses of the system's model can be analyzed.

Let $Y_{...}$ represent the sum of all the responses obtained during the simulations, $Y_{...} = \sum y = \sum_{i=1}^{a} \sum_{j=1}^{b} \cdots \sum_{w=1}^{m} \sum_{s=1}^{l} y_{ij...ws}$. Let $Y_{i...}$ represent, the sum of all the responses with the factor A at level i, $Y_{i...} = \sum_{j=1}^{b} \cdots \sum_{w=1}^{m} \sum_{s=1}^{l} y_{ij...ws}$.

Considering the values of $Y_{i...}$ normalized with the number of experiments, $n = b \times \cdots \times m \times l$, for which the variable A is at level i, we compute the average value of the performance for A at level i:

$$C_{A_i} = \frac{Y_{i...}}{n} \tag{15}$$

The values of the C_{A_i} plotted against the objectives values provide the so-called linear graphs. Besides showing the trend of the objectives to the variation of a single discrete variable (with all the other variables effects averaged out), they also show the eventual presence of higher order effects, if more than two levels per factor are available from the sampling procedure. In case of ordinal discrete variables, e.g., the number of batteries in a satellite, the higher-order effects may have a certain significance indicating that the performance is not linear with increasing value of that factor. In case of categorical variables instead, e.g., the type of batteries to be implemented in the power subsystem or the type of launcher to be used for the mission, the higher-order effects are not so significant *per se* since there is no *increasing* or *decreasing* direction. This aspect have an implication on the type of matrix design selected for sampling the sub-space formed by the discrete variables only. In principle all the combinations of categorical design factors shall be experimented. Each one of these combinations represent a different system architecture that needs to be explicitly assessed. For the ordinal design factors instead, fractional-factorial designs may suffice to compute their effect on the output due to the fact that these types of variables usually have monotonic trends. However, this does not always have to be the case thus accurate matrix-design selection have to be made by the engineering team depending on the type of problem at hand.

The interaction between two discrete variables can be computed using an approach similar to that used before. For the interaction between factor A and factor B, for instance, a matrix with

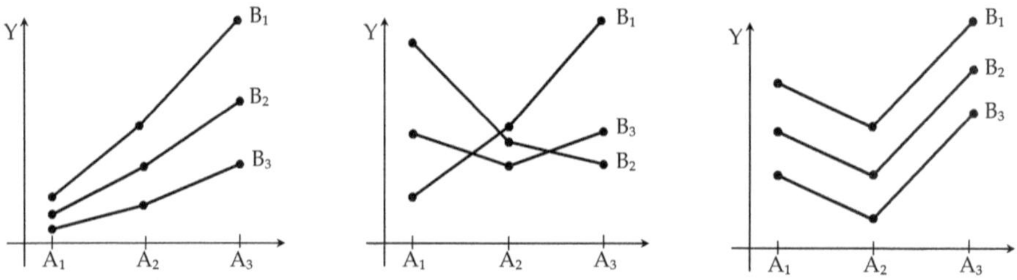

(a) Synergistic Interaction (b) Anti-synergistic Interaction (c) No interaction

Fig. 8. Interaction graphs with 2 discrete variables at 3 levels. Adapted from (Phadke, 1989).

dimensions equal to $a \times b$ is filled with the following coefficients:

$$C_{A_i B_j} = \frac{Y_{ij\cdots}}{r} \tag{16}$$

In this case $Y_{ij...}$ indicates the sum of the $r = c \times \cdots \times m \times l$ responses with the factor A at level i and factor B at level j. For each level of A, for instance, b average performance can be plotted against the objectives values, providing the so-called interaction graphs, see Fig. 8. When the lines of an interaction graph are not parallel it indicates the presence of synergistic or anti-synergistic effects, i.e., interactions. A synergistic effect is present when the improvement of a performance given the variation of a factor is enhanced by the variation of another one. An anti-synergistic effect is the exact opposite, (Phadke, 1989). In Fig. 8, the higher-order behavior of the objective to the variation of the variable levels is indicated by the fact that the lines are not perfectly straight over the three levels of variable A, for instance.

The interactions between continuous and discrete variables, eventually detected by sensitivity analysis, can be graphically presented using a mix of contour plots, or single-variable trends, and linear graphs as will be shown in the following subsection.

The synergistic utilization of the results from sensitivity analysis with the RSM and linear and interaction graphs allows the engineering team to study only on the most relevant trends, identified with the sensitivity analysis, and to more effectively select the best combination of design-variable levels. The purpose of this methodology is to support the engineering team and the decision-makers design process and trade-off analysis, and we believe that with this combination of mathematical techniques and graphical results the initial goal is accomplished. However, at this stage of the methodology, the surrogate model could also be used with automatic optimization techniques to provide an optimum (in case of single objective) or a Pareto front of optima (in case of multiple objectives) solutions. A discussion on single- or multiple-objectives optimization techniques is beyond the scope of this chapter. A vast amount of literature dealing with this topic can be found by the interested readers. Coello Coello et al. (2007) and Back et al. (2000), for instance, provide a broad overview and many references.

6.3 The Earth-observation mission, visualization of the design region

The results obtained with the sensitivity analysis in the previous section suggested that some variables influence the objectives and the constraints more than others. This allowed us to reduce the number of important graphs and to focus the attention on only a few of them. Indeed, the graphs in Fig. 9 are an alternative, more detailed and more focused, way of looking at the same data used to compute the sensitivity analysis.

In the interaction graph of Fig. 9(a) the two discrete variables related to the orbit of the satellite are considered. For all the levels of (A) and (B) the average (as previously discussed in this section) value of the equatorial coverage is plotted. The number of days for a repeating ground-track and the total number of orbits in that time period have a synergistic effect on the coverage. In particular, as expected with a higher orbit (e.g., 13 orbits in 1 day and $H = 1258.6$ km) the average equatorial coverage is larger compared to a case with a lower orbit (e.g., 29 orbits in 2 days and $H = 725.2$ km). The combinations of factors levels A1-B3 (i.e., 15 orbits in 1 day), A2-B3 (i.e., 30 orbits in 2 days), and A3-B3 (i.e., 45 orbits in 3 days) lead to the same configuration since the altitude of the orbit is the same, $H = 567.5$ km. The comparison between the performances of an A1-B1 configuration and A3-B2 configuration on the resolution at swath-edge, and on the equatorial coverage, as a function also of the minimum elevation angle and the instrument aperture diameter (factor C) is presented in Fig. 9(b). The light-gray area represents the revisit time constraint for the A3-B2 configuration, set as 100% of equatorial coverage in 24 h. The dark-gray area represents the same constraint for the A1-B1 configuration. A higher orbit (dashed lines in Fig. 9(b)) allows to meet the re-visit constraint with a larger minimum elevation angle thus also improving the resolution performance at the edge of the swath. For the A3-B2 configuration, with $\epsilon = 30^\circ$ and the instrument aperture diameter equal to 0.7 m the resolution at the edge of the swath is 12.7 m/pixel, and 1.26 m/pixel at subsatellite point. For the A1-B1 configuration, instead, the resolution at subsatellite point is slightly worse, i.e., 2.2 m/pixel, but at the edge of the swath a resolution of 7 m/pixel can be obtained. Further, for an A1-B1 configuration, the fact that the minimum elevation angle can be up to 30° gives the satellite the possibility to actually observe over the entire geometrical swath width with the maximum possible slewing angle, i.e., $(E) = 50^\circ$, and at a higher resolution than a A3-B2 configuration. The aperture diameter of the instrument, paradoxically, plays a more relevant role in the determination of the data rate, thus on the down-link margin than on the actual resolution, as demonstrated by the sensitivity analysis. Indeed, in Fig. 9(d) the down-link margin constraint is plotted as a function of the instrument aperture diameter and the minimum elevation angle, for the configuration A1-B1 and with $(H) = 30$ W and $(I) = 1$ m. An A3-B2 configuration would push the coverage constraint down, with the side result of allowing less flexibility in selecting the instrument aperture diameter. The effect on the cost is plotted in Fig. 9(c). The assumption is that a higher orbit, would require less manouvers for pointing the instrument of the satellite in one particular direction and the effect is in a reduced cost (difference between the full and the dashed lines). The constraint on the launcher mass availability is mainly driven by the instrument aperture diameter. Indeed the mass and power consumption of the payload is scaled with the diameter, and so does the mass of the satellite and its cost. The *Delta II* class of launchers allows for enough flexibility until the payload aperture diameter of about 0.9 m.

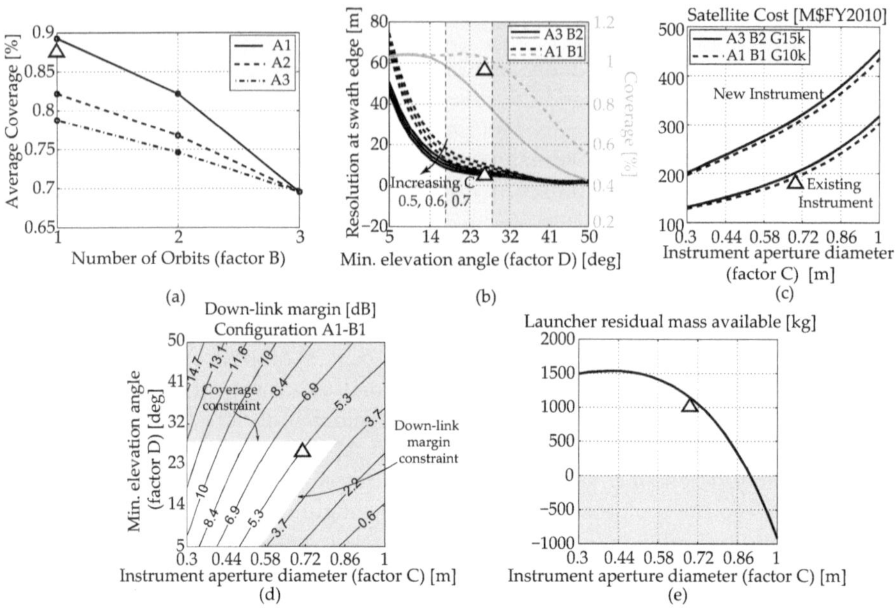

Fig. 9. Analysis main results. Δ is a tentative selected baseline. The light-gray area of (b) represents the revisit time constraint for the A3-B2 configuration, set as 100% of equatorial coverage in 24 h. The dark-gray area of (b) represents the same constraint for the A1-B1 configuration.

The triangles in Fig. 9 represent a tentative selection of the baseline. In particular an A1-B1 architecture has been selected, with $(C) = 0.7$ m, $(D) = 30°$, $(E) = 50°$, $(F) = 120$ s, $(G) = 10000$, $(H) = 30$ W, $(I) = 1$ m, $(J) = 2$, $(K) = 2$, $(L) = 1$. With these settings of the design variables a confirmation experiment was performed on the model. The simulation yield to a cost of the satellite of 188 $M\$FY2010$, a mass of 1330 kg and an overall power consumption of 1 kW. The resolution at the edge of the swath is 7.3 m/pixel and 2.2 m/pixel at sub-satellite point. The equatorial coverage after 24 h is 100% and the down-link margin is 4.1 dB. The results from the verification experiment are very close to the values that can be read from the graphs in Fig. 9. This indicates that the sampling technique and the regression analysis provided reliable results. Sensitivity analysis and graphical support in the form of contour plots, variable trends and interaction graphs enabled a thorough reasoning on the phenomena involved. This allowed us to quickly select a system baseline that meets the constraints balancing the objectives under analysis.

7. Uncertainty analysis and robust design

Uncertainty analysis and robust design are often considered complementary design activities implemented for determining the performances of the system under uncertain operating conditions. In particular, uncertainty analysis is the study of the uncertain distribution characteristics of the model output under the influence of the uncertainty distributions of the model inputs. With these settings, the purpose of uncertainty analysis is to *simply* propagate the uncertainty trough the model. When the analysis presents both controllable and uncontrollable factors, the latter being intrinsically uncertain parameters (e.g., operating environmental conditions), the purpose of the uncertainty analysis is to obtain settings of the controllable design variables that optimize the performances while at the same time minimize the impact of the uncertainties on the system. In this case we talk about robust design.

In general, uncertainty can be classified in two types, stochastic and epistemic. The stochastic or aleatory uncertainty describes the inherent variability associated with a certain phenomenon. It is usually modeled by stochastic processes when there is enough information to determine the probability distributions of the variables. The epistemic uncertainty is characterized instead by the lack of knowledge about a specific characteristic of the system. If seen in this perspective, the value of the controllable design variables and its related uncertainty can be classified as epistemic and, as discussed in the previous section, these variables are modeled as uniformly distributed between a minimum and a maximum value. However, epistemic uncertainty can also be related to uncontrollable factors for which there is too little information for determining a proper probability distribution. In this case the use of uniform distributions to characterize their uncertainty has been criticized for the main reason that a phenomenon for which there is lack of knowledge cannot be represented by any specific probability distribution (Helton et al., 2006).

For the design of a complex system, in case of both epistemic and stochastic uncertainty, probability theory alone is considered to be insufficient for a complete representation of the implications of the uncertainties on the performances. Therefore, in the following subsection we introduce a unified method for propagating the uncertainty through the model, in the presence of stochastic and epistemic uncertain factors. The main design questions we will try to answer in this section are the following:

In case of uncertainties of any type, how do they propagate through the model of the system? What are the factors that are mostly responsible for performance dispersion? How robust is the design?

7.1 The unified sampling method

In this subsection we introduce a modified implementation of the Sobol sampling technique. A Sobol sequence only allows to sample uniformly on the design space. Uniform distributions are the only necessary distributions of the design variables when the purpose of the analysis is to select a certain baseline that optimize the performances, as discussed in the previous sections. The unified sampling technique, instead, allows to cope with any type of epistemic and stochastic distributions of the uncertain factors, typical when the focus of the analysis is that of propagating the uncertainty throughout the model.

The problem of determining the probability distribution of the output, given the probability distributions of the inputs of a model is related to the computation of a multi-dimensional integral. A direct numerical integration or the analytical solution of the integral can become practically infeasible with already few uncertain variables. Therefore, the direct Monte-Carlo simulation is amongst the most widely adopted methods for uncertainty analysis, since it does not require any type of *manipulation* of the model. When it comes to long-running models, as is usually the case for complex space systems in a collaborative environment, the method of Monte Carlo, using random-sampling techniques, has the recognized disadvantage of

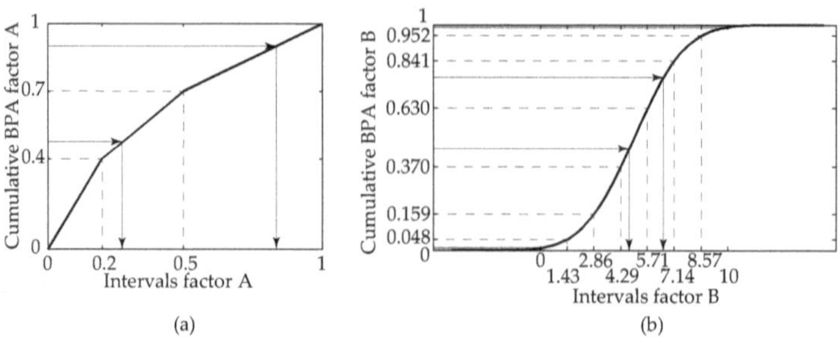

Fig. 10. Representation of the cumulative distributions of (a) the epistemic uncertain variable, and (b) the stochastic (normal) uncertain variable. The dashed lines connect the BPAs to the relative uncertainty intervals. The arrows represent the projection of the sample points from the BPSs domain to the uncertainty-intervals domain.

being computationally expensive, since it generally requires a large number of simulations to compute the mean, the variance and a precise distribution of the response (Rubinstein, 1981). Helton & Davis (2003) compare LHS with a random sampling technique for the propagation of uncertainty into mathematical models. Their analysis corroborates the original results obtained by McKay et al. (1979), and demonstrates that stratified sampling provides more stable Cumulative Distribution Functions (CDFs) of the output than random sampling, with the result that less samples are required for a given accuracy in the determination of the CDFs.

As discussed previously, also epistemic uncertainty must be considered for the design of a complex system. Thus, for the development of the unified sampling technique presented in this section we inherit some ideas and some nomenclature from the evidence theory derived from the initial work of Dempster (1967; 1968) and Shafer (1976). When lack of knowledge about a certain system behavior is present, and when the available historical and statistical sources become sparse, the engineering team is forced to evaluate and combine disparate data sources not perfectly tailored to the purpose at hand based on judgmental elements. Structured expert judgment is increasingly accepted as scientific input in quantitative models, and it is dealt in a number of publications, see for instance (Cooke, 1991) and (O 'Hagan & Oakley, 2004). The result of the combination of experts judgments on the uncertainty of a specific phenomenon leads to the creation, for every single uncertain factor, of so-called *Basic Probability Assignments*, BPAs. The BPAs represent the level of confidence that the engineering team has on the fact that the value of the factor of interest lies in a certain interval of possible values. The uncertainty interval is divided into n subsets and for each of them a certain belief, or probability, that the actual value of the uncertain parameter will lie within that subset is assigned. The set of the n beliefs form the BPA for the factor under analysis.

Consider for instance the epistemic uncertain factor (A) in Figure 10(a). The uncertainty interval of factor A is equal to $[0, 1]$, divided into 3 subsets $[0, 0.2] \cup [0.2, 0.5] \cup [0.5, 1]$. Suppose that the judgment of the engineering team-members on the uncertainty structure of factor A leads to the conclusion that the actual value of A will lie in the subset $[0, 0.2]$ with a probability equal to 0.4, in the subset $[0.2, 0.5]$ with a probability equal to 0.3 and in the subset $[0.5, 1]$ with a probability of 0.3. Thus the BPA of factor A is equal to $[0.4, 0.3, 0.3]$ and its cumulative function is reported on the y axis of Figure 10(a). The idea is to extend

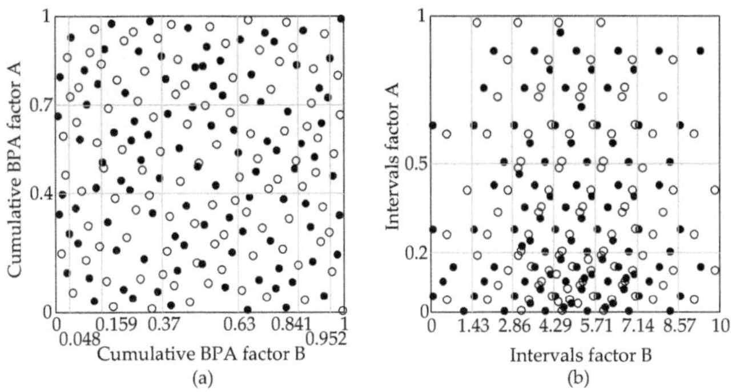

Fig. 11. Unified sampling method. Representation of (a) the uniform sampling in the BPAs domain, and (b) the corresponding sample points in the uncertainty-intervals domain.

the concept of the BPA also to the stochastic variables in such a way to obtain a unique representation of the uncertainty structure of the inputs. For a stochastic variable the cumulative distribution function is continuous. If the uncertainty interval of the stochastic factor is discretized in m subsets, then the discretized CDF can be expressed in the form of Basic Probability Assignments as in case of the epistemic uncertain factors. Consider the normally distributed uncertain factor (B) of Figure 10(b). Its uncertainty interval is equal to $[0, 10]$, divided in 7 subsets, for instance, as reported on the x axis of Figure 10(b). According to the CDF of the normal distribution, the BPAs associated to this discretization are the following $[0.0480, 0.1110, 0.2106, 0.2608, 0.2106, 0.1110, 0.0480]$. The cumulative BPAs are reported in the y axis of Figure 10(b). In the case of stochastic uncertainty, there is the possibility of having infinite tails of the distributions, as in the case of the normal one. However, if the minimum and the maximum values of the uncertainty intervals represent a high percentile, e.g., 0.95 and 0.05, or 0.99 and 0.01 (as in the case of factor A), the error is acceptably small in most of the cases. In Figure 10(b) the gray areas represent the error that arise when considering a truncated normal distribution. The probabilities of the first and the last intervals are overestimated of a quantity equal to the smallest truncation percentile (0.01 in this case).

The unified sampling method is executed in two steps. First, a uniform sampling on the space formed by the cumulative values of the BPAs is executed, Fig. 11(a). Then, each sample point within each interval in the BPA domain is mapped to the corresponding point in the uncertainty interval domain, Fig. 11(b). This passage from the BPAs domain to the uncertainty intervals domain is also represented by the arrows in Figure 10. With this procedure the final sample is collected according to the aleatory/epistemic probability distribution of the factors.

Experience and common sense tells that the more the BPA intervals, the better the approximation of the output probability distribution. However, in the case of epistemic-factor uncertainty the number of BPA intervals depends on the degree of knowledge of the engineering team on the behavior of the factors themselves. If the initial uniform sampling is performed according to a stratified technique, the resulting response CDF will be more stable than what could be obtained by using a random technique, as demonstrated by Helton & Davis (2003) and McKay et al. (1979). Further, if a Sobol sequence is implemented all the advantages already discussed in the previous chapters would still hold. This is particularly true if seen in the perspective of computing the sensitivity analysis using the RBSA, which is directly applicable if the unified sampling method is used. The computation of sensitivity analysis under uncertainty settings allows to identify the contribution of the inputs to the uncertainty in the analysis output, so to drive the effort in better describing the uncertainty of only the most relevant factors.

7.2 The Earth-observation mission, uncertainty analysis for cost, mass and power budgets

In the traditional system engineering process, design margins are used to account for technical budget uncertainties, e.g., typically for cost, mass and power. A certain percentage of the baseline's performance is added to account for both uncertainties in the model and uncertainties about eventual assumptions made at a preliminary phase that will likely be modified in advanced phases, due to an increased level of detail and knowledge. For instance, the results presented in the previous section were obtained with a 15% margin on the total satellite mass, total power consumption and propellant stored on board. The results without margins would be different. In particular, the satellite mass would be equal to 1048 kg, the power consumption euqual to 830 W and with a cost saving of 15 $M\$FY2010$. The unified sampling method allows the engineering team to obtain more insight in the uncertainty structure of the solution by focussing on every single source of uncertainty. This will enable a more informed decision-making process on the allocation of the budgets to each subsystem and each element.

In the case of the Earth-observation mission we considered the uncertain parameters and the uncertainty structure presented in Table 3. A mix of normal, log-normal and epistemic distributions has been considered. The normal and the log-normal uncertain variables are centered on the values needed to obtain the results presented before. The epistemic uncertain intervals and BPAs are determined in such a way that the value of the factors needed to obtain the previous results is at the center of the first epistemic interval. Using the unified sampling method, with 200 samples, we obtained the results shown in Fig. 12. In Fig. 12(a,b,c) the probability density estimates of the performances are presented. The histograms are plotted with an adjusted scale, so to obtain a total area of the bars equal to 1. The black and gray arrows are positioned in correspondence to the values of the performance previously computed with and without margins, respectively. It is clear that the margins approach does not provide the same insight provided by the PDFs and the histograms on the performances of the system under uncertain factors. In particular, the PDF and CDF trends shown in Fig. 12 allow the engineering team to better understand the behavior of the system under analysis, bringing two main advantages. First, the uncertainty can be allocated to single subsystems and single elements more effectively using the unified sampling method. Second, the final performance can be precisely assessed according to the desired confidence level. Further, having a precise distribution of the performances allows for more effective budget-allocation management for subsequent phases of the design process. In Fig. 12(d,e,f) the empirical cumulative distribution functions of the performances are presented. The CDF estimate, computed with 2000 samples using a random sampling method, is also represented. The fact that the empirical CDF and the CDF estimate are very close to each other corroborates the initial statement that the unified sampling method, being a stratified sampling method, is able to provide accurate results with a reduced computational effort.

Uncertain Variables	Intervals Min	Intervals Max	Distribution
Margin δV [%]	0	0.25	Epistemic[a]
Specific Impulse [s]	280	320	Normal[d]
Thrusters inert mass fraction [%]	0.2	0.4	Epistemic[b]
ADCS sens. mass [Kg]	58	70	Log-normal[e]
ADCS sens. power [W]	33	45	Log-normal[e]
Antenna mass density [Kg/m^2]	9	11.5	Normal[d]
Solar cells η [%]	0.17	0.23	Normal[d]
Solar array power dens. [W/Kg]	90	110	Normal[d]
Batteries energy dens. [$W-h/Kg$]	25	75	Normal[d]
PCU mass [Kg]	27	50	Log-normal[e]
Regulators mass [Kg]	33	55	Log-normal[e]
Thermal subs. mass [Kg]	20	50	Log-normal[e]
Struct. mass margin [%]	0	1	Epistemic[c]

Table 3. Settings of the design variables.[a]Intervals $[0, 0.04, 0.1, 0.17, 0.25]$, BPA $[0.4, 0.3, 0.2, 0.1]$. [b]Intervals $[0.2, 0.25, 0.3, 0.4]$, BPA $[0.4, 0.35, 0.25]$. [c]Intervals $[0, 0.25, 0.5, 0.75, 1]$, BPA $[0.4, 0.3, 0.2, 0.1]$.[d]$\mu = 0$ $\sigma = 1$, Min and Max are the 0.01 and 0.99 percentile respectively.[e]$\sigma = 1$, Max is the 0.99 percentile, Min corresponds to $X = 0$.

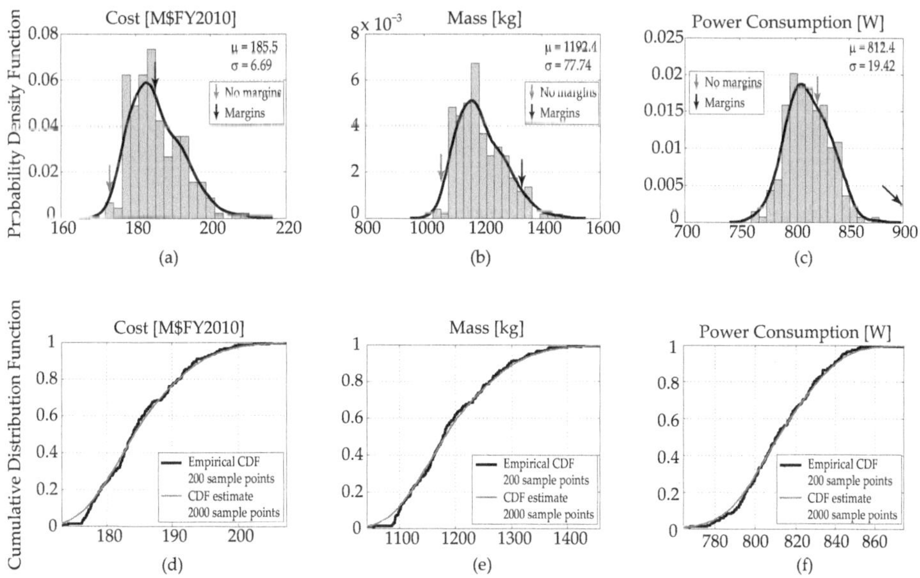

Fig. 12. Uncertainty-analysis results

7.3 Robust design and the augmented mixed hypercube approach

Robustness is a concept that can be seen from two different perspectives, at least according to the discussion so far. One can define robustness of the system with respect to the effect of uncontrollable factors (aleatory and/or epistemic) and, if interested in obtaining a robust design, one can select that combination of controllable design-factor values that minimizes the variance while optimizing the performance. This concept was already expressed in the previous section, and it is the most common way of thinking of robust design. However, robustness can also be defined as the insensitivity of a certain design baseline to modification of the design variables in subsequent phases of the design process, thus providing an intrinsic design-baseline robustness figure. The modification of the levels of the design variables is likely to happen, especially when the baseline is at an early stage of the design process (phase 0/A). In this sense, robustness can be linked to the programmatic risk encountered when modifying a set of design parameters at later stages of the design process. In the first case, instead, robustness is more related to the operational-life risk of the system (if the uncertainties derive from the operational environment, for instance).

In both cases the Mixed Hypercube approach and the unified sampling method, and the utilization of the design techniques proposed in this chapter, provide a valuable tool in the hand of the engineering team. The sampling approaches described in this chapter are summarized in Fig. 6 and Fig. 13. When the purpose of the analysis is to study the best settings of the controllable design variables to optimize the performances while meeting the constraints, the mixed hypercube approach (see Fig. 6 in conjunction with RBSA, response surfaces and linear and interaction graphs) provide a way to answer many of the most common design questions. When the purpose of the analysis is to obtain a robust design, thus studying the settings of the controllable design factors that optimize the performances while keeping the system insensitive to uncertain factors, then the Augmented Mixed Hypercube, AMH, approach shall be used, see Fig. 13(a). For every combination of the levels of the controllable design variables, an uncertainty analysis can be executed using the unified sampling method to obtain the performance of the system, and the relative statistics, due to uncertain factors. When, instead, the effect of the modification of the controllable design variables in later stages of the design process is under investigation, the general case presented in Fig. 13(b) can be implemented. The variables used to determine the baseline can be studied in perspective of their uncertain future variation. The continuous variables are more likely to be modified, since the discrete ones commonly represent different architectures of the system (whose change usually bring more radical modifications of the design, thus most likely high costs). However, in general a figure of robustness can be computed for each combination of discrete-factor levels. The approach in Fig. 13(b), without architectural variables, was also used for the budget-margins analysis presented in the previous section.

One last remark regards the possibility to use the Augmented Mixed Hypercube for a wider search. The analysis performed with the AMH, as presented in this chapter, is restricted to the portion of the design space delimited by the variability ranges of the design variables. Sometimes a single hypercube is sufficient to entirely cover the design space, sometimes instead a narrower hypercube might be needed to avoid major lack-of-fit conditions. In this case more than one hypercube may be implemented to study different regions of the design space as different alternative baselines of the system. In this case, the methodologies presented

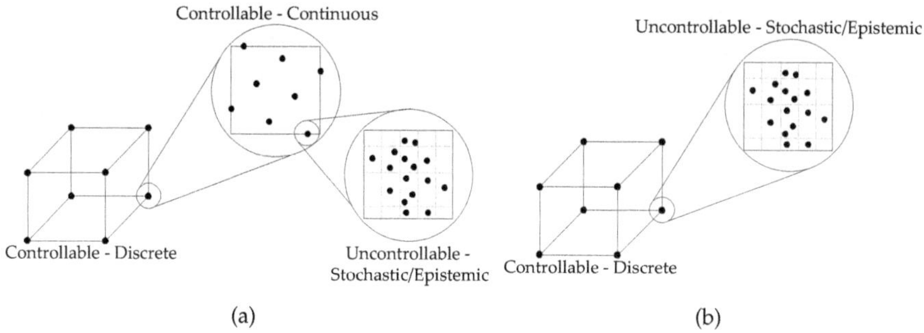

Fig. 13. Augmented Mixed Hypercube sampling procedure for robust design.

in this chapter will not only support the engineering team in selecting the best configuration for each single baseline, but will also allow to compare and trade between the baselines based on their performances, constraint-violation conditions and robustness.

8. Conclusions

Design-space exploration is the fundamental activity with which the model of a complex system is analyzed to understand the effect of the design choices on the performance(s) and to set the values of the variables in such a way that the final product will perform as required by the customer(s). This activity often involves many stakeholders, with many objectives to be balanced, many constraints and many design variables, thus posing the problem to be extremely difficult to solve with a non-structured approach. The purpose of this chapter was to discuss on subsequent analysis steps and synthesis methodologies that could serve as a guideline for exploring the design space of complex models in a standardized and possibly more efficient way. In particular, the goal was to bring fundamental analysis techniques from the discipline domain level to the system domain level. This is done to support the decision-making process and provide a unified design approach that could be implemented in a collaborative environment, in the presence of long-running models with limited time available for the design process. For this motivation, all the proposed techniques do not require any kind of manipulation of the original model and they are developed pursuing the reduction of the required computational effort as much as possible.

The Augmented Mixed Hypercube approach developed and presented step-by-step in this chapter demonstrated to be a flexible sampling method with which many fundamental design questions could be answered. The AMH is slightly more elaborated than other conventional sampling techniques but it allows the engineering team to gain a great deal of insight in the problem at hand with continuous and discrete, controllable and uncontrollable design variables with one unified method. The final baseline of the Earth-observing satellite, for instance, was selected according to a non-conventional mission architecture for an observation satellite, i.e., quite a high orbit altitude. This choice was mostly driven by the need to balance the coverage requirement and the resolution performance, while keeping the cost down. The *risk* of obtaining conventional design baselines is behind the corner when non-structured, expert-judgment driven, approaches are implemented. However, very often, especially in preliminary design phases, expert judgment is a fundamental ingredient to a good system baseline. In fact the AMH also allows to take expert-judgment into account with a unified epistemic-stochastic sampling approach. The Regression Based Sensitivity Analysis presented in this chapter, coupled with the AMH, allows to compute global variance-based sensitivity indices with a reduced computational effort if compared to other global sensitivity analysis methods. The great advantage of sensitivity analysis performed already at an early stage of the design process, as demonstrated with the Earth-observation mission, is that it could speed up the process itself providing the engineering team with an *X-ray machine* that allows to efficiently understand the effect of their design choices on the system.

We would like to close this chapter with few final thoughts on possible implementation of the methods proposed here. In case of low-complexity systems, when few variables are under analysis, and when previous experience on similar systems is present, these methods could be used as a confirmation of the expected trends, or as a proof for the analysis underlying assumptions. For complex and new systems, the implementation of the methods could reduce the engineering-team effort in exploring different solutions and architectures. In the cases where very experienced specialists are present within the engineering team (that would probably have already a clear picture of the priorities of the factors for the inherent problem), the standardized graphical approach could be a valid tool for them to explain thoughts and solutions. However, understanding performance trends in the presence of constraints and multiple objectives beforehand could also for them be a non-trivial task. On the other hand, the less experienced team members could benefit from the tool even with easy problems and expected behaviors, thus improving the overall design process, quality and effectiveness.

The contribution of the human factor is fundamental for obtaining a final product with a high cost/effectiveness value. With the integrated design approach presented in this chapter we do not mean to substitute the humans in the process of designing but, quite on the contrary, to better support their activities.

9. Acknowledgments

The authors are grateful to Ron Noomen (Section Astrodynamics and Space Missions, Faculty of Aerospace Engineering, Delft University of Technology) for many useful discussions on the test case presented in this chapter.

10. References

Antonov, I. & Saleev, V. (1979). An economic method of computing lpt sequences., *USSR Computational Mathematics and Mathematical Physics* 19(1): 252–256.

Back, T., Fogel, D. & Michalewicz, Z. (2000). *Evolutionary Computation.*, Vol. 1-2, Institute of Physics Publishing, Bristol.

Box, G., Hunter, W. & Hunter, J. (1979). *Statistics for Experimenters. An Introduction to Design, Data Analysis and Model Building*, Wiley, New York.

Box, G. & Wilson, K. (1951). On the experimental attainment of optimum conditions., *Journal of the Royal Statistical Society* 13(B): 1–45.

Bratley, P. & Fox, B. (1988). Implementing sobol's quasirandom sequence generator, *ACM Transactions on Mathematical Software* 14(1): 88–100.

Coello Coello, C., G.B, L. & Van Veldhuizen, D. (2007). *Evolutionary Algorithms for Solving Multi-Objective Problems.*, Springer Science and Business Media, New York, NY. Genetic and Evolutionary Computation Series.

Cooke, R. (1991). *Experts in Uncertainty.*, Oxford University Press, New York, NY.

Cramer, E. J., J. E. Dennis, J., Frank, P. D., Lewis, R. M. & Shubin, G. R. (1993). Problem formulation for multidisciplinary optimization, *AIAA Symposium on Multidisciplinary Design Optimization*, Houston, TX. CRPC-TR 9334.

Cukier, R., Levine, H. & Shuler, K. (1978). Nonlinear sensitivity analysis of multiparameter model systems, *Journal of Computational Physics* 1(26): 1–42.

Dempster, A. (1967). Upper and lower probability inferences based on a sample from a finite univariate population., *Biometrika* 2-3(54): 515–28.

Dempster, A. (1968). A generalization of bayesian inference., *Journal of the Royal Statistical Society* (30): 205–47.

Helton, J. & Davis, F. (2003). Latin hypercube sampling and the propagation of uncertainty in analyses of complex systems., *Reliability Engineering and System Safety* 1(81): 23–69.

Helton, J., Johnson, J., Oberkampf, W. & Sallaberry, C. (2006). Sensitivity analysis in conjunction with evidence theory representations of epistemic uncertainty., *Reliability Engineering and System Safety* 1(91): 1414–1434.

Iman, R. & Conover, W. (1982). A distribution-free approach to inducing rank correlation among input variables., *Commun. Statist. - Simula. Computa.* 3(B11): 311–334.

International Council on Systems Engineering [INCOSE] (2007). Systems engineering vision 2020, *Technical Report INCOSE-TP-2004-004-02*. Available from `http://incose.org/ProductsPubs/products/sevision2020.aspx`.

Kuri, A. & Cornell, J. (1996). *Response Surfaces. Design and Analyses*, Marcel Dekker, Inc., New York. Second Edition, Revised and Expanded.

McKay, M., Beckmank, R. & W.J., C. (1979). A comparison of three methods for selecting values of input variables from a computer code, *Technometrics* 21: 239–245.

Mistree, F., Lautenshlager, U., Erikstad, S. & Allen, J. (1994). Simulation reduction using the taguchi method., *NASA Contractor Report* (CR-93-4542): 252–256.

Montgomery, D. (2001). *Design and Analysis of Experiments*, John Wiley & Sons, New York. Fifth Edition.

Morris, M. (1991). Factorial sampling plans for preliminary computational experiments, *Technometrics* 33(2): 161–174.

O 'Hagan, A. & Oakley, J. (2004). Probability is perfect, but we can 't elicit it perfectly., *Reliability Engineering and System Safety* (85): 239–248.

Phadke, M. (1989). *Quality Engineering Using Robust Design.*, Prentice-Hall, Englewood Cliffs, NJ.

Press, W., Teukolsky, S., Vetterling, W. & Flannery, B. (2007). *Numerical Recipes. The Art of Scientific Computing*, Cambridge University Press, New York. Third Edition.

Rubinstein, R. (1981). *Simulation and the Monte Carlo method.*, Wiley, New York, NY.

Saltelli, A., Tarantola, S., Campolongo, F. & Ratto, M. (2004). *Sensitivity Analysis in Practice*, John Wiley & Sons Ltd, Chichester, West Sussex, England.

Shafer, G. (1976). *A mathematical theory of evidence*, Princeton University Press, Princeton, NJ.

Sobieszczanski-Sobieski, J. (1989). Multidisciplinary optimization for engineering systems: Achievements and potential, *Technical Report NASA Technical Memorandum 101566*, NASA Langley Research Center.

Sobieszczanski-Sobieski, J. & Haftka, R. T. (1995). Multidisciplinary aerospace design optimization: Survey of recent developments, *34th Aerospace Sciences Meeting and Exhibit*, Reno, Nevada. AIAA 96-0711.

Sobol, I. (1979). On the systematic search in a hypercube., *SIAM Journal on Numerical Analysis* 16(5): 790–793.

Sobol, I. M. (1993). Sensitivity analysis for nonlinear mathematical models, *Mathematical Modeling and Computational Experiment* 1: 407–414.

Sobol, I. M. (2001). Global sensitivity indices for nonlinear mathematical models and their monte carlo estimates, *Mathematics and Computers in Simulation* 55: 271–280.

Taguchi, G. (1987). *System of Experimental Design. Engineering Method to Optimize Quality and Minimize Costs.*, UNIPUB/Kraus International Publications, New York.

Tedford, N. P. & Martins, J. R. (2006). On the common structure of MDO problems: A comparison of architectures, *11th AIAA/ISSMO Multidisciplinary Analysis and Optimization Conference*, Portsmouth, VA. AIAA 2006-7080.

Viana, F., Venter, G. & Balabanov, V. (2010). An algorithm for fast optimal latin hypercube design of experiments, *Int. J. Numer. Meth. Engng* (82): 135–156.

Wertz, J. & Larson, W. (1999). *Space Mission Analysis and Design*, Springer, New York.

Yi, S. I., Shin, J. K. & Park, G. J. (2008). Comparison of MDO methods with mathematical examples, *Structural and Multidisciplinary Optimization* 35: 391–402.

4

Life Cycle Cost Considerations for Complex Systems

John V. Farr
United States Military Academy
USA

1. Introduction

Because of complexity and technology, the upfront costing of complex systems has become a tremendous challenge. We understand how to cost hardware and to a lesser extent software. However, we are still developing tools and processes for costing the integration and interfaces of complex systems. As we scale to larger and more complex systems, system-of-systems (SoS), and enterprises our ability to determine costs becomes less relevant and reliable. Our estimates can be off by an order of magnitude. Unfortunately, this often the result of requirements creep as much as it is our inability to translate requirements to products.

Cost estimation techniques can be divided into three categories: parametric costs estimates or PCEs, analogies, and detailed engineering builds. Figure 1 shows their applicability throughout a typical product life cycle. We chose to ignore accounting in the chapter. However, capturing expenses in a formal manner is certainly the best way to ascertain costs. Obviously, developing true costing amounts and utilizing good cost management requires good accounting practices and the tracking of expenses using activity based costing techniques. Table 1 summarizes the advantages and disadvantges of these various techniques.

In this chapter we present some of the methods, processes, tools (MPTs) and other considerations for conducting analysis, estimation and managing the life cycle costs (LCCs) of complex systems.

2. Life cycle considerations

In today's global business environment, engineers, information technology professionals and practitioners, and other related product development professionals integrate hardware, software, people, and interfaces (i.e., complex systems) to produce economically viable and innovative applications while ensuring that all pieces of the enterprise are working together. No product or services are immune from cost, performance, schedule, quality, and risks and tradeoffs. Yet engineers spend most of their formal education focused on performance and most of their professional careers worrying about resources and schedule. Too often we become fixated on the technical performance to meet the customer's expectations without worrying about the downstream costs that contribute to the total LCCs of a system. Unfortunately, in many cases the LCCs or total ownership costs (TOCs) are ignored because either the total costs would make the project untenable (especially for large government projects) or the increased acquisition costs needed to reduce the LCCs would make the project unacceptable.

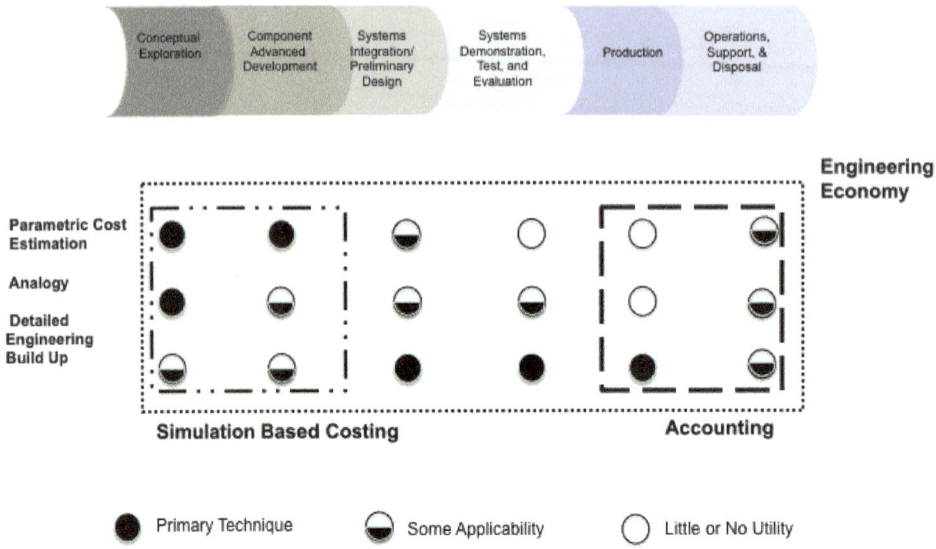

Fig. 1. Cost estimation techniques throughout the life cycle (modified from NASA, 2008)

Technique	Description	Advantages	Disadvantages
Actual Costs/ Extrapolation	Use costs spent during prototyping, hardware engineering development models and early production items to project future costs for the identical system	• Could provide detailed estimate • Reliance of actual development data	• Development data may not reflect cost correctly • Higher uncertainty • Often mistakenly use contract prices to substitute for actual cost • Various levels of detail involvement • Require existing actual production data
Analogy/ Comparative/ Case-based Reasoning	Compare available data from similar project previously completed and adjust estimates for the proposed project	• Reliance of historical data • Less complex than other methods • Save time	• Subjective/bias may be involved • Limited to mature technologies • Reliance of single data point • Hard to identify appropriate analog • Software and hardware often do not scale linearly • Not always possible to find programs of similar scope and complexity

Cost Accounting	Formulate based on the expenditures of reliability, maintainability, and decomposed component cost characteristics	• Reliance of detailed data collection	• Accounting ethics (i.e. cook the books) • Post-production phase strongly preferred • Requires of large and complex data collections • Labor intensive
Detailed Engineering Builds/Bottom-Up	Estimate directly at the decomposed component level that leads to a total combined estimate	• Most detailed at the component level through work breakdown structures • Systemic oriented • Highly accurate • High visibility of cost drivers	• Resource-intensive (time and labor) • May overlook system integration costs • Reliance of stable systems architectures and technical knowledge • Highly prone to double-counting • Lacks ability to capture economies of scale
Expert Judgment/ Delphi Method	Use human experts' knowledge and experience via iterative processes and feedbacks to general consent estimates	• Available when there are insufficient data, parametric cost relationships, or unstable system architectures	• Subjective/bias • Detail cost influence/driver may not be identified • Program complexities can make estimates less reliable • Human experience and knowledge required
Parametric/ Cost Estimating Relationship	Use mathematical expressions and historical data to generate cost relationships models via statistical and regression analysis	• Statistical predictors provide information on expected value and confidence of prediction • Less reliance of systems architectures • Less subjective	• Heavy reliance of historical data • Attributes within data may be too complex to understand • Possibly resource intensive (time and labor) • Difficult to collect data and generate correct cost relationships during cost model development • Limited by data and independent variables
Top-Down	Use the overall project characteristics as the base and generate estimates by decomposing into lower level components and life cycle phases.	• Fast and easy deployment • Minimal project detail required • Systemic oriented	• Less accurate than others • Tend to overlook lower level component details or major cost drivers • Limited detail available for justification

Table 1. Summary of LCCs estimating techniques (from Young et al., 2010)

We have an extensive array of economic techniques and tools at our disposal to predict and monitor LCCs and schedules yet overruns are commonplace and in general are the rule and not the exception; especially for large software enabled systems. Figure 2 shows some of the external and internal factors that we must tackle in conducting cost analysis and then must be addressed when managing the program in the most effective manner.

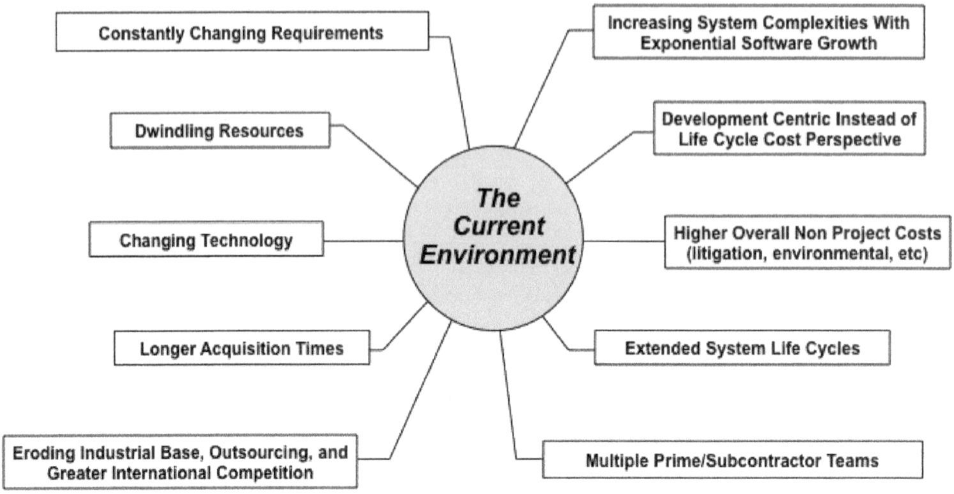

Fig. 2. Some of the factors that can affect the cost of a system (modified from Stevens Institute of Technology, 2008)

The specific purposes utilizing a LCCs perspective in acquisition management, product development, product upgrades, etc., includes:

- Estimate the TOCs to the stakeholder,
- Reduce/capture TOCs through using LCCs tradeoffs in the systems engineering/product development process,
- Control cost through using LCCs contractual provisions in procurements,
- Assist in day-to-day procurement decisions, and
- Understanding TOCs implications to determine whether to proceed to next development phase.

3. Issues surrounding complex systems

Figure 3 shows cost incurred and the ability to influence LCCs over a typical systems life cycle. The figure clearly shows the importance of upfront systems engineering and managing requirements. Because we do not allocate sufficient resources early in a program/project we often make bad engineering decisions that lead to unplanned downstream costs.

From a LCCs perspective what is even more critical is that while developing products and programming and committing funds when we simply do not have the techniques to estimate costs to a high degree of accuracy. The top down tools we used to estimate costs early in the product development cycle are gross rules of thumb at best. When combined with requirements creep, unstable funding, etc., cost estimates of ± 100% are to be expected. As shown in Figure 4 many factors can contribute to cost and schedule overruns.

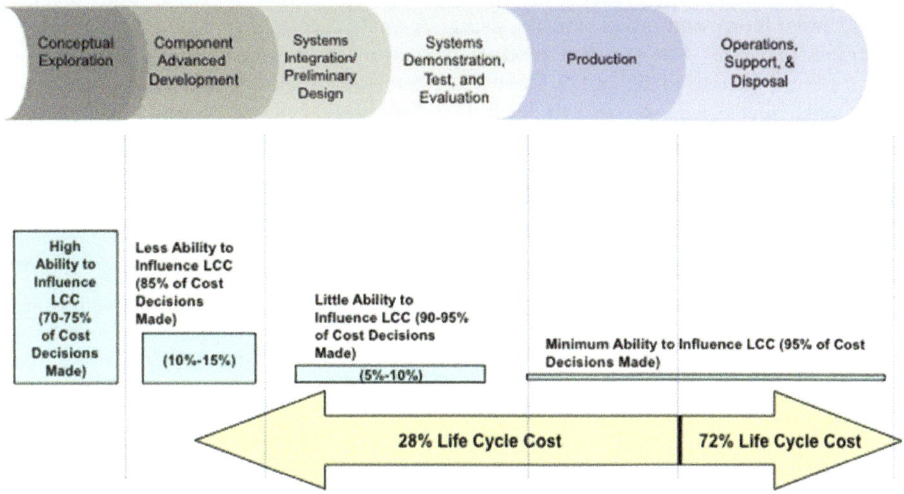

Fig. 3. Costs incurred and committed during our systems life cycle acquisition process (modified from Andrews, 2003)

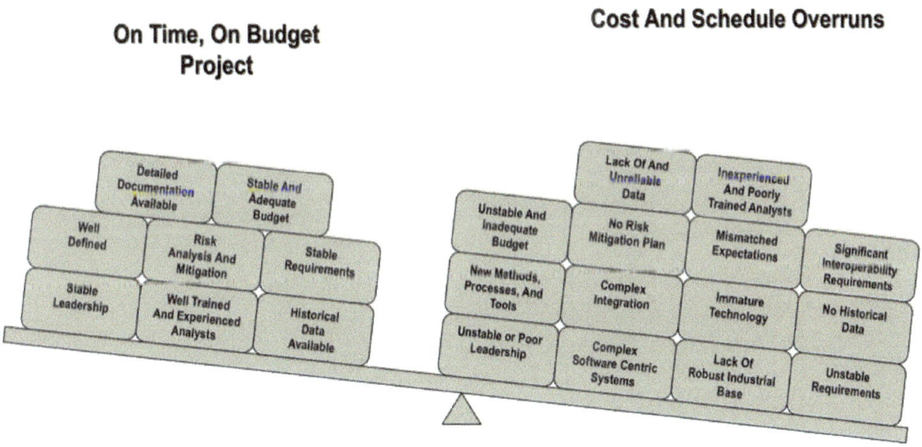

Fig. 4. Challenges cost estimators typically face (modified GAO, 2009)

The techniques for estimating systems costs vary depending upon where we are in the life cycle. Taking our seven-phase model of conceptual exploration, component advanced development; systems integration and preliminary design, systems demonstration and test and evaluation, production, and operations support and disposal, different techniques might be used to estimate costs. For example, early in conceptual exploration the only technique that might be satisfactory is some type of parametric cost estimation techniques such as Constructive Systems Engineering Cost Model (COSYSMO), which will be explained later in detail. As move further into the product development cycle (say at the end of preliminary design) estimating will be conducted using bottoms up approach/engineering build of the system. Finally, as we enter into production, we will modify our engineering bottoms up model to more accurately reflect the final design elements of hard, software, and interfaces/integration and track costs using formal accounting techniques. Table 2 demonstrates that very early in the product development cycle we simply do not know enough about the system to accurately develop costs. Unfortunately this is when budgets are allocated, bids developed, etc. In order for LCCs to become more accurate we most use software and other formal engineering tools sooner in the design.

Baseline Created	Technical Work Products From Which Estimates Are Developed	Methodologies Used to Develop Cost Estimates
Customer	Customer Requirements • Capabilities • Characteristics Concept of Operations or CONOPS	**Top Down** • Based Upon Number/Complexity of Requirements • Based Upon Number/Complexity of Scenarios • Based Upon Number/Complexity of External Interfaces **Analogous** • Estimates Based Upon Complexity of Technical Work Products Compared to Similar Complexity of Similar Projects *Estimates Are Based On Experience And Historical Data With A ±75% Accuracy*
System	System Requirements Preliminary Architecture	**Top Down** • Based Upon Number/Complexity of Requirements • Based Upon Number/Complexity of Scenarios • Based Upon Technology Maturity • Based Upon Architecture Complexity **Analogous** • Estimate Based on Complexity of Technical Work Products Against Known Projects **Bottom Up** • Estimates Based Upon Architecture *Estimates Are Based On Experience And Formal Design And Systems Engineering (SE) Tools With A ±50% Accuracy*
Component (HW, SW, Process)	Component Requirements • Hardware (HW) and Software (SW) Systems Architecture • Document All HW, SW, Processes, and Interfaces Test Architecture	**Bottoms Up** • Estimates Based Upon Architecture, Technologies Selected, Testing Plan, etc. *Estimates Are Based On Formal Design (Work Breakdown Structure, COCOMO, COSYSMO, Function Point, etc) And SE Tools With A ±10% Accuracy*
Design, Test, and Production	System Into Production HW, SW, and Processes Design and Test Strategy Service Agreements	**Bottoms Up** • Estimates Based Upon Detailed Design, Test Schedules, Implementation Details, and Other Technical Work Products • Delivered Solution Architecture *Estimates Are Detailed Bottoms Up Based Upon All Technical Work Products*

Table 2. Cost and schedule estimates as a function technical baseline work products (modified from Barker, 2008)

4. Hardware, software, systems engineering and management costs

4.1 Hardware costs

If we use a hierarchal approach (a system of systems/enterprise is composed of systems, systems are composed of subsystems, and subsystems are composed of components) any of these levels will be the building block of a bottoms-up estimate. In its simple form, hardware can be separated into physical component that comprise these building blocks plus the labor for estimating purposes. We can think if this as levels of our work breakdown structure or WBS. Note that when developing LCCs for any component of systems is to correctly develop the WBS and assigning hardware (HW), software (SW), integration, etc., for every phase.

As a first cut and if the WBS is developed correctly, we could use these categories as a way to classify costs. Unfortunately, depending upon where you are in the product life cycle we will need to adjust costs to account for technology maturity which might include readiness levels (Technology Readiness Levels or TRLs, Systems Readiness Levels or SRLs, Integration Readiness Levels or IRLs), learning curve issues, etc. NASA (2011) presents a tutorial on TRLs.

As you transition from a top down cost estimating relationship such as COSYSMO, you could use rough relations to estimate these costs over the product life cycle and refine them as the design becomes more final. The WBS and cost models developed must evolve as you move further down the life cycle.

4.2 Software

Software dominates most complex systems. The COnstructive COst Model or COCOMO family of models (see the Center for Systems and Software Engineering, 2011a) are the most widely used software estimation tools in industry. Most developers have validated models for translating lines of code in costs. The challenge for estimating software costs is translating requirements to some type of architecture/requirements to lines of code. Without experience in developing the product software and integrations costs are impossible to develop. The GAO (2009) presents a good overview of the challanges and technqiues for estimating and costing software.

4.3 Interfaces/Integration at the system level

No overarching methodology exists for costing the integration of hardware, software, and developing the interfaces. Interfaces/integration challenges are the key reason why the costs of systems scale non linearly. We know from the DoD, NASA, and other developers of large SoS problems that we do not know how to estimate their costs. The GAO (2009) summarized current major DoD procurements all had experienced significant cost and schedule growth.

4.4 Systems engineering/project management costs

One area that has received significant attention because it is often underfunded and has been connected to major cost overruns is systems engineering and project management (SE/PM). Figure 5 shows some of the SE/PM functions that comprise this category. Stem, et al (2006) reported that the average SE/PM costs for major aircraft programs had increase from 8% in the 1960s to about 16% in the 1990s of the total development costs. The SE/PM components are significant to controlling costs, schedule, and quality during product design. However, what are the SE/PM concerns post production? These also are significant for upgrades and supportability issues.

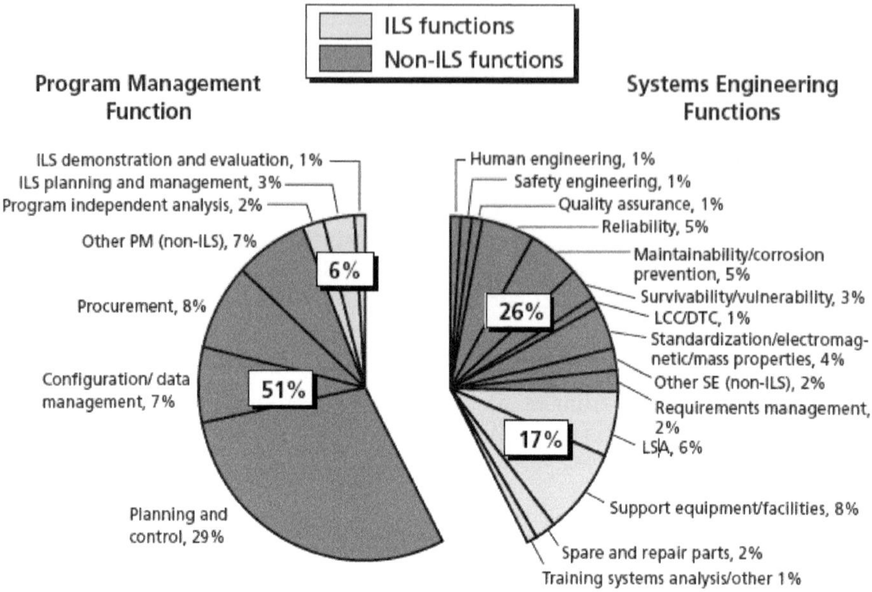

Fig. 5. SE/PM as a function of Integrated Logistics Support (ILS) for a typical Air Force program (from Stem, et al., 2006)

According to Stem et al. (2006) of Rand there is about roughly a 50/50 split of systems engineering and project management costs for most large defense programs. An as shown in Figure 6, these costs can be significant and depending upon maturity, oversight, complexity, etc., can account for about 20% of the development costs. This figure uses lot numbers across product line. Unfortunately, COSYSMO only provides a technique for estimating systems engineering cost during the development phase. Research is underway to identify quantitative means for estimating project management costs from a top down perspective (Young et al, 2011). For services based costing (SBC) to evolve this will be needed.

The COSYSMO is a model that can help people reason about the economic implications of systems engineering on projects. Similar to its predecessor, COCOMO II (Center for Systems and Software Engineering, 2011b), it was developed at the University of Southern California as a research project with the help of BAE Systems, General Dynamics, Lockheed Martin, Northrop Grumman, Raytheon, and SAIC. COSYSMO follows a parametric modeling approach used to estimate the quantity of systems engineering labor, in terms of person months, required for the conceptualization, design, test, and deployment of large-scale software and hardware projects. User objectives include the ability to make Proposal estimates, investment decisions, budget planning, project tracking, tradeoffs, risk management, strategy planning, and process improvement measurement (see Valerdi, 2005 and 2006).

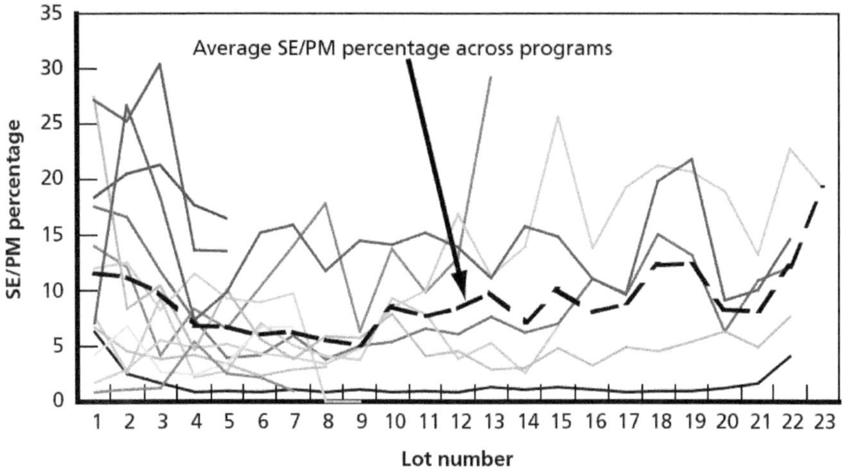

Fig. 6. Average systems engineering and project management costs for 22 major Air Force programs (from Stem et al, 2006)

Each parameter in the COSYSMO Algorithm is part of a Cost Estimating Relationships (CERs) that was defined by systems engineering experts. COSYSMO is typically expressed as (Valerdi, 2005, 2006)

$$PM_{NS} = A\left(\sum_{k}\left(\omega_{e,k}\Phi_{e,k} + \omega_{n,k}\Phi_{n,k} + \omega_{d,k}\Phi_{d,k}\right)\right)^{E} \prod_{j=1}^{14} EM_j \quad (1)$$

where:

- PM_{NS} = effort in Person Months (Nominal Schedule)
- A = calibration constant derived from historical project data
- E = represents diseconomies of scale
- k = {REQ, IF, ALG, SCN}
- w_k = weight for "easy", "nominal", or "difficult" size driver
- Φ_k = quantity of "k" size driver
- EM = effort multiplier for the *j*th cost driver. The geometric product results in an overall effort adjustment factor to the nominal effort.

The size of the system is the weighted sum of the system requirements (REQ), system interfaces (IF), algorithms (ALG), and operational scenarios (SCN) parameters and represents the additive part of the model while the EM factor is the product of the 14 effort multipliers.

Obviously there are some shortcomings to this type of approach that would be inherent in any top down model develop early in the life cycle and would include:

- The model is developed on historical data - unless you have significant experience in that domain the model should not be used; and
- Requirements are difficult to use for estimating in that it is difficult to correlate requirements and effort. COSYSMO does recognize this implicitly by distinguishing between pure and equivalent requirements.

5. Methods and tools

5.1 Engineering economy

Engineering economics/economy, is a subset of economics for application to engineering projects. Engineering economics uses relatively simple mathematical techniques to make decisions about capital projects by making comparison of various alternatives. Engineering economy techniques allows for comparisons by accounting for the time value of money. Most engineers are trained in engineering economy and it is the predominate collection of techniques that are used in support of LCCs analysis of complex systems.

Spreadsheets have dramatically changed how we conduct economic analysis of alternatives. What once involved manipulation of equations and tables can now modeled in a spreadsheet using only a few basic commands. The use of spreadsheets are ideal because

- Most problems repetitive calculations that can be expressed as simple formulas as a function of time. Note that Excel® has built in functions for most engineering economy equations.
- Sensitivity analysis is key to conducting good analysis and by properly designing a spreadsheet the parameters can be changed and plots easily developed.
- Complex models can be rapidly and easily built and are for the most part self documenting.
- The user can develop professional reports and plots using the functionality in most spreadsheets.

5.2 Simulation based costing

Systems and enterprises at the most basic level are an integrated composition of elements or sub systems governed by processes that provide a capability to satisfy a stated need or objective. Thus, simulation is an ideal way to analyze these systems. To develop a system or enterprise successfully you must first define the problem that exists, identify the mission requirements (or business drivers) of the organization(s) needing the problem to be solved, evaluate high-level CONOPS for solving the problem, select the concept that makes the most sense in light of the product or mission requirements, develop an operational concept around the selected concept, create architectures and derived requirements for the subsystems, components, and configuration items consistent with the decomposition of the system, design the integration, test and evaluation process for the parts of the system, conduct the integration and test process for the parts of the system, manufacture/assemble the parts of the system, deploy the system, train operators and maintainers, operate/maintain the system, refine the system, and finally retire the system. Simulation can play a key role during each of these phases to assess risk for operational analysis and LCCs. Simulation can be used to prototype the systems, evaluate CONOPS, and used in determining the cost and associated risk.

For simulation based costing (SBC) analysis constructive simulations are the primary analysis tool. Simulation is important for cost analysis because

- the system can be prototyped,
- the only method to model the complex interactions of sub-systems and components,
- conduct CONOPS and "what if" trade space studies, and
- using a combination of the above assess the variability/risk of a LCCs estimate.

Figure 7 demonstrates how simulation can be used throughout the life cycle to assess risk. Note how the distribution of the cost estimate (Y axis) and in the input (triangles on the X axis) both have less variability as the product/project becomes more mature and defined.

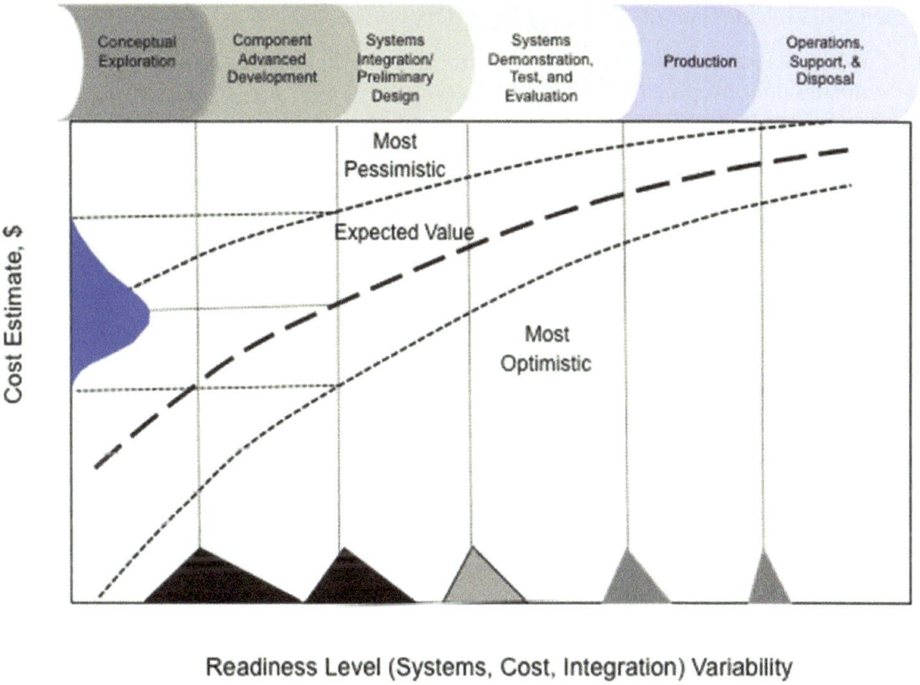

Fig. 7. Cost risk as a function of product life cycle phases

5.3 Parametric cost estimation

The following definitions are used to describe parametric cost estimation (modified from NASA, 2008 and DoD, 1995):

- Parametric Cost Estimates or PCEs - Estimate derived from statistical correlation of historic system costs with performance and/or physical attributes of the system.
- Parametric Cost Model - A mathematical representation of parametric cost estimating relationships that provides a logical and predictable correlation between the physical or functional characteristics of a system, and the resultant cost of the system. A parametric cost model is an estimating system comprising of CERs and other parametric estimating

functions, e.g., cost quantity relationships, inflation factors, staff skills, schedules, etc. Parametric cost models yield product or service costs at designated levels and may provide departmentalized breakdown of generic cost elements. A parametric cost model provides a logical and repeatable relationship between input variables and resultant costs.

- Cost Estimating Relationship or CERs - An algorithm relating the cost of an element to physical or functional characteristics of that cost element or a separate cost element; or relating the cost of one cost element to the cost of another element. CERs can be a functional relationship between one variable and another and may represent a statistical relationship between some well-defined program element and some specific cost, etc. Many costs can be related to other costs or non-cost variables in some fashion but not all such relationships can be turned into CERs.

PCEs utilizes CERs and associated mathematical algorithms, logic, processes to establish cost estimates and are probably the most widely used tool to capture experience. Figure 8 shows a process that can be used for developing CERs for PCEs. Like any mathematical based process, it should only be used for the range described by the "relationship" data.

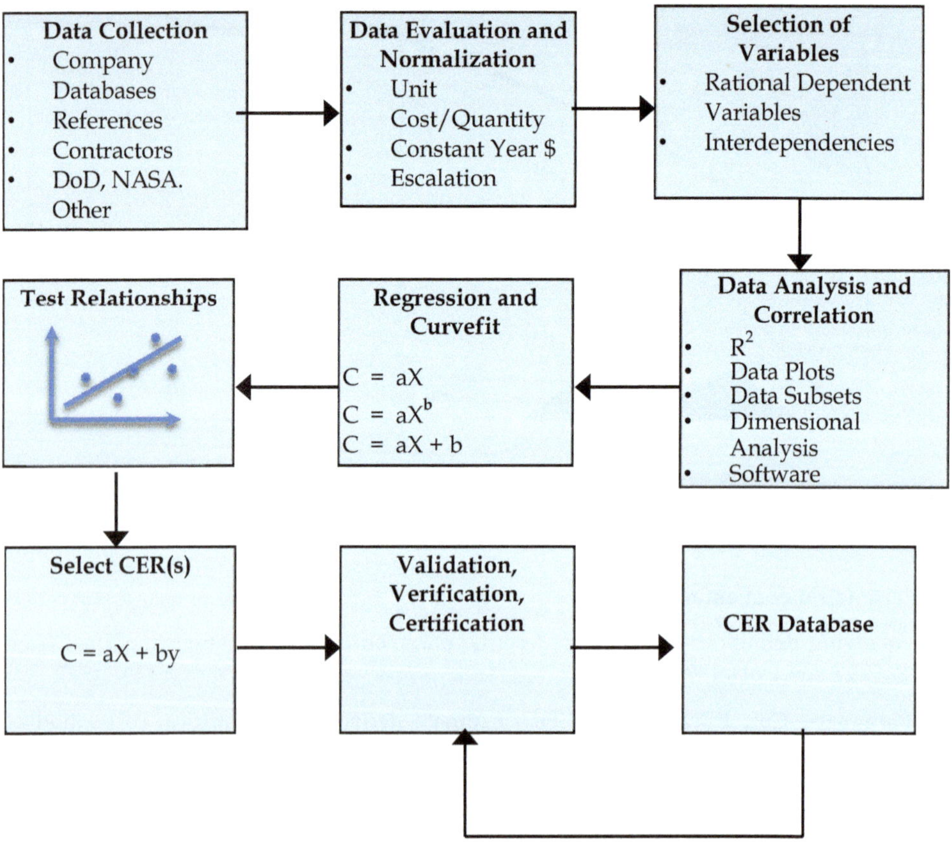

Fig. 8. Process for determining parametric cost estimates (modified from DoD, 1995)

The techniques used in estimating software are much more mature than systems. At best the tools commonly used are estimates and analogies and have little mathematical basis. Whether purely a service's centric or a physical system, most products now have significant software element. The methodology for estimating software has been around for over 30 years and can be classed as PCEs tool. However, because of new languages, hardware/software integration challenges, computer aided software tools, etc., techniques/algorithms must be continually updated. Software estimating is still dominated by experience supplement with quantitative techniques. NASA (2002) has an online handbook describing indepth parametric cost estimating.

5.4 Analogy

Analogy estimates are performed on the basis of comparison and extrapolation using like items or efforts. In many instances this can be accomplished using simple relationships or equations representative of detailed engineering builds of past projects. Obviously, this is the preferred means to conduct a cost estimate based upon past programs that is technically representative of the program to be estimated. Cost data is then subjectively adjusted upward or downward, depending upon whether the subject system is felt to be more or less complex than the analogous program (from NASA, 2008).

5.5 Engineering build or bottom up methodology

The engineering build or bottom up methodology rolls up individual estimates for each element/item/component into the overall cost estimate. This can be accomplished at the WBS element or at the component level. This costing methodology involves the computation of the cost of a WBS element by estimating at the lowest level of detail and computing quantities and levels effort to determine the total system cost. Obviously, this is the most accurate means to develop a cost estimate. The challenge is early in the systems development that a bottom's up approach cannot be utilized because the systems haven't been fully designed. Ideally, you would like to take bottom-up estimates and scale based upon experience. In order to imporve our cost estimates we must conduct bottoms-up estimating soon in the product life cycle. This requires good systems engineering to translate requirements to physical architecture.

6. From requirements to architectures

From a set a system requirements or CONOPs a functional description is developed where the system level requirements or "whats" are translated to "hows" using tools such as functional block diagrams. This functional hierarchy process and interdependencies are shown in Figure 9. The functional description provides the basis for either a physical architecture or a WBS.

7. Costing software

Almost every aspect of our modern society is controlled by software. You can look no further than the defense industry to see how dramatic and persuasive software has become. Consider the following military examples the

- F4 fighter had no digital computer and software (Early 70's),
- F16A fighter had 50 digital processors and 135 thousands of lines of code or KLOC (Late 70's),
- F16D fighter had 300 digital processors and 236 KLOC (Late 80's),
- B-2 bomber has over 200 digital processors and 5,000 KLOC (Late 90's), and
- The US Army's Future Combat Systems (FCS) will have over 16,000 to 50,000 KLOC (Late 00's).

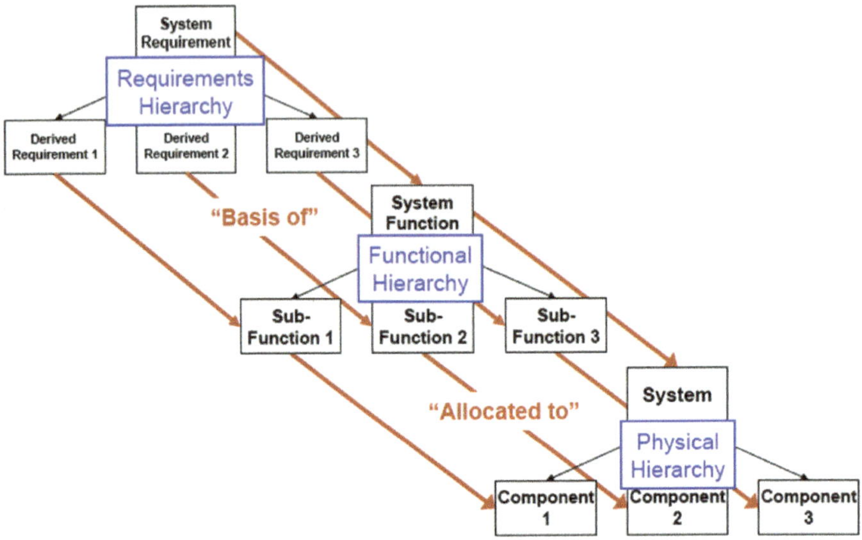

Fig. 9. Role of functional and physical views of a system (from Stevens Institute of Technology, 2009)

Software requirements growth (% of functionality provided by software) has grown from less than 10% in the 1980s to 80% in our current world (National Research Council, 2008).

Software is also redefining the consumer's world. Microprocessors embedded in today's automobiles require software to run, permitting major improvements in their performance, safety, reliability, maintainability, and fuel economy. According to Elektrobit (2007), today's high-end automobiles contain up to 70 electronic control units that control the vehicle's major functions. The average car in 1990 had one million lines of code; by 2010, the average car is expected to have up to 100 million lines of code with software and electronics contributing to over one-third of the cost of a car. New devices in the consumer electronics sector have dramatically changed how we play and manage music and conduct personal computing to extend that we manage our daily activities. As software becomes more deeply embedded in most goods and services, creating reliable and robust software is becoming an even more important challenge. Despite the pervasive use of software, and partly because of its relative immaturity especially with regards to integrating complex hardware and software applications, understanding the economics of software presents an extraordinary challenge.

Engineers typically know how to estimate hardware - we can simply count up the components. However, software and integration/interfaces continue to be the challenge in costing complex systems. Thus, we wrote this chapter to expose readers to the myriad of methods to estimate software. As you will see, historical analysis dominates software cost estimation.

Probably the most important tool in developing a software (or any) cost estimate is to develop some type of functional representation to capture all elements in the life cycle. This includes (modified from DoD, 2005):

A product-oriented family tree composed of hardware, software, services, data, and facilities. The family tree results from systems engineering efforts during the acquisition of a defense materiel item.

A WBS displays and defines the product, or products, to be developed and/or produced. It relates the elements of work to be accomplished to each other and to the end product. A WBS can be expressed down to any level of interest. However the top three levels are as far as any program or contract need go unless the items identified are high cost or high risk. Then, and only then, is it important to take the work breakdown structure to a lower level of definition.

Most models are a mix of expertise based and hybrid because of the subject nature of many of the inputs and algorithms. Expertise is nothing more than subjective human estimating combined with some simple heuristics. One large defense contractor uses the expertise and algorithm to estimate software costs:

- Estimate the number of function points based upon requirements, like projects, etc;
- Use Intermediate or COCOMO II (see Boehm et al, 2000) to estimate the resources required; and then
- Multiple the software development time by 175% to estimate costs.

This is one example of an experienced based algorithm combined with a mathematical model to produce a hybrid technique. Most companies use "rules of thumb" with hybrid techniques to estimate software development costs.

The original COCOMO is an algorithm-based model developed by Boehm (1981) and is used predicts the effort and schedule for a software product development. The model is based on inputs relating to the size of the software and a number of cost drivers that affect productivity COCOMO and drew on a study of about sixty projects with software ranging in size from 2,000 to 100,000 lines of code. Most companies even today use a modified version of one of the COCOMO family of models to estimate software development times and efforts.

The original COCOMO consists of a hierarchy of three increasingly detailed versions (modified from NASA, 2008):

- Basic COCOMO computes software development effort (and cost) as a function of program is good for quick, early, rough order of magnitude estimates of software costs;
- Intermediate COCOMO (Boehm et al, 2000) computes software development effort as function of program size and a set of "cost drivers" that include subjective assessment of product, hardware, personnel and project attributes; and

- Detailed COCOMO incorporates all characteristics of the intermediate version with an assessment of the cost driver's impact on each step (analysis, design, etc.) of the software engineering process engineering.

The basic COCOMO, which is also referred to as COCOMO 81 (Boehm, 1981), is a static model that utilizes a non-linear single valued input equation to compute software development effort (and cost) as a function of software program size. The main input into the model is estimated KDSI. The model takes the form:

$$E = aS^b \tag{2}$$

where: E = effort in person-months,
S = size of the software development in KDSI, and
a, b = values dependent on the development mode

Note that all models that COSYMO and other COCOMO based models all use this type of exponential model. Typically they all follow the form presented in Equation 2 with additional multiplicative factors.

8. Cost management

8.1 Introduction

Engineering cost management can be defined as the process to identify, allocate, and track resources needed to meet the stakeholder's requirements. An integrated, process-centered, all backed with quantifiable data and documented processes provides real and tangible benefits to all stakeholders. Engineering cost management can best be described as an integrated, process-centered, measurable, and disciplined approach to LCCs and management to make the tradeoffs between cost, performance, schedule, and risk. Good cost management practices, supported by sound analysis, can lead to (modified from NASA, 2008):

- Complete, unambiguous, and documented functional requirements in order to meet LCCs goals;
- Bounded and clearly defined product functional expectations and acceptance criteria, understood and agreed to by all stakeholders;
- More accurate, credible, and defensible scope, cost, and schedule estimates with realistic assessments of risk;
- More complete and timely risk identification, leading to more effective risk mitigation;
- A basis for properly quantifying, evaluating, and controlling the acceptance and timing of changes to requirements (i.e., precluding "scope creep");
- Final products that deliver better reliability, adaptability, usability, performance, maintainability, supportability, and functionality -- in short, higher quality and value;
- Insight into near, mid and long term technology, design, infrastructure and operational investment needs as they relate to different effects on the phases and trade-offs within the life-cycle;
- Earlier and more consistent visibility to problems (fewer surprises);
- Understanding the costs for each step in the development process;
- More efficient project management; and
- Organizational credibility and reputation.

Engineers play a critical role in corporate or business planning. Engineers are involved in cost management from top-level corporate planning to costing components and sub systems. All require the same basic understanding of time value of money, risk, and life cycle perspective.

Engineering cost management is employed as a means of balancing a project's scope and expectations of risk, quality, and technical performance to ensure that the most cost effective solution is delivered and consists of three steps:

1. Define the requirements, level of quality desired, and the budget,
2. Ensure that the risk, scope, and quality are aligned with the budget, and
3. Monitor and manage the balance of these four components throughout the life of the project by using sound engineering techniques.

The ability to use analysis techniques such as those discussed allow an engineer to conduct defendable and rigorous analysis that can not only provide representative costs but can help scope a technical problem.

One important technique to help manage costs is cost as an independent variable (CAIV). Though mainly a technique that is used solely by government, its underlying principles have utility in the commercial sector. The challenges of managing the costs of open source and off the shelve technology presents a unique costing challenge because integration not development is the key cost driver. The complexity, especially given the amount of software in most modern systems, Lastly, formal tracking using project management techniques to estimate, track, and manage costs. This is beyond the scope of this chapter but is an important for managing costs and are commonly used.

8.2 Cost as an independent variable

Cost as an Independent Variable (CAIV) is a formal methodology for reducing TOCs while maintaining performance and schedule objectives. It involves developing, setting, and refining cost objectives in a systematic method while meeting owner/user requirements. CAIV entails setting aggressive, realistic cost objectives for acquiring systems and managing program risks to obtain those objectives. Cost objectives must balance against market and budget realties with projected out-year resources, taking into account existing technologies as well as the high-confidence matriculation of new technologies (from Kaye, et al, 2000). In essence the CAIV concept means that, once the system performance and objective costs are decided (on the basis of cost-performance trade-offs), then the acquisition process will make cost more of a constraint, and less of a variable, while obtaining the needed capability of the system. Figure 10 shows this graphically.

CAIV is founded upon two primary principles. First, LCCs are constrained. Unfortunately, this is all to often limited to development and production costs. Whereas some programs do obtain additional funding when needed, such funding is often at the expense of other business units, programs, or future modernization. Second, "trade space" is the foundation for smart decisions. Trade space is the range of alternatives available to the buyers. It is four-dimensional, comprising performance, TOCs, schedule, and risk impacts (from Kaye, et al., 2000). Many of the methods presented such as SBC can be used for this trade space analysis.

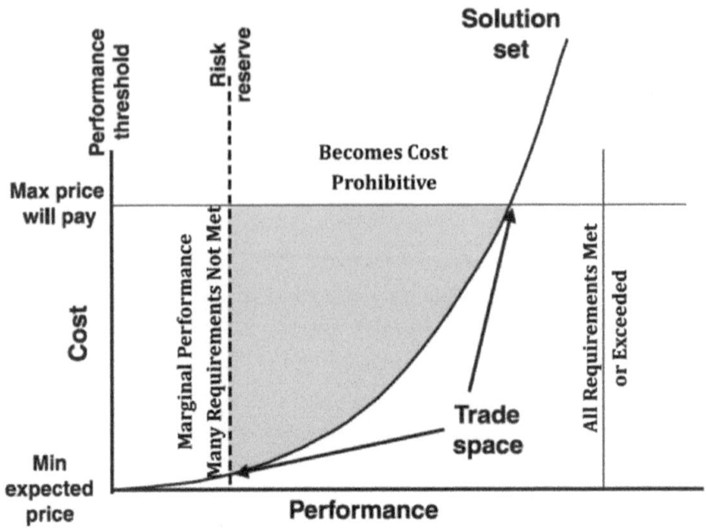

Fig. 10. CAIV representation (modified from Kaye, et al., 2000)

8.3 Formal cost accounting

Cost accounting is obviously the best way to track costs. The emergence of activity based costing techniques have made the engineers job easier when trying to ascertain true costs. Activity Based Costing (ABC) tracks costs—both direct and indirect—to their source. While traditional accounting practices have concentrated on evaluating inventory for asset based reporting, ABC links the resources consumed to the activities performed and then links these activities directly to their products. As a result, ABC provides a basis for strategic product and service pricing by capturing the direct relationships between costs, activities, and products. This is particularly useful when the primary cost factors are directly traceable to individual products or traditional direct costs. Most costs in industrial companies today are indirect, resulting, when indirect costs are uniformly allocated across products, in invalid management support information. This is particularly true in a service organization—commercial or government—attempt to use traditional inventory accounting techniques for management support will inevitably lead to inappropriate decisions.

All engineers need to understand the basics of cost accounting. As systems become more complex, the role of the engineer has diminished in terms of developing detail proposals. Most engineers now develop LCCs estimates for the system. Unless you are working at the senior management level, you do not need an in depth accounting background.

9. Summary

Costing systems is complex and consists of a variety of techniques to include analogies, PCEs, and detailed bottom-ups modeling. Unlike the mature knowledge encompassed by the traditional engineering disciplines, the techniques and tools for costing and managing complex systems are rapidly evolving and being driven mainly by the commercial sector. Also, the MPTs and techniques are often not presented in the open literature because of the competitive advantage afforded any company that can accurately estimate the LCCs of a product. Thus much of the MPTs presented were gleamed from government sources especially the DoD the National Aeronautical and Space Administration. Fortunately, the DoD and NASA are in many ways the intellectual thought leader on costing and estimating of complex systems because of the sheer size and complexity of their projects/programs. There is probably no one size fits or collect of MPTs, and certainty no substitution for experience, that are repeatable for LCCs estimation. However, much research, especially for techniques applicable early in the life cycle, is needed to better ascertain true LCCs.

Good engineers follow a disciplined and structured approach when developing a product/system. Costing hardware, software, and integration requires an understanding of many MPTs and terminology that few engineers have received formal training. Once technical characteristics have been ascertained from the requirements, selecting the right MPTs is critical to accurately determining costs early in the development cycle and estimating realistic LCCs.

In the evaluation and reengineering of existing systems, the functional analysis serves as a basis for developing WBS or CBS leading to the collection of costs by functional area. Unfortunately, if you can develop architectures/WBS you have a well-understood system suitable for realistic costs estimates which is often long after a budget has been establish.

10. References

Andrews, Richard, "An Overview of Acquisition Logistics," Fort Belvoir, VA: Defense Acquisition University, 2003, available online at https://acc.dau.mil/CommunityBrowser.aspx?id=32720, accessed on April 2, 2007

Barker, Bruce, personal note, April, 2008

Boehm, Barry, Software Engineering Economics, Prentice-Hall, 1981

Boehm, Barry, Abts, Chris, A. Brown, Winsor, Chulani, Sunita, Clark, Bradford K., Horowitz, Ellis, Madachy, Ray, Reifer, Donald J. and Steece, Bert, Software Cost Estimation with COCOMO II, Englewood Prentice-Hall, 2000

Center for Systems and Software Engineering, COCOMO Tools, University of Southern California, 2011, available onine at http://csse.usc.edu/csse/tools/, accessed on 19 August 2011a

Center for Systems and Software Engineering, COCOMO II, University of Southern California, 2011, available online at http://sunset.usc.edu/csse/research/COCOMOII/cocomo_main.html, accessed on 19 August 2011b

Department of Defense, "Parametric Cost Estimating Handbook," Joint Government/Industry Initiative, Fall, 1995

Department of Defense Handbook, "Work Breakdown Structure for Defense Material Items," available onine at http://www.acq.osd.mil/pm/currentpolicy/wbs/MIL_HDBK-881A/MILHDBK881A/WebHelp3/MIL-HDBK-881A%20FOR%20PUBLICATION%20FINAL% 2009AUG05.pdf, accessed on 30 July, 2005

Elektrobit, "Freescale and Elektrobit IntroduceAutosar Sfotware Bundle for Automotive Microcontrollers," available onine at http://www.elektrobit.com/news-987-195-freescale_and_elektrobit_introduce_autosar_software_bundle_for_automotive_microcontrollers, 10 October 2007, accessed on 19 August 2011

Government Accounting Office (GAO), "Cost Estimating and Assessment Guide Best Practices for Developing and Managing Capital Program Costs," GAO-09-3SP, March 2009, available online at http://www.gao.gov/new.items/d093sp.pdf, accessed 19 August 2011

IBM, "Software Estimation, Enterprise-Wide, Part I: Reasons and Means," available onine at http://www.ibm.com/developerworks/rational/library/jun07/temnenco/index.html, accessed November 18, 2008

Kaye, M. A., Sobota, M. S., Graham, D. R., and Gotwald, A. L., 2000, "Cost as an Independent Variable: Principles and Implementation," Available onine at http://www.dau.mil/pubs/arq/2000arq/kaye.pdf, from the Acquisition Review Quarterly, Fall, 2000, accessed January 2008

National Aeronautics Space Administration, "Cost Estimating Handbook," Available onine at www.nasa.gov/ceh_2008/2008.htm, accessed 28 January 2010

National Aeronautics Space Administration, "Parametric Cost Estimating Handbook," Available onine at http://cost.jsc.nasa.gov/PCEHHTML/pceh.htm, 2009, accessed 19 August 2011

National Aeronautics Space Administration, Technology Readiness Levels Demystified , Available online at http://www.nasa.gov/topics/aeronautics/features/trl_demystified.html, 20 August 2010, accessed 19 August 2011

National Research Council of the National Academies, Air For"Pre-Milestone A Systems Engineering - A Retrospective Review and Benefits for Future Air Force Systems Acquisition," Air Force Studies Board, 2008

Stem, David E., Boito, Michael and Younossi, Obaid, "Systems Engineering and Program Management - Trends and Costs for Aircraft and Guided Weapons Programs," ISBN 0-8330-3872-9, Published by the RAND Corporation, Santa Monica, CA, 2006

Stevens Institute of Technology, "SYS 625 Fundamentals of Systems Engineering Class Notes," 2008

Stevens Institute of Technology, "SYS 650 System Architecture and Design," Course Notes, 2009

Valerdi, Ricardo, "The Constructive Systems Engineering Costing Model (COSYSMO)," A Dissertation Submitted in Partial Fulfillment of the Requirements for the Degree of Doctor of Philosophy, University of Southern California, August 2005, Available onine at http://csse.usc.edu/csse/TECHRPTS/PhD_Dissertations/files/Valerdi_Dissertation.pdf, accessed 19 August 2011

Valerdi, Ricardo, "Academic COSYSMO User Manual - A Practical Guide for Industry and Government," Version 1.1, MIT Lean Aerospace Initiative, September 2006

Young, Leone Z., Farr, John V., Valerdi, Ricardo, and Kwak, Young Hoon, "A Framework for Evaluating Life Cycle Project Management Costs on Systems Centric Projects,"31th Annual American Society of Engineering Management Conference, Rogers, AK, October, 2010

Young, Leone Z., Wade, Jon, Farr, John V., Valerdi, Ricardo, and Kwak, Young Hoon, "An Approach to Estimate the Life Cycle Cost and Effort of Project Management for Systems Centric Projects," International Society of Parametric Analysts (ISPA) and the Society of Cost Estimating and Analysis (SCEA)2011 ISPA/SCEA Joint Annual Conference and Training Workshop, Albuquerque, New Mexico, June 7 - 10, 2011

5

Integrated Product Service Engineering – Factors Influencing Environmental Performance

Sofia Lingegård, Tomohiko Sakao* and Mattias Lindahl
Department of Management and Engineering, Linköping University
Sweden

1. Introduction

In society today there is increased awareness about environmental problems, e.g. climate change and pollution. This, in combination with concern about future shortages of natural resources, has resulted in increased pressure to find innovative strategies that can tackle these problems. Simply put, the main reasons for these problems are tied to society's use of products, and in general caused by:

- Number of products used - the growing population poses a need for an increasing number of products.
- Time products are used - the average time a product is used before it is scrapped has decreased. There are several reasons for this, e.g. quickly-changing needs and poor quality.
- How materials and energy are consumed for a product - in general, the material and energy invested for a product is not re-used or is used in an inefficient way.

Clearly, strategies for tackling these problems need to be investigated. During the last two decades, industry and academia have proposed and tried to implement several strategies and solutions. From academia, these include Functional Economy (Stahel 1994) and the Integrated Product Service Engineering (IPSE) concept, also often called Product/Service Systems (PSS) (e.g. (Mont 2002; Tukker and Tischner 2006; Sakao and Lindahl 2009)). PSS is defined, for instance, as "a marketable set of products and services capable of jointly fulfilling a user's needs" (Goedkoop, van Halen et al. 1999). Service in this chapter includes operation, maintenance, repair, upgrade, take-back, and consultation. In addition to this definition, other authors (Tukker and Tischner 2006) regard PSS as a value proposition, one including its network and infrastructure. Another concept, named Total Care Products (Functional Products), has been developed as well with some connection to PSS. It comprises "combinations of hardware and support services". The economically efficient functioning of this concept should be achieved by the proposition of an "intimate business relationship" between the service provider and the customer. As a result, both the provider and the customer obtain benefits through sharing existing business risks (Alonso-Rasgado, Thompson et al. 2004; Alonso-Rasgado and Thompson 2006). Furthermore, the proposal of a "life cycle-oriented design" (Aurich, Fuchs et al. 2006) highlights an important step for the

*Corresponding Author

"product and technical service design processes" integration. It is also interesting that Aurich et al. address designing products and services based on life cycle thinking. Furthermore, some specific engineering procedures and computer tools have been developed and validated with industrial cases (e.g. (Sakao and Shimomura 2007; Sakao, Birkhofer et al. 2009; Sakao, Shimomura et al. 2009)).

However, the research in this area is still in its infancy and a number of questions remain unanswered. Specifically, a general weakness in existing literature is that even though a large number of authors have stressed PSS' environmental and economic potential (e.g. (Roy 2000; Mont, Singhal et al. 2006)), very few studies have proved PSS' potential for changing environmental performance.

In the manufacturing industry, the trend of servicizing has been evident regardless of the environmental concern or the academic debate (e.g. (Sakao, Napolitano et al. 2008)). In much of the manufacturing industry today, numerous companies' business offerings are a combination of physical products and services. In fact, over 50% of the companies in the USA and Finland provide both physical products and services (Neely 2007). Some manufacturing firms are even strategically shifting from being a "product seller" towards becoming a "service provider" (Oliva and Kallenberg 2003). Namely, the industry possesses a driver for service integration, something which should be seen as an interesting opportunity for academia (Isaksson, Larsson et al. 2009).

As explained above, PSS is a likely solution for environmental problems from the theoretical and practical viewpoints. However, little is known scientifically about PSS' impact on environmental performance. It is the research community who should respond to this lack of knowledge, and this is the overall subject of this chapter.

There are two main questions to consider. One is under which conditions PSS is a suitable offering, since it is a prerequisite for PSS to work in business practice in order to realize its influence on environmental performance. In general, PSS approaches seem to work well if any of the following conditions apply (Tukker and Tischner 2006):

- products with high costs to operate and/or maintain;
- complex products that require special competencies to design, operate, manage and/or maintain;
- products with considerable consequences or costs if not used correctly or appropriately;
- products where operational failure or downtime is not tolerated;
- products with long life; or
- products with only a few major customers on the market.

In addition, recent research has reported on characteristics of products suitable for PSS. For instance, (Lay, Copani et al. 2010) argue that the innovativeness of products has positive influences on the integration of product and service. Theoretical investigation has also begun: For instance, property rights (Furubotn and Pejovich 1972) have gained attention as a key for PSS to be meaningful (Hockerts 2008; Dill, Birkhofer et al. 2011). Yet, all these literature are insufficient, especially from scientific viewpoints.

The other main question is which PSS factors influence the environmental performance in comparison with traditional product-sales type business. (Tukker 2004) is one of very few who have attempted to analyze the relation between PSS types and their influence on environmental impact, yet he fails to present a thorough background and reasons.

In sum, thus far there has been growing interest in PSS. Among other things, there has been relatively more work with the analytical approach (e.g. (Mont 2002)), and less work with PSS synthesis (e.g. (Sakao and Lindahl 2009)). Even with relatively more work available on analysis, there is analysis to be conducted as to PSS' factors making PSS meaningful as a business and influential on environmental impacts. This PSS with a certain level of complexity is believed to be a good example of areas where Systems Engineering (Lindemann 2011) can contribute.

2. Objective and method

This chapter endeavours to lead the scientific discussion regarding which IPSE factors are expected to, in theory, lower the environmental impact of a life cycle compared to a traditional product sales business. To do so, the IPSE concept is introduced, first with an emphasis on engineering processes rather than an object such as PSS. In the following sections, four aspects from theory will be discussed: product development, information asymmetry, economies of scale, and risk. These sections discuss how environmental impacts are influenced from a product life cycle perspective, and highlight crucial factors theoretically. They are followed by an overall discussion and an examination of some promising future work. The chapter provides the research community with a first theoretical cornerstone regarding environmental performance by IPSE. To practitioners, it will be an eye opener for how they engineer.

3. Redefining IPSE

Our research group at Linköping University and KTH (The Royal Institute of Technology) in Sweden has developed what is termed Integrated Product Service Engineering (IPSE) (Lindahl, Sundin et al. 2006). IPSE has the following characteristics in relation to other existing concepts. First, and in common with PSS, IPSE looks at combinations of products and services. Second, IPSE is a type of engineering, which is different from PSS per se. In addition, it attempts holistic optimization from the environmental and economic perspectives throughout the life cycle. Third, IPSE consists not only of design as the most influential activity, but possibly other engineering activities such as maintenance, upgrade, remanufacturing, etc. Therefore, IPSE has to deal with the time dimension of the life cycle. Figure 1 depicts different interesting processes for IPSE, obviously showing various disciplines and different aspects to be addressed.

This section reveals additional characteristics of IPSE. An IPSO (Integrated Product Service Offering) is an offering that consists of a combination of products and services that, based on a life cycle perspective, have been integrated to fit targeted customer needs. Further, IPSO means that products and services have been developed in parallel and are mutually adapted to operate well together. This contrasts with the traditional product sale, where the provider transfers control and responsibility to the customer at the point of sales. An IPSO often creates close contact between the supplier and customer, leading e.g. to offers being customized and improved to better suit the customer. In many cases, the service provider retains responsibility for the physical products in the IPSO during the use phase. One example is when a client does not own the machines installed by the supplier, but only uses them and pays for the manufactured volumes; then, when the customer does not need them anymore, the supplier takes back the machines. Such cases increase the provider's interest to ensure that the customer uses machines installed as long as possible and that any disturbances, such as the need for repairs, are reduced. The increased responsibility by the IPSO supplier also potentially facilitates improvements identified and implemented in comparison to traditional sales. This could lead to a product lifetime extension.

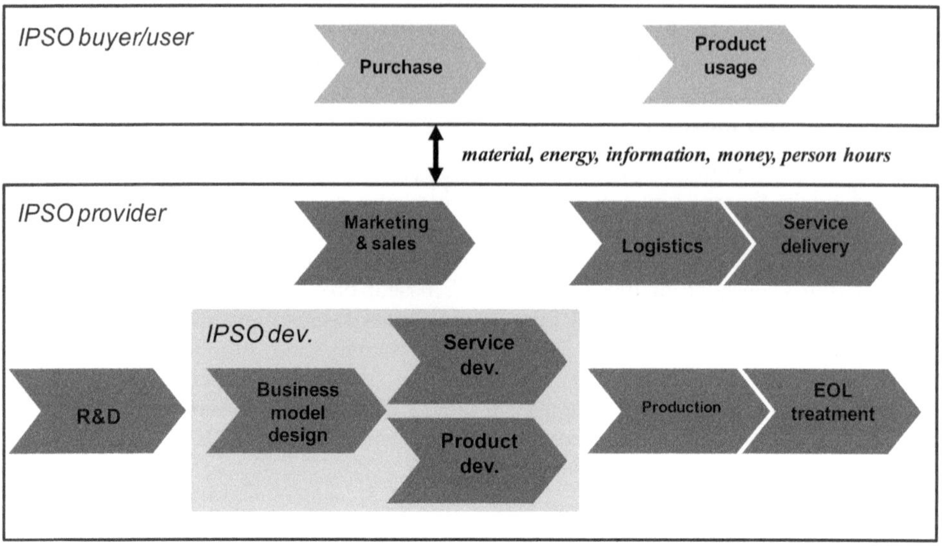

Note: IPSO; Integrated Product Service Offering. EOL; end of life.

Fig. 1. Processes of IPSE's interest (Sakao, Berggren et al. 2011)

Based on (Sakao 2009), IPSE is explained in comparison to Ecodesign (environmentally conscious design) due to some commonality with Figure 2 (a) and (b), where different types of engineering activities are put on the identical graph. The graph depicts the environmental impact of a certain type of product with high impact from its usage phase, which holds true in many cases. The horizontal axis represents the time dimension on the life cycle. Bars represent the environmental impact from each phase such as production and usage (scaled with the left vertical axis). A dotted line represents the accumulated influence of the activity at each phase of the life cycle's environmental impact. It is shown that the design phase has by far the highest ratio (some 80%), which is generally known.

As seen by the dotted line, Ecodesign is obviously crucial, since it is the design activity with the dominant influence. However, is Ecodesign sufficient? The answer is no, since it leaves out control after the design phase. This is why IPSE is more effective, including the possible employment of other engineering activities such as maintenance. Naturally, company management must be committed if they are to carry out IPSE. IPSE includes a business issue, e.g. how to sell services.

What characteristics of IPSE are to be paid particular attention to in this chapter? The first is its length on the time dimension. It can be as long as 20 - 30 years in the case of an investment machine (e.g. aircraft engine) or facility (e.g. railway). Therefore, IPSE has to address much of this dimension with the fact that the earlier a certain action is taken the more effective its outcome is in general. It is actually realized by effective design. Thus, design is naturally a core of IPSE.

Then, what is design? A seminal work by (Pahl and Beitz 1996) states "design is an engineering activity that … provides the prerequisites for the physical realization of solution ideas" (originally in (Martyrer 1960)). It has a lot to do with the processing of information - information about needs and wants from stakeholders and through the product life cycle, as well as about function and structure of the product. Effective processing of information plays a central role in IPSE - this is the second characteristic.

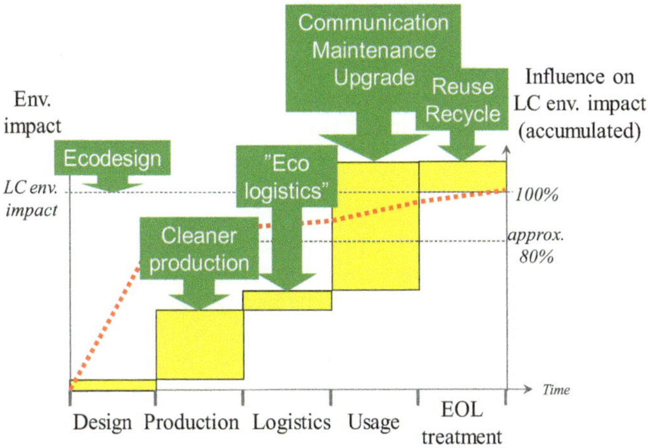

(a) Various Eco-activities

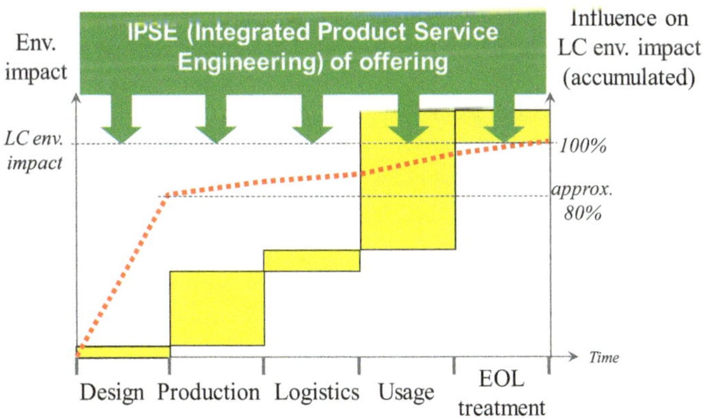

(b) IPSE

Note: The environmental impact (shown by bars) is a rough estimation of active products. EOL and LC stand for end-of-life and life cycle, respectively.

Fig. 2. Comparison of IPSE and other activities.

Then, design of what? This is the next relevant question as discussed in (Cantamessa 2011), which points out an artefact, i.e. an object to be designed, is today "integrated and systemic product-services linked in a high-level user experience". Also acknowledging co-creation of value by a provider and a customer/user is a strong idea behind the servicizing (see e.g. (Vargo and Lusch 2004)), a provider cannot get rid of influence from its customer/user to create the intended value. Thus, a provider can design something contributing to its value, but cannot design the value itself. This means that control of risks of the value creation process is crucial. Thus, this risk is the third characteristics.

In sum, IPSE can be defined as an engineering activity controlling risks of value creation through dealing with information originating from a wide window on the time dimension. These three characteristics are discussed in the following sections with their relevant theories: time dimension and design with the theory of product development, information processing with theory about information asymmetry, and risk. In addition to these, economies of scale are also discussed since it is vital to business activities in general.

4. Product development

According to ENDREA† (ENDREA 2001), product development is defined as: "all activities in a company aiming at bringing a new product to the market. It normally involves design, marketing and manufacturing functions in the company". A product can in this context be both physical and non-physical. As is well known, when developing new products, designers typically follow a general procedure (sequence of activities), a so-called product development model. A product development model normally involves design, marketing and manufacturing activities. The current business model for many products, to get the customer to buy the product, implies that the focus is normally on cutting down the cost for manufacturing the product and delivering it to the customer. This is done in order to get a price that is accepted by the customer. It also implies that little focus is placed on later phases of the product's life cycle, e.g. the use phase (with activities such as use of energy and consumables, service and maintenance, and upgrading) and end-of-life. At the same time, life cycle cost studies and life cycle assessments have shown that for many products, it is during the use-phase (in reality often the longest phase of a product's life) and its related activities where the major costs and environmental impact for the product occur. Figure 2 shows, in a basic way (different products have different profiles), the environmental impact accumulation over the product's life cycle.

When developing IPSO, the basic principal is to consider all life cycle phases in order to optimize the offering from a life cycle perspective. The idea is to get the lowest total cost for the offering possible, not only to get the lowest cost for product. This generates new conditions for the product development. Since the focus is expanded to cover more life cycle phases, e.g. the use phase, it implies that the number of potential offering solutions

† Engineering Research and Education Agenda (ENDREA). ENDREA was a joint effort between four of the major Swedish institutes of technology: Chalmers University of Technology in Göteborg, the Royal Institute of Technology in Stockholm, Linköping Institute of Technology in Linköping and Luleå University of Technology in Luleå. Funding came from the Swedish board for strategic research, SSF, industry and the participating universities. The main idea behind ENDREA was to create a national cooperation in creating a new type of research in the engineering design area.

increases, which is good from an optimizing perspective. At the same time, costs are often associated with the use of materials and energy, which in turn provides a negative environmental impact, implying that more cost-optimized products usually have less environmental impact.

Figure 2 also illustrates the different phase's impact on the total environmental impact and how important the design phase is, especially the early part of it. This is at the same time logical, since it is in the early phases of product development that the product specification is defined, i.e. what parameters must/should be focused on. Examples of parameters are: how it will be used; how long it will work; what type of power it will use; what type and amount of consumables will be used during the normal use phase; what spare parts will be needed; and what is the lifetime of the product. Today, many companies' main concern in their product specifications is how to optimize and improve the production of their products, and how to develop products that are not too durable. This is important, since the predominate way of earning money is by selling products to customers.

At the same time, the initial product specification sets up boundaries for potential actions in the later phases. This is a well-known fact for people working with product development, often referred to as the "design paradox". When a new design project starts, very little is known about the final product, especially if the product is a new one for the designers. As the work on the product progresses, knowledge is increased. At the same time, the scope of freedom of action decreases for every product decision step taken, since time and cost drive most projects. Costs for later changes increase rapidly, since earlier work must be redone (Ullman 2002). The paradox is that when the general design information is needed, it is not accessible, and when it is accessible, the information is usually not needed.

Figure 3 shows the principal relation between freedom of action, product knowledge and modification cost‡. The figure is the author's further development of three figures: the design paradox (Ullman 2002), costs allocated early but used late in the project (Andreasen 1987) and the cost for design changes as a function of time during the planning and production process (Bergman and Klefsjö 2003).

Figures 2 and 3 illustrate the importance of the design phase as well as getting in relevant requirements as early as possible in the development process. It also shows the problem with traditional product development. Often, little care is taken in product development (and in its specification) for future services, maintenance, and end-of-life-treatment. Traditionally, the initial focus is on developing the physical product; once that is done, a possible service (intangible product) is developed, but this is hindered by the limitations set up from the physical product. When developing IPSO, the development is accomplished in an integrated and parallel approach.

The rate of market and technological changes has accelerated in the past decade. This implies that companies must be pro-active in the sense that they must be able to rapidly respond to fluctuations in demand (Collaine, Lutz et al. 2002). Central to competitive success in the present highly-turbulent environment is: the company's capability to develop new products (Gonzalez and Palacios 2002); to improve, further develop and optimize old

‡ This figure can also be found in the author's licentiate thesis Lindahl, M. (2000). Environmental Effect Analysis - an approach to design for environment Licentiate Thesis, Royal Institute of Technology.

products; and to do so faster than competitors (Stalk and Hout 1990). Designers must develop and proceed faster, while at the same time covering an increased number of different demands on the product. A way to handle these challenges is to do more of the product development in a more parallel and concurrent way in order to e.g. shorten the calendar time (from start to stop) and increase the collaboration over competence disciplines. One concept in line with this is Integrated Product Development§ (IPD), whose basic idea is to increase the efficiency in product development by more parallel activities and a higher degree of co-operation between functions, levels and individuals in an enterprise (Olsson 1976; Andreasen 1980). Norell (1999) characterizes the performance of IPD as follows: parallel activities; cross-functional collaboration by multifunctional teams; structured processes; and front-loaded development. The four characteristics above are in line with what (Wheelwright and Clark 1992), (Cooper, Edgett et al. 1998), and (Wilson, Kennedy et al. 1995) regard as important features for successful product development.

However, if a business model is changed from selling products to providing a function via IPSO, this also changes the conditions for development. When selling products, there is a need to constantly sell new ones in order to survive. In order to do so, the company must constantly come out with new models and/or features, and do so at an increased speed to keep competitors out. This also implies that a company should not want to offer all potential technical improvements in new products, but rather split them up over several versions in order to be able to sell more products over time.

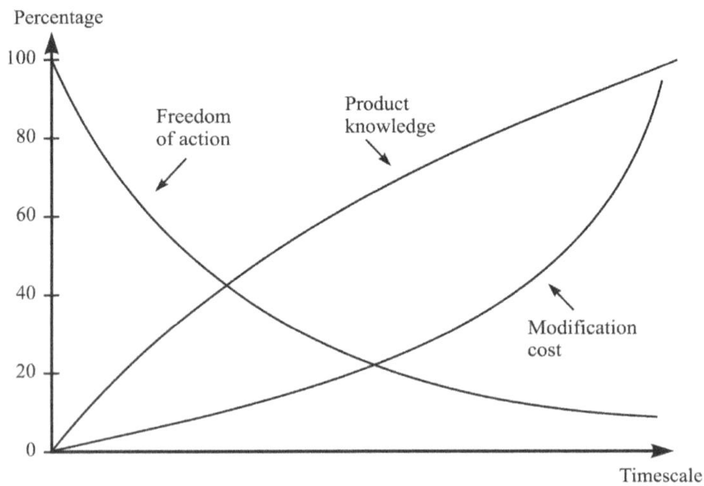

Fig. 3. The relation between "Freedom of action", "Product knowledge" and "Modification cost" is shown (Lindahl and Tingström 2000).

However, if a company sells IPSO, this is changed since the focus is not on selling products but rather on selling functionality to the customer. In principal, once an IPSO is sold to a customer, the company wants him/her to use it for as long a time as it is economically

§ Other similar common terms which correspond to this concept are Concurrent Engineering (Söderved, 1991), (Prasad, 1997) and Lean Product Development (Mynott, 2001).

interesting. If a company has technology that can e.g. cut down the energy consumption during use, it will implement the best technique at once instead of taking it in steps. Instead of spending time on developing different versions of a product, with IPSO the company in principal has more time for developing more optimized offerings - offerings that are more cost-efficient and effective, and therefore in general give a lower negative environmental impact. Nevertheless, it will still be relevant for shortening the calendar time (from start to stop).

5. Information asymmetric between a provider and a user

In general, environmental impact of a product life cycle is determined by product characteristics themselves and processes on the product. The former includes the type and amount of materials in a product, while the latter includes how to treat the product at EOL (end of life). Thus, the environmental impact of a product can be decreased by changing either its characteristics or its processes. However, one has to own and apply appropriate information to do so. There are different types of such information about a product itself or processes along the life cycle phases such as design, manufacturing, usage, and EOL. In addition, the information may not be documented in such a way that it is easily transferrable to another actor as depicted in Figure 4.

Who owns the information on how to improve the environmental aspect of the product and processes at different stages of the life cycle? Information asymmetry exists in many cases between the OEM, who in many cases designs a product, and the user. For instance, how the substances contained in a product are toxic is not necessarily known to a user but is to a designer. In addition, how to attain the best energy performance for the product in practice may be more hidden to a user than to a designer - the user simply does not know how to operate the given product for the best performance, or the provider has more knowledge of the best available technologies at the moment. There can be various reasons for this, such as a lack of user education in spite of the existence of the necessary information, or the strategy of a user as a company not to get the competence.

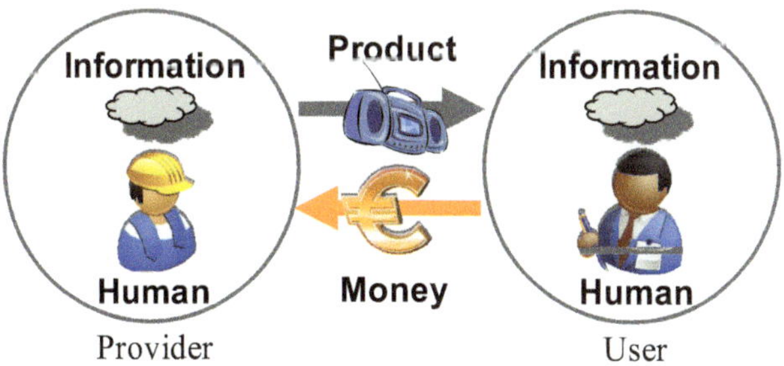

Fig. 4. General illustration of information owned by provider and user

Note that information asymmetry in the "market for lemons" addressed by (Akerlof 1970) is not the main issue of this chapter. In that case, the information possessed by a provider is about a product at a point of sale and is unchanged after the sale of the product, as it is based on a product-sales type business and the provider has no access to the product afterwards. This is shown with gray lines in Figure 5: the information of a user about the product increases along time and can surpass that of a provider. Note that variation of speed of the increase along time is not considered in this graph. In IPSE, on the other hand, a provider can obtain more information with access to the product during usage, and could maintain superiority regarding product information over the user. This is drawn as Cases 1 and 2 in Figure 5, to refer to the same and a higher speed as compared to the user, respectively. In Case 3, due to the lower speed than the user, the provider is surpassed by the user.

Information asymmetry can be a weapon for a provider to obtain payment in IPSE and makes IPSE meaningful as a business. For example, in the case where an OEM owns more information about usage or EOL of a product, there is potential for the OEM to provide IPSO so that the environmental impact is less than would be for product sales. It is also often reasonable for an OEM to be able to provide maintenance or upgrade service of its product. From the viewpoint of environmental performance, on the other hand, information asymmetry is a hindrance to improvement, since it is costly to transfer information to an actor who needs it.

Some regulations are effective so as to diminish the information asymmetry - a simple example is a symbol of "no to be put it in a dustbin" attached to an electronic product by the WEEE (Waste Electrical and Electronic Equipment Directive) (EU 2003). This symbol conveys effective information from a provider to a user: this product should not be disposed of in a regular dustbin from an environmental viewpoint. As is explained by Cerin (Cerin 2006), this type of information flow has potential to decrease the environmental impact. However, everything is not covered by regulations. A user may be willing to pay for information that contributes to the environmental performance of the product. This is where business opportunities for an OEM as an IPSO provider can be found.

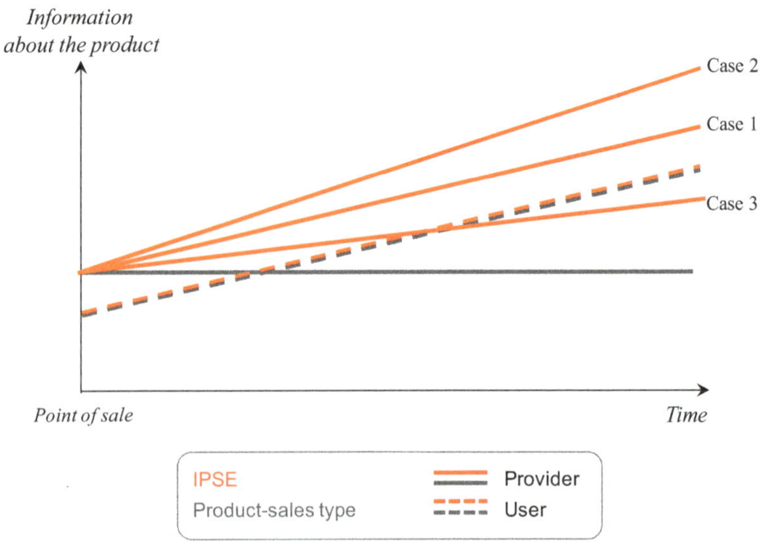

Fig. 5. Transitions of amount of information about a product after sales

Summarizing the discussion above, three levels of information asymmetry are assumed to exist in this context. If there is no (or too little) information asymmetry, there will be no gain in environmental performance through IPSE and no IPSE activities. On the other hand, in case there is a high level of information asymmetry, i.e. enough to make IPSE meaningful, there would be economic activities as well as environmental gain. The rest is an intermediate level, where there are no IPSE activities and thus loss of environmental performance. Note that this discussion focuses on a single parameter, information asymmetry; there can be other influential parameters if IPSE is meaningful.

6. Economies of scale

Economies of scale are the result of an increased number of units produced or distributed, making it possible for the unit price to decrease (Chandler 2001; Cook, Bhamra et al. 2006). An IPSE provider has the possibility to attain economies of scale through several different aspects. To provide IPSE is, in some cases, equal to being responsible for all the life cycle costs of the offering, which provide incentives to optimize the total cost as well as to realize economic development, and potentially environmental development (Lindahl, Sundin et al. 2006; Tukker and Tischner 2006). The provider would be able to gain economies of scale for both the products and the services. Leverage in production and administration could be created by offering the same services to different customers (Morey and Pacheco 2003). Another way of decreasing costs and achieving economies of scale could be realized when answering customers' demands by constantly configuring the same technology and skills in different ways (Cook, Bhamra et al. 2006). For a certain industry the market capacity is limited, which means that a single company may not reach its scale of economy since its market share is relatively fixed for a certain period of time. It is not possible to realize large-scale effects with only a few customers, since much information is needed before, during and after the delivery which results in high transaction costs (Arnold 2000). If a number of companies outsourced their processes to one organization, this would aggregate the volume and the production efficiency would increase (Gao, Yao et al. 2009). This would also bring down the transaction costs, since they were created when transferring goods and services (Chandler 2001). If the transactions occur frequently they are better handled within one single organization, since hierarchical governance facilitates administrative control and coordinated adaptability (Toffel 2008). Furthermore, customers want to benefit from the knowledge of the supplier, and are reluctant to do business with several suppliers if they want an integrated and global offering (Mathieu 2001). However, the number of actors should be enough to make sure all the components of the offer are delivered by experts (Mont 2004).

Reduced transaction costs are not the only costs to consider. New costs for complementary products may also appear for the provider in the beginning, but will benefit from economies of scale after the transition (Toffel 2008). Even though IPSE offerings imply customized solutions to achieve economies of scale, they have to be combined with well-defined modular structures at the component level (Windahl, Andersson et al. 2004). If a company wants to profit from economies of scale, this standardization of components is to be the first step (Arnold 2000). This could also be useful when considering remanufacturing, since parts that are worn out quickly or require frequent upgrading should be placed in an accessible way (Sundin and Bras 2005). Considering the remanufacturing, this process could also benefit from an economies of scale perspective. The IPSE approach would provide the manufacturer with the knowledge of how many products that are entering the process, as well as when they would do so, which would provide the IPSE provider with a remanufacturing plan that is easier to manage (Sundin and Bras 2005).

When it comes to other steps in the life cycle of the offering, the IPSE provider can economically afford a high level of specialization and technological features due to economies of scale, and can thereby optimize resource consumption and waste production, leading to better eco-efficiency for the company. The provider also often gains a competitive advantage over the customer when it comes to experience and knowledge concerning the product. With this information, the provider can optimize maintenance routines and thereby minimize the cost (Toffel 2008). Furthermore, the provider can benefit from scale effects when observing how the equipment is repaired across their whole customer base and use this knowledge (Toffel 2008). Further increased knowledge and understanding will result in increased availability and reduced product failures (Alonso-Rasgado, Thompson et al. 2004). Economies of scale can also emerge when the provider is in charge of the operations at the site of the customer, when the expertise of the provider in running the equipment can provide reduction in lead time and scale affects (Lay, Schroeter et al. 2009).

In sum, there are economies of scale in IPSE as well. Major positive factors include carrying out similar services so that an organization can learn from one service and apply it to another. In the case of IPSE, in contrast to the case of selling physical products, exactly the same offering does not exist, since a customer or user is involved in the service. This difference means that IPSE requires more involvement of staffs of a provider learning to gain economies of scale. Another factor is a market capacity, and it is necessary to take into account transaction cost and complementary product cost. Needs addressed by IPSE differ slightly from one offering to another. Therefore, modularization is a key to gain economies of scale, but service modularization needs more research than product modularization (e.g. (Simpson, Siddique et al. 2006)).

7. Risk

There are various types of risk, namely possible negative consequences from the environmental viewpoint. Reasons for this include an actor's lack of necessary information due to another actor's possession of the information, which was already discussed in the section on information asymmetry. There is another reason as well - non-existence of information.

Whether a product is better from an environmental standpoint for a given need is not necessarily certain at the time the product is first used. Different factors for this originate from the environment (not in the meaning of sustainability) and users. The former includes the speed of progress of the technology used in the product (or product generations) (see e.g. (Deng and Williams 2011)). If a new product is more energy efficient than the original one, and it becomes available before the end of usage, it may be better environmentally to switch to the new product. The user factor includes his/her discontinuity with the need for the chosen product (see different classical reasons for this in (Hanson 1980)). For instance, a change in demand causing a user to stop using a product after a short time, and owning another product in addition, generates additional environmental impact.

How can these different types of uncertainty be better handled? A provider could do this. If a provider promises a user in a contract that the "best" available technology is provided within the contract period, the user can avoid the uncertainty of the technology progress. For the user's discontinuity of the need, a provider could give an option to a user so that the user can return the product to the provider after a certain period of time. By doing so, a user can shorten the time of holding that risk. The "trick" behind this is scale of economy that enables a provider to cancel different types of risks arising from its users. Thus, variety of the needs by a group of many customers is cancelled.

In sum, there are different types of uncertainty, due to unavailable information. In the case of product sales, they generate risks of producing higher environmental impact than if this uncertainty and risk is managed through IPSE. Note that this is not merely an actor's lack of information; rather, the information is not available in spite of a willingness to get it. This is where business opportunities for IPSO exist, and existing research has not approached with that viewpoint. For instance, uncertainty in PSS has been researched as an object to be reduced for more accurate cost estimation (Erkoyuncu, Roy et al. 2011). Note that e.g. leasing by itself does not improve EOL management of leased products (Lifset and Lindhqvist 1999). If there is a high degree of uncertainty of technological progress or demand discontinuity, and if the risk can be cancelled by an OEM, IPSO has potential to decrease environmental impact.

8. Concluding discussion

This chapter endeavoured to lead theoretical discussion regarding which IPSE factors are expected to increase environmental performance of a life cycle compared to a traditional product sales business. Four aspects from theory were discussed and their relevance was pointed out. In the theory of product development, information about a product is pointed out to be a crucial parameter, although the theory is to be adapted according to the nature of the offering - IPSO as opposed to a physical, traditional product. Then, asymmetry of the information about a product between a provider and a user was identified as a key for IPSE to be meaningful also through comparison with the product sales type business. Economies of scale were brought into the discussion and this remains to be an important issue for IPSE but with different characteristics from the product sales type business. Finally, risk was discussed and pointed out to be a crucial parameter to be controlled after sale and economies of scale were shown to be an enabler to control the risk in a better way. As shown in these four sections, these aspects are interlinked with each other (see Figure 6) and need to be further investigated. Nevertheless, the chapter has provided a first theoretical cornerstone regarding conditions for IPSE to be a meaningful business style and IPSE's influential factors on environmental performance.

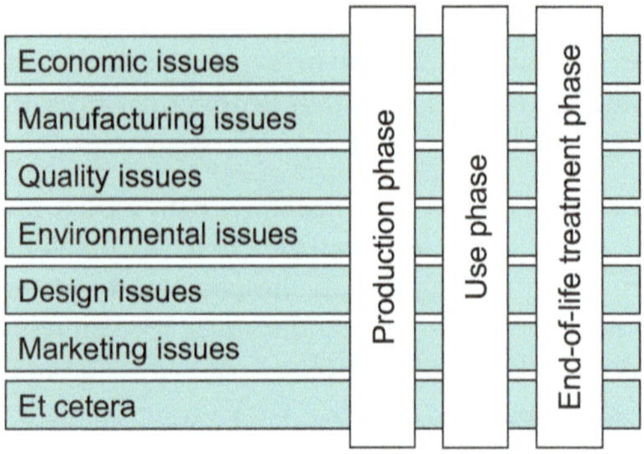

Fig. 6. Relations between different issues at each phase of a life cycle

9. Acknowledgment

This research was partially supported by a Trafikverket (the Swedish Transport Administration)-funded project "Integrated Product Service Offerings of the Railway Infrastructure System".

10. References

Akerlof, G. (1970). "The market for lemons: quality uncertainty and the market mechanism." Quarterly Journal of Economics 84: 488-500.

Alonso-Rasgado, T. and G. Thompson (2006). "A rapid design process for Total Care Product creation." Journal of Engineering Design 17(6): 509 - 531.

Alonso-Rasgado, T., G. Thompson, et al. (2004). "The design of functional (total care) products." Journal of Engineering Design 15(6): 515-540.

Andreasen, M. (1987). Integrated Product Development. Berlin, Springer.

Andreasen, M. M. (1980). Machine Design Methods Based on a Systematic Approach. Lund, University of Lund. Ph.D. Thesis.

Arnold, U. (2000). "New dimensions of outsourcing: a combination of transaction cost economics and the core competencies concept." European Journal of Purchasing & Supply Management 6(1): 23-29.

Aurich, J. C., C. Fuchs, et al. (2006). "Life cycle oriented design of technical Product-Service Systems." Journal of Cleaner Production 14(17): 1480-1494.

Bergman, B. and B. Klefsjö (2003). Quality from Customer Needs to Customer Satisfaction. Lund, Studentlitteratur AB.

Cantamessa, M. (2011). Design ... but of What. The Future of Design Methodology. H. Birkhofer. London, Springer: 229-237.

Cerin, P. (2006). "Bringing economic opportunity into line with environmental influence: A discussion on the Coase theorem and the Porter and van der Linde hypothesis." Ecological Economics 56 209– 225.

Chandler, A. D. (2001). Scale and Scope: The Dynamics of Industrial Capitalism. Cambridge, The Belknap Press of Harvard University Press.

Collaine, A., P. Lutz, et al. (2002). "A method for assessing the impact of product development on the company." International Journal of Production Research 40(14): 3311 - 3336.

Cook, M. B., T. A. Bhamra, et al. (2006). "The transfer and application of Product Service Systems: from academia to UK manufacturing firms." Journal of Cleaner Production 14(17): 1455-1465

Cooper, R. G., S. J. Edgett, et al. (1998). Portfolio Management for New Products. Reading, MA, Perseus Books.

Deng, L. and E. D. Williams (2011). "Functionality Versus 'Typical Product' Measures of Technological Progress." Journal of Industrial Ecology 15(1): 108-121.

Dill, A. K., H. Birkhofer, et al. (2011). Property Rights Theory as a Key Aspect in Product Service Engineering. International Conference on Engineering Design, Copenhagen.

ENDREA (2001). ENDREA nomenclature. Linköping, ENDREA - Engineering Research and Education Agenda.

Erkoyuncu, J. A., R. Roy, et al. (2011). "Understanding service uncertainties in Industrial Product-Service System cost estimation." International Journal of Advanced Manufacturing Technology 52(9-12): 1223-1238.

EU (2003). "Directive 2002/96/EC of the European Parliament and of the Council of 27 January 2003 on waste electrical and electronic equipment (WEEE)." Official Journal of the European Union L 37: 24-39.

Furubotn, E. G. and S. Pejovich (1972). "Property Rights and Economic Theory: A Survey of Recent Literature." Journal of Economic Literature 10: 1137-1162.

Gao, J., Y. Yao, et al. (2009). "Service-oriented manufacturing: a new product pattern and manufacturing paradigm." Journal Intelligent Manufacturing 22(3): 435-446.

Goedkoop, M. J., C. J. van Halen, et al. (1999). Product Service Systems, Ecological and Economic Basics. Hague, Dutch Ministry of Housing, Spatial Planning and the Environment.

Gonzalez, F. J. M. and T. M. B. Palacios (2002). "The effect of new product development techniques on new product success in Spanish firms." Industrial Marketing Management 31(3): 261-271.

Hanson, J. (1980). "A proposed paradigm for consumer product disposition processes." Journal of Consumer Affairs 14: 49–67.

Hockerts, K. (2008). Property Rights as a Predictor for Eco-Efficiency of Product-Service Systems. CBS Working Paper Series. Frederiksberg, Copenhagen Business School.

Isaksson, O., T. C. Larsson, et al. (2009). "Development of product-service systems: challenges and opportunities for the manufacturing firm." Journal of Engineering Design 20(4): 329 - 348.

Lay, G., G. Copani, et al. (2010). "The relevance of service in European manufacturing industries." Journal of Service Management 21(5): 715-726.

Lay, G., M. Schroeter, et al. (2009). "Service-Based Business Concepts: A Typology for Business-to-Business Markets." European Management Journal 27(6): 442-455.

Lifset, R. and T. Lindhqvist (1999). "Does Leasing Improve End of Product Life Management?" Journal of Industrial Ecology 3(4): 10-13.

Lindahl, M., E. Sundin, et al. (2006). Integrated Product and Service Engineering - the IPSE project. Changes to Sustainable Consumption, Workshop of the Sustainable Consumption Research Exchange (SCORE!) Network, supported by the EU's 6th Framework Programme, Copenhagen , Denmark.

Lindahl, M. and J. Tingström (2000). A Small Textbook on Environmental Effect Analysis. Kalmar, Department of Technology, University of Kalmar.

Lindemann, U. (2011). Systems Engineering versus Design Methodology. The Future of Design Methodology. H. Birkhofer. London, Springer: 157-167.

Martyrer, E. (1960). "Der Ingenieur und das Konstruieren." Konstruktion 12: 1-4.

Mathieu, V. (2001). "Service strategies within the manufacturing sector: benefits, costs and partnership." International Journal of Service Industry Management 12(5): 451-475.

Mont, O. (2004). "Institutionalisation of sustainable consumption patterns based on shared use." Ecological Economics 50(1-2): 135-153.

Mont, O., P. Singhal, et al. (2006). "Chemical Management Services in Sweden and Europe: Lessons for the Future." Journal of Industrial Ecology 10(1/2): 279-292.

Mont, O. K. (2002). "Clarifying the concept of product–service system." Journal of Cleaner Production 10(3): 237-245.

Morey, E. and D. Pacheco (2003). "Prouct service systems: Exploring the potential for economic and environmental efficiency."

Mynott, C. (2001). Lean product development: the manager's guide to organising, running and controlling the complete business process of developing products. Northampton, UK, Westfield Publ.

Neely, A. (2007). The servitization of manufacturing: an analysis of global trends. 14th EurOMA Conference, Ankara.

Oliva, R. and R. Kallenberg (2003). "Managing the transition from products to services." International Journal of Service Industry Management 14(2): 160-172.

Olsson, F. (1976). Systematic Design. Lund, University of Lund. Doctoral Thesis.

Pahl, G. and W. Beitz (1996). Engineering Design: A Systematic Approach. London, Springer-Verlag: 1.

Prasad, B. (1997). Concurrent Engineering Fundamentals - Integrated Product Development - Volume 2. Upper Saddle River, New Jersey, Prentice-Hall.

Roy, R. (2000). "Sustainable product-service systems." Futures 32: 289–299.

Sakao, T. (2009). A View of Service, Quality, and Value for Sustainability. 12th International QMOD Conference, Verona.

Sakao, T., C. Berggren, et al. (2011). Research on Services in the Manufacturing Industry based on a Holistic Viewpoint and Interdisciplinary Approach. CIRP International Conference on Industrial Product-Service Systems, Braunschweig.

Sakao, T., H. Birkhofer, et al. (2009). "An Effective and Efficient Method to Design Services: Empirical Study for Services by an Investment-machine Manufacturer." International Journal of Internet Manufacturing and Services 2(1): 95-110.

Sakao, T. and M. Lindahl, Eds. (2009). Introduction to Product/Service-System Design. Springer's global publishing programme in engineering and management. London, Springer.

Söderved, H. (1991). Concurrent Engineering - ett arbetssätt för effektiv produktframtagning (in Swedish only). Stockholm, Sveriges Mekanförbund.

Sakao, T., N. Napolitano, et al. (2008). "How Are Product-Service Combined Offers Provided in Germany and Italy? – Analysis with Company Sizes and Countries -." Journal of Systems Science and Systems Engineering 17(3): 367–381.

Sakao, T. and Y. Shimomura (2007). "Service Engineering: A Novel Engineering Discipline for Producers to Increase Value Combining Service and Product." Journal of Cleaner Production 15(6): 590-604.

Sakao, T., Y. Shimomura, et al. (2009). "Modeling Design Objects in CAD System for Service/Product Engineering." Computer-Aided Design 41(3): 197-213.

Simpson, T. W., Z. Siddique, et al. (2006). Product Platform and Product Family Design: Methods and Applications. New York, Springer.

Stahel, W. R. (1994). The Utilization-Focused Service Economy: Resource Efficiency and Product-Life Extension. The Greening of Industrial Ecosystems. Washinton DC, National Academy Press: 178-190.

Stalk, G. J. and T. M. Hout (1990). Competing Against Time - How Time-Based Competition is Reshaping the Global Markets. New York, The Free Press, A Division of Macmillan Inc.

Sundin, E. and B. Bras (2005). "Making functional sales environmentally and economically beneficial through product remanufacturing." Journal of Cleaner Production 13(9): 913-925.

Toffel, W. M. (2008). Contracting for Servicizing.

Tukker, A. (2004). "Eight Types of Product-Service System: Eight Ways to Sustainability? Experiences from Suspronet." Business Strategy and the Environment 13. 246 - 260.

Tukker, A. and U. Tischner (2006). New Business for Old Europe. Sheffield, Greenleaf Publishing.

Ullman, D., G. (2002). The Mechanical Design Process. New York, McGraw-Hill Higher Education.

Vargo, S. L. and R. F. Lusch (2004). "Evolving to a New Dominant Logic for Marketing." Journal of Marketing 68(1): 1-17.

Wheelwright, S. C. and K. B. Clark (1992). Revolutionizing Product Development: Quantum Leaps in Speed, Efficiency, and Quality. New York, Free Press.

Wilson, C. C., M. E. Kennedy, et al. (1995). Superior Product Development: Managing the Process for Innovative Products: A Product Management Book for Engineering and Business Professionals, Blackwell Publishers.

Windahl, C., P. Andersson, et al. (2004). "Manufacturing firms and integrated solutions: characteristics and implications " European Journal of Innovation Management 7(3): 218-228.

6

A Safety Engineering Perspective

Derek Fowler and Ronald Pierce
JDF Consultancy LLP
UK

1. Introduction

Safety is a viewpoint. By this we mean that safety is not in itself an attribute of a system but is a property that depends on other attributes and on the context in which the system is used. The question that arises (and which we will attempt to answer in some detail) is which attributes of a system determine whether it is safe or not, in its context of use.

Throughout this chapter, the term *system* is used in the widest sense - ie it includes not just the technical elements (equipment) but also all other elements - eg the human-operator and operational procedures - that necessarily make up the complete, end-to-end system.

We will start by attempting to dispel what seems to be a not-infrequent misconception (also reflected in some safety standards) that safety is mainly dependent on reliability (and / or integrity, depending on one's definition of the terms - see section 3.1 below). This we feel is important for those readers who may have had some previous exposure to areas of safety engineering in which this view is held and will lead to the inescapable conclusion that we need a broader view of system safety than is sometimes taken.

Next we will establish some basic safety concepts firstly by defining key terms, and then by considering two distinct categories of safety-related system and seeing how the system properties determine safety in each case.

Finally, and for the most part of this Chapter, we will explain how the broader approach to safety works and show that it is closely linked with (not 'special and different' from) systems engineering in general.

2. "Safety is reliability" – Dispelling a myth

[Leveson, 2001] in a review of major software-related accidents and the implication for software reliability, presents compelling evidence that software reliability had never been the cause of such disasters - on the contrary, in every case investigated, the software had performed in exactly the manner that it was designed to. The problem was that the software was designed to do the wrong thing for the circumstances under which it "failed" (or, as in the case of Ariane V, for example) was used for a purpose (ie in a context - see above) different from that for which it was originally designed. Professor Leveson quite rightly, therefore, poses the question as to why, in most software safety standards, so much emphasis is placed on processes to improve software reliability whilst not ensuring also that the resulting systems actually perform the intended function – ie allowing them to be what one might call "reliably unsafe". This same misconception is prevalent also at the system level in, for example, European Air Traffic Management (ATM) - see [Fowler, 2007].

We can illustrate the problem by considering the simple, everyday example of a car airbag for which, for the sake of this discussion, we wish to make a case that it would be safe.

If we were to simply follow a failure-based process - ie focus on how reliable the airbag needs to be in order to be 'safe' - we would start (at the wrong point, as we will see shortly) by identifying the hazards[1] presented by the airbag. Such hazards are those caused by the airbag's two main failure modes: failure to operate when required, and operating when not required. We would then use a risk classification scheme to derive safety requirements that specify the maximum frequency with which those failures could be allowed to occur and from that we would deduce more detailed safety requirements which limit the frequency with which the causes of the hazards could be allowed to occur.

Even if the results were valid, they would lead us only to:

- an understanding of how reliable the airbag needs to be - so that it operates when required; this would, however, not give any assurance that, when it did operate, the airbag would actually protect the front-seat occupants from death or serious injury in the event of a collision, and
- the totally **irrational** conclusion that putting an airbag in a car would only increase the risk of death or serious injury to the front-seat occupants, because of the finite (albeit small) possibility that it would operate when not intended to!

Of course, what is missing is any evidence of a positive safety contribution from the airbag - only the possibility of actually being killed / seriously injured by it - without which we would have no case for fitting one.

If instead we were to take a more **rational** view, we would start from the position that in the event of, say, a head-on collision without an airbag there is a very high risk of death or serious injury to the driver (and other front-seat occupant(s)) of a car. This risk we can call *pre-existing* because, by definition, it is inherent in driving and has nothing whatsoever to do with the airbag – indeed it is to mitigate this risk that we are intending to fit the airbag in the first place. So, the more rational approach would be to:

- firstly assess how effective the airbag would be when it did work – ie by how much the *pre-existing* risk from driving would be reduced by the airbag - and what properties of the airbag determine the amount of this reduction; and then
- assess the *system-generated* risk, induced by airbag failure.

Thus, given the correct set of functional properties – eg shape, location, strength, compressibility, sensitivity to 'g' forces, speed of deployment etc – as well as adequate reliability and integrity, our safety case should show that the airbag would make a positive contribution to the reduction in the identified pre-existing risk that is very much greater than the system-generated risk due to airbag failure. This would be a much more balanced, and rational conclusion than what emerged above from considering only airbag failure.

---

[1] A state of a system that could lead to an accident - see section 3.1.

3. System safety – Basic concepts

3.1 Definitions

This section defines the safety terms that are used in the rest of this chapter. In most cases, as there is no single, universally accepted definition, the ones given below have been adapted from those in Part 4 of international functional-safety standard IEC 61508 [IEC, 2010] and, if not actually used by, should at least be understood in, any safety-related sector.

- *Harm* death or serious physical injury or damage to the health of people, or serious damage to property or the environment
- *Hazard* a situation, state or event that may result in harm[2]
- *Risk* combination of the frequency of occurrence and severity of harm
- *Safety* freedom from unacceptable risk
- *Reliability* the ability of a system / element to provide a long period of continuous delivery of an intended service (or function) without failure
- *Integrity* the ability of a system, under all stated conditions, to provide all the services (or functions) required by the users, with no unintended or un-commanded services (or functions)
- *Failure* termination of the ability of a functional unit to provide a required function or operation of a functional unit in any way other than as required

3.2 Risk acceptability

3.2.1 The ALARP principle

Risk may, in extremis, be either so great that it would be intolerable under any circumstances or so small as to be insignificant and therefore may be discounted altogether. In practice, however, risk will usually fall somewhere between these two extremes and the ALARP principle requires that any risk shall be reduced to a level that is as low as reasonably practicable, bearing in mind two factors: the benefits resulting from its acceptance, and the costs of any further reduction. ALARP is described in more detail in IEC 61508 [IEC, 2010], Part 5, Annex C; other standards and practices use different acronyms such as ALARA (USA) and SFAIRP (Aus). In the UK, the ALARP principle has a specific legal connotation and expert advice should be sought before applying it! [Ladkin, 2008].

A practical way of specifying what is *tolerable* risk, and in some cases applying the ALARP principle, either qualitatively or quantitatively, is the so-called Risk Classification Scheme (also known as a Hazard-Risk Index).

3.2.2 Risk Classification Schemes

Risk Classification Schemes (RCSs) are used in a number of industry sectors. Their form is as variable as their usage but a typical example, from the ATM sector [EUROCONTROL 2010], is shown in Figure 1.

---

[2] Whether or not a hazardous event results in harm depends on whether people, property or the environment are exposed to the consequence of the hazardous event and, in the case of harm to people, whether any such exposed people can escape the consequences of the event after it has occurred

An RCS is typically set out as a matrix, in which the severity of possible outcomes is mapped against the frequency with which the outcomes might occur.

		Severity Class			
Probability per flt hr		1	2	3	4
Frequent	$P>10^{-4}$	A	A	A	B
Probable	$P<10^{-4}$	A	A	A	C
Occasional	$P<10^{-5}$	A	A	B	D
Remote	$P<10^{-6}$	A	B	C	D
Improbable	$P<10^{-7}$	A	C	D	D
Extremely Improbable	$P<10^{-8}$	B	D	D	D

Fig. 1. A Risk Classification Scheme

In this ATM example, the **severity** of outcome ranges from Class 1 (an accident involving death and/or serious injury[3]) to Class 4 (the lowest level of safety-significant incident) and, in practice, would be explained by detailed descriptions and illustrative examples. The **frequency** of outcome is shown both qualitatively and quantitatively, as the probability per flight hour (or per operating hour). The grid is populated with the tolerability / acceptability of the risk and, in this example, includes the ALARP principle[4] as follows:

- Risk Class A is defined as intolerable
- Risk Class B is tolerable only if risk reduction is impracticable or cost grossly disproportionate to improvement
- Risk Class C is tolerable if cost of risk reduction would exceed the benefit of improvement
- Risk Class D is defined as broadly acceptable

An RCS should be tailored to the purpose and function of the system or service concerned and the risks and benefits involved. It would normally be published in the Safety Management System for the organisation responsible for the operation, and be approved by the relevant safety-regulatory authority. It can be used in one of two ways:

- for safety monitoring of an on-going operation (see section 6.5 below) - in this case the achieved risk can be compared with what is tolerable / acceptable according to the RCS, and / or
- for *a priori* safety assessment - in this case the severity of each potential outcome is assessed and the required maximum frequency of occurrence, in order to achieve an acceptable (or at least tolerable) level of risk, is obtained from the RCS.

---

[3] In European ATM, it is not usual practice for accidents to be 'graded' by the number of people killed and/or seriously injured.

[4] If the ALARP principle were not applied to the RCS, the red boxes (labelled 'A') might remain the same as in Figure 2 but the rest of the grid would show only what was tolerable.

One of the main attractions of RCSs is that they are relatively simple to use - and therein lies a potential problem, unless each user it careful to check the following:

- at what level in the system hierarchy, and within what scope, the values apply
- where the probability / frequency values used in the scheme came from and whether they are appropriate and (still) valid
- how the aggregate risk can be deduced from analysis of individual hazards, in restricted segments of the total system
- what allowance needs to be made for *pre-existing* risk - see also section 3.3.3 below.

The significance of some, if not all, of these 'health warnings' should become more apparent in the subsequent sections of this chapter.

3.3 Safety-related systems and their properties

3.3.1 Safety-related systems – General

Consider the two types of safety-related system (SRS) shown in Figure 2. Case (a) is a system - say, a complete nuclear power plant - which simply provides a service into its operational environment. Because the service in the case of a nuclear power plant is the provision of electrical power then, from a purely safety viewpoint, we do not care whether the service is provided or not. What we do care about are the hazards (eg radiation leakage), and the related level of risk, that a failure internal to the system might present to its operational environment (and to the people therein).

Case (b) is quite different. Here we have, first of all, a set of hazards that already exist in the operational environment. If, for example, the System was our car airbag (see section 2 above) then these hazards (and associated risks) would be those (*pre-existing*) hazards / risks inherent in driving a car, and the operational environment (from the airbag's perspective) would be the whole car.

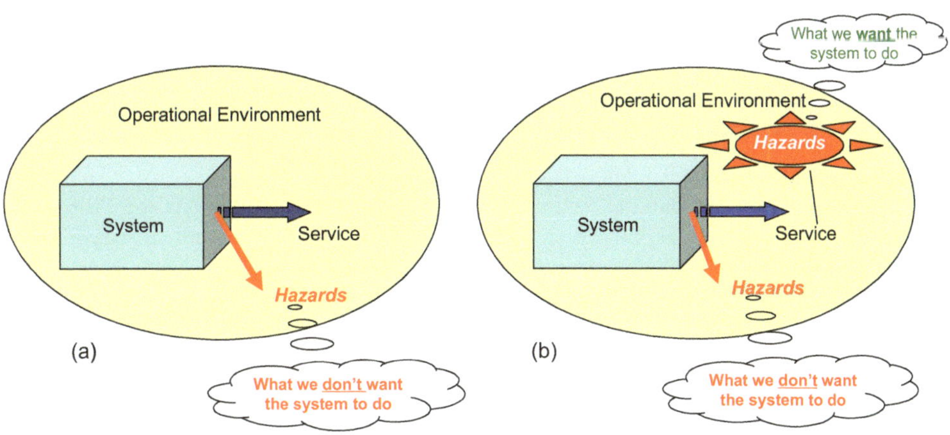

Fig. 2. Two types of Safety related System

As we will see in more detail in section 3.3.3 below, what is very important about Figure 2 is that for case (b) the mitigation of *pre-existing* hazards / risks (ie what we want the system to do) and the inevitable introduction of *system-generated* hazards / risks (ie what we don't want the system to do) depend on entirely different properties of the system.

Before that, however, we will introduce IEC 61508 [IEC, 2010], probably the most widely accepted, international standard on the functional safety of systems.

3.3.2 IEC 61508

IEC 61508 has a particular model of how SRSs influence the real world that is based on the concept of the *Equipment Under Control* (EUC) which itself is regarded as the hazard-creating system[5] and for which SRS are designed in order to mitigate those hazards [Pierce & Fowler, 2010]. Since IEC 61508 is a generic standard, to be adapted for application to a wide range of specific industry sectors, it has no particular view on the nature of the EUC, which could be a nuclear reactor, a chemical plant, an oil rig, a train, a car, or an aircraft etc[6].

The standard then defines two types of SRS that are intended to mitigate the hazards and risks associated with the EUC:

- *Control Systems* (eg a railway signalling system) which provide *Safety Functions* that are designed to maintain continuously a tolerable level of risk for the EUC, and
- *Protection Systems* (eg an automatic train protection (ATP) system or car airbag) which provide *Safety Functions* that are designed to intervene when they detect a hazardous state developing within the EUC and/or its Control System(s), and put the EUC / its Control System(s) into a safe, or at least safer, state[7].

As far as a Control System is concerned, the hazards and risks associated with its EUC are clearly *pre-existing*, since they are caused by the latter not the former. Similarly, the hazards and risks associated with the combination of an EUC and its Control System(s) are pre-existing as far as a Protection System is concerned.

With this concept, an SRS (Control and/or Protection system) is put in place to reduce the pre-existing risks to an acceptable level. IEC 61508 refers to this as *Necessary Risk Reduction* but does not actually stipulate what is "acceptable", this being left to local or national considerations, including legal frameworks, for the applicable industry sector.

As we will see in section 3.3.3 below, safety integrity requirements on Control Systems are usually expressed as probability of failure per operating hour, whereas for Protection Systems they are usually expressed as probability of failure on demand. In either case, the target probability will of course depend on the level of the pre-existing risk. The

[5] Equivalent to Figure 2 case (a).

[6] In these examples, the EUC is very tangible and for these it is probably a better term than the equivalent term "operational environment" used in Figure 3(b). However, in some cases - eg air traffic management and a railway level crossing (see section 4) - the EUC is much less tangible and "operational environment" might be better. Whatever term is used, the principles are exactly the same and the concept of pre-existing risk is paramount!

[7] Eg, an ATP system is designed to stop a train if it passes a signal at danger.

objective of the SRS for Control and Protection systems is *risk control* and *risk reduction* respectively[8].

In both cases, IEC 61508 is quite clear that the safety functional requirements (specifying functionality and performance of the Safety Functions) must be completely and correctly identified before the SRS can be designed. This requires hazard and risk analysis of the EUC not (initially at least) hazard and risk analysis of the SRS(s) themselves. Once the safety functionality and performance requirements of the Safety Functions have been identified, the tolerable failure rates of the Safety Functions can then be identified, and the Safety Integrity Level (SIL) for each Safety Function established[9].

3.3.3 Safety properties

We can build on our simple example of a car airbag to explain more generically, and develop, the above principles since it can readily be seen that an airbag fits Figure 2 case (b), and the IEC concept of a Protection System, very well. Figure 3 shows the risk, in the Operational Environment (or EUC), with and without an SRS - ie R_U and R_A respectively. As we saw for the airbag in section 2 above, the safety case for the SRS depends on its making a (much) bigger positive contribution to safety when operating as intended (ie the *success* case as represented by the green, right-to-left arrow) than any negative contribution caused by its failure or incorrect / spurious operation (ie the *failure* case as represented by the solid red, left-to-right arrow).

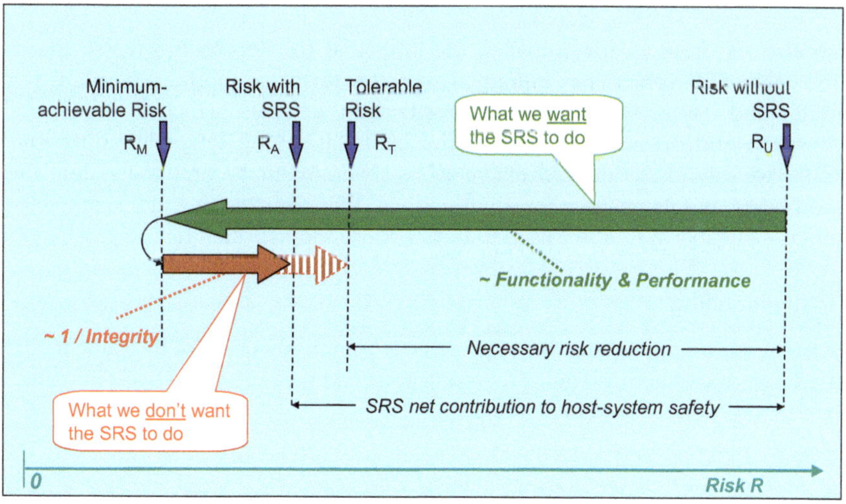

Fig. 3. Risk Graph for a Safety-related System

[8] Version 2 of IEC 61508, issued in 2010, is much clearer about the application of risk reduction to protection systems and risk control to continuously-operating control systems, than was the earlier (2000) version.

[9] This is exactly what we said for the airbag example in section 2, but is not always followed in some industry sectors!

There are a number of very important points to note about this diagram:

- R_U has nothing to do with the SRS – ie it is the *pre-existing* risk, as above
- R_M is the theoretical minimum risk that would exist in the complete absence of failure / spurious operation of the SRS – it is not zero, because there usually are some accident scenarios for which an SRS cannot provide mitigation[10]
- since R_M is defined as the risk in the absence of failure, it must be determined only by the *functionality & performance* of the SRS, as explained in section 2 above
- the risk increase R_A-R_M is caused entirely by failure / spurious operation of the SRS - thus it is the *system-generated* risk and is determined primarily[11] by the *reliability & integrity* of the SRS
- the safety case for the SRS is based on showing that $R_A << R_U$
- if we now introduce R_T, the maximum tolerable level of risk, then an interesting conclusion emerges: given that R_T is fixed (eg by a regulatory body), then the maximum tolerable failure rate of the SRS - ie a function of the length of the extended red (l-r) arrow (R_T-R_M) - depends on the length of the green (r-l) arrow (R_U-R_M); in other words, the tolerable failure rate depends on how successful the SRS is in reducing the pre-existing risk in the first place
- overall, R_U-R_T fits the IEC 61508 definition of Necessary Risk Reduction
- if, as we desire, R_A-R_M<<R_U-R_M, then the overall risk actually achieved (ie R_A) is much more sensitive to changes in the length of the green (r-l) arrow (ie to changes in functionality and performance) than to proportionate changes in the length of the red (l-r) arrow (ie to changes in reliability & integrity).

We can also see from Figure 3 that in the limit that R_M approaches R_T, so the integrity required of the SRS approaches infinity! This raises further important questions regarding the origins and use of traditional risk-classification schemes, which are often are based entirely on R_T and do not take any account of R_M in setting tolerable failure rates for a system. As we saw in section 3.2.2 above, RCSs generally model only the system's negative effects on safety, not its positive contributions and, therefore, to get a more complete picture of where risks lie in a system we need to turn to more sophisticated forms of risk modelling.

3.3.4 Risk modelling

One of the systems engineering techniques that is commonly used for the static modelling of risk in a safety assessment is Fault Tree Analysis (FTA) [IEC, 2006b]. This is illustrated, for a very basic Protection System, in Figure 4.

An accident would occur if one, or both, of two conditions occurs, as follows:

- firstly, the (pre-existing) hazard occurs and the consequences of the hazard are not mitigated; in this case, if the hazard were never mitigated, then the accident rate would be the same as the hazard occurrence rate – ie the hazard occurrence rate would be the

[10] Eg for an airbag these include fire, or being hit from behind by a vehicle with high relative velocity.
[11] The word "primarily" is used here because (as is more generally the case) it is may be possible to provide additional functionality to mitigate some of the causes and / consequences of system-generated hazards .

pre-existing risk (R_U) defined in Figure 3. The situation that the hazard is not mitigated could arise because Protection System either: operates but is not effective; or fails to operate.

- or secondly, the Protection System operates spuriously (eg when it is not supposed to, or in a way different from what was required) and the consequences of this (system-generated) hazard are not mitigated.

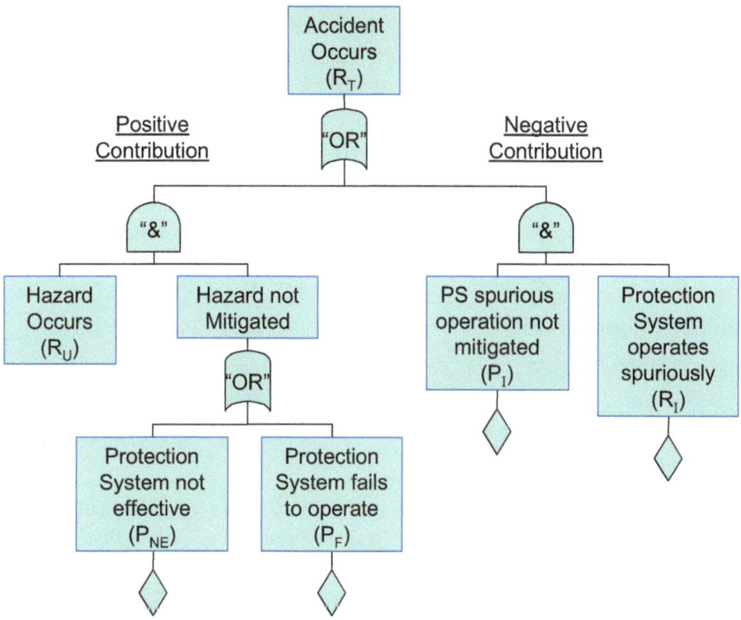

Fig. 4. Simple Accident Fault Tree - Protection System

It is the presence of the external input R_U that distinguishes Figure 4 from Fault Trees in general, and it is this that enables the computation of both the positive and negative contributions to safety.

It should be noted that a simple failure (to operate) of the Protection System is shown on the *success* side of the model rather than on the *failure* side - corresponding to shortening the green arrow on Figure 3 rather than lengthening the red arrow. This would be valid for our airbag if driver behaviour were not affected by the knowledge that the car had an airbag - because the risk of airbag failure would simply be the same as not having an airbag at all for the period of the failure. However, if drivers drove less carefully and / or faster in expectation that the airbag would always protect them in the event of a head-on collision then the consequences (and therefore the risk) from failure of the airbag to operate when required would be correspondingly greater - in this case the effect of the failure would better be shown on the failure side of the model. The latter case is an example of what is very common in safety-related environments, especially where humans are involved, and requires extra care to be taken when incorporating such loss-type failures in a risk model.

In practice, risk models are much more complex than the simple illustration in Figure 4. Their uses range from a discrete safety assessment of part of an overall system up to developing a risk model for an entire operation. An example of the latter use, from the ATM field, is EUROCONTROL's Integrated Risk Picture (IRP) [Perrin & Kirwan, 2009]. As a complete model of both positive and negative contributions to safety, the IRP has proved to be a much more powerful tool, in the management of functional safety, than a simple RCS (see section 3.2.2 above) and can be used in many different ways including *a priori* safety assessment, safety monitoring and safety-strategy formulation. Such models are used also in some parts of the nuclear and rail industries but, to the authors' knowledge, not in other industry sectors at the moment.

4. Requirements engineering – The key to safety assessment

Capturing, and then satisfying, a complete and correct set of safety requirements is as fundamental to any *a priori* safety assessment as requirements engineering is to systems engineering in general, as explained below.

4.1 Requirements capture

Some crucial issues regarding requirements capture can be expressed through the simple, but rigorous, requirements-engineering (RE) model shown in Figure 5. This model has been adapted from [Jackson, 1995], in the introduction to which Dr Jackson sums up the requirements-capture problem perfectly, as follows:

"We are concerned both with the world, ***in which the machine serves a useful purpose****, and with the machine itself. The competing demands and attractions of these two concerns must be appropriately balanced. Failure to balance them harms our work".*

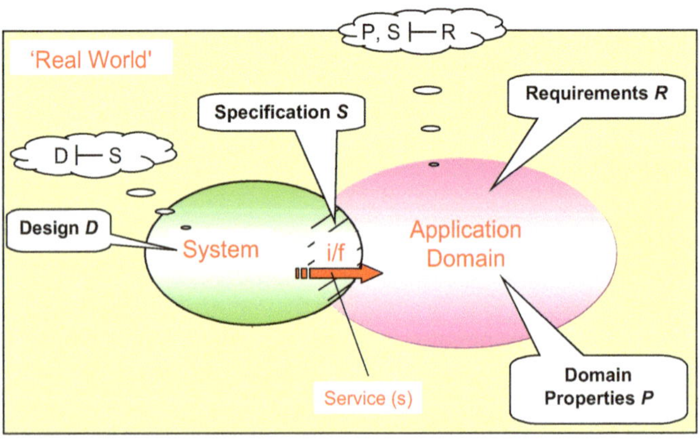

Fig. 5. Jackson Requirements-capture Model - General Form

In this context, what has been said in section 2 above about the lack of a success approach in safety assessments is an example of a pre-occupation with the machine itself at the expense of considering its useful purpose (ie to reduce pre-existing risk). Figure 5 helps clear our thinking as follows.

In the Jackson model, the *system* exists in the *real world*. The part (ie subset) of the real world that influences the system, and into which the system provides a *service* through an interface (*i/f*), is known as the *application domain*. *Requirements* are what we want to make happen in the application domain[12] and are defined in that domain - not in the system.

A *specification* is what the system has to do across the interface in order that the *requirements* can be satisfied - ie a specification takes an external, or "black-box", view of the system. Another way of thinking about a specification is that it contains all the shared properties between the service provider and the service user - therefore it might include things that the service user has to do, not just what the system has to do.

Design, on the other hand, describes what the system itself is actually like and includes all those characteristics that are not directly required by the users but are implicitly necessary in order for the system to fulfil its specification and thereby satisfy the user requirements. Design is essentially an internal, or "white-box", view of the system.

The formal notation in the "bubbles" in Figure 5 defines two relationships that must be shown to be true in requirements capture:

1. that the specification *S* satisfies the requirements *R*. However, this can be true only for a given set of properties *P* of the application domain; therefore, if any one of these three sets of parameters is changed then satisfaction demonstration is invalidated until one of the other sets is also changed
2. that the design *D* satisfies the specification S

The distinction, and relationship, between requirements, specifications, application-domain properties, and design are not merely academic niceties; rather, they provide the essential foundations for developing systems that do, and can be shown to do, everything required.

What is described above in this section applies, of course, to systems engineering in general. However, if the assertion at the beginning of section 1 is correct then it should be possible to apply the same principles to the safety perspective. By comparing Figure 5 with Figure 2 (b) we can see that there is a direct equivalence for safety, as shown in Figure 6.

The main differences from Figure 5 are limited to:

- the Application Domain is now known as the Operational Environment / EUC - merely a change in terminology
- the aspects of the Requirements of particular interest are the Safety Criteria - ie the level of safety that has to be achieved in the Operational Environment / EUC
- for safety-related systems of the type shown in Figure 2 (b) above, the *pre-existing hazards* exist in the Operational Environment / EUC [13] - therefore, the primary Requirements are for the service(s) provided by the system to reduce the associated pre-existing risks to the level defined by the Safety Criteria.

Otherwise, everything that is said in section 4.1.1 above applies to safety.

---

[12] Since the service users are in the Application Domain these requirements are sometimes called *User* Requirements

[13] Indeed they are the most important properties of the operational environment / EUC!

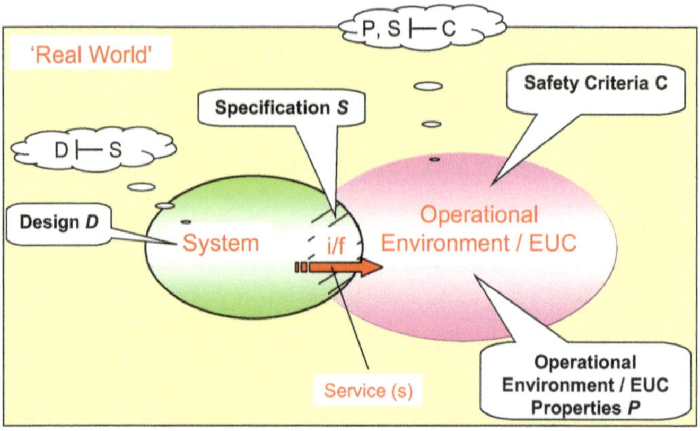

Fig. 6. Safety Engineering form of the Jackson Requirements-capture Model

4.2 Safety requirements satisfaction

Implementation of the design, in the built and integrated system, involves a **third** relationship that must be shown to be true:

1. the implementation *I* satisfies the design *D*

The validity of this relationship requires two objectives to be satisfied in implementation of the design - ie showing that:

- the required properties (functionality, performance, reliability and integrity) of the built system satisfy the requirements established for the design, and
- no emergent properties (eg common-cause failures) and unwanted functionality have been introduced inadvertently such that they could adversely affect the ability of the built system to satisfy the (safety) requirements established for the design.

Because these two objectives are generic - ie apply to all properties of a system - there is no difference in principle between the satisfaction of safety requirements and the satisfaction of design requirements in general. That said, there is usually a difference in degree, in that safety requirements require a higher level of assurance that they have been captured, and then satisfied, completely and correctly.

5. Safety assurance and safety cases

5.1 Safety assurance

Safety assurance, like systems assurance in general, relies on planned, systematic activities to provide the necessary confidence that a service or functional system satisfies its requirements (which are themselves complete and correct), in its intended environment[14]. *Assurance activities* are systematic in that they specify how the *assurance objectives* (ie what has to be demonstrated) are to be achieved, as indicated in Figure 7.

[14] From a safety perspective, this would mean achieving an acceptable or tolerable level of safety - see definition of safety assurance in [European Commission, 2005]

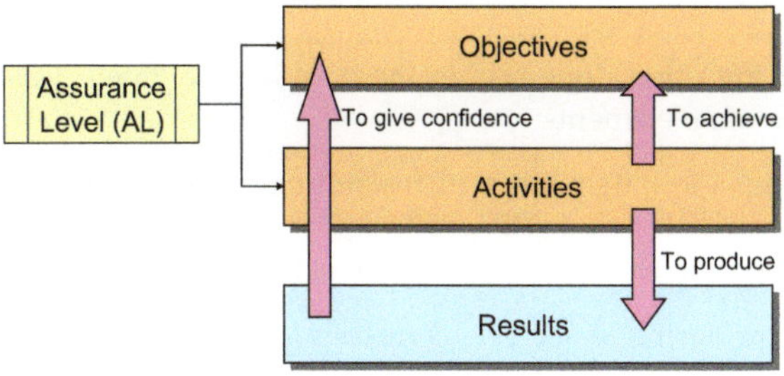

Fig. 7. Safety Assurance - Basic Elements

Which assurance objectives have to be achieved, and the rigour with which they have to be achieved, are often determined by assurance levels (ALs), which are based upon the potential consequences of the anomalous behaviour of the system element concerned, as determined by the system safety assessment process.

The AL implies that the level of effort recommended for showing compliance with safety requirements increases with both the severity of the end effect of the element failure. and the probability / likelihood of occurrence of that end effect, given that the failure has occurred [Mana et al, 2007][15]. The results (outputs) of the activities are then used to show that the assurance objectives have been achieved.

For high-integrity systems in particular, there is a further issue that safety assurance is often used to address and concerns the safety-integrity of system elements - software functions and human tasks, in particular. Whereas it may be necessary to specify Safety Integrity Requirements for all elements of a system in order to show compliance with a numerical Safety Criterion, it is usually very difficult to show in a direct way - through, for example, test results - that such requirements are actually satisfied in implementation. In such situations, it becomes necessary to adopt a more indirect, assurance-based approach which uses the rigour of the development processes to give confidence that the requirements are likely to be / have been satisfied. This is reflected in, for example, airborne software standard DOD 178B [RTCA, 1992] and IEC 61508 both of which are assurance based.

The problem with safety standards is that their use can become highly *proceduralized*, leaving open two important questions:

- where did the full set of assurance objectives come from in the first place?
- how was it decided which objectives and activities have to be done (or may be ignored) for a given AL, and how to we know that this would be appropriate for a particular application?

We can address this problem by putting safety assurance into an *argument* framework but in order to understand this we first need to look have a brief look at Safety Cases.

---

[15] In some standards, the likelihood of occurrence of the failure is also taken into account - ie the assurance is based on the risk associated with a failure, not just the consequence thereof. This is the case with IEC 61508 and related standards, in which the term SIL (Safety Integrity Level) is used instead of AL.

5.2 Safety cases

Safety assessments are often done within the context of a Safety Case which, like a legal case, comprises two main elements:

- a set of *arguments* - ie statements which claim that something is true (or false), together with
- supporting *evidence* to show that the argument is actually true.

Safety arguments are normally set out hierarchically; this is shown in Figure 8 using and adapted form of goal-structuring notation (GSN). In safety work [Kelly & Weaver, 2004], GSN is simply a graphical representation of an argument / evidence structure and usually starts with the top-level claim (Arg 0) that something is (or will be) acceptably (or tolerably) safe; this is then decomposed such that it is true only if, and only if, the next-level argument statements (in this case Arg 1 to 4) are all true. The *strategy* text should explain the rationale for that decomposition.

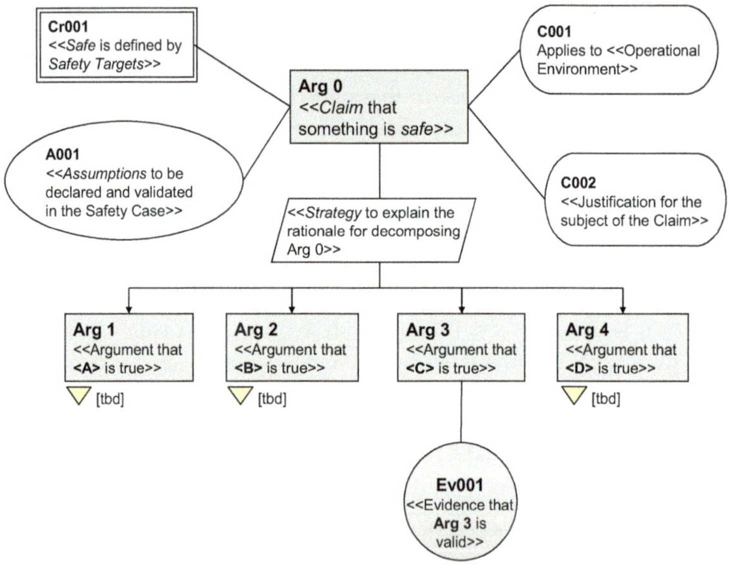

Fig. 8. A generic high-level Safety Argument

The *claim* is supported by vital contextual information, as follows:

- what is meant by *acceptably safe* is defined by means of *safety criteria*
- the *context* for the claim must include a description of the operational environment for which the claim is being made - we can deduce from section 4.1 above how critical this is to the validity of the claim
- *assumptions* are usually facts on which the claim depends and over which the organisation responsible for the safety case has no managerial control - eg driver behaviour, in the case of our airbag
- if the claim relates to a major change to a safety-related system, it is good practice to provide a *justification* for that change.

The arguments would then be further sub-divided until a level is reached at which a piece of documented evidence, of a manageable size, could be produced to show that the corresponding argument statement is valid. The question is how to ensure that a safety argument is complete and rigorous - for this, we use the three formal relationships derived in section 4, as follows.

- **Arg 1** - the system has been **specified** to be safe - ie meets the appropriate safety criteria - in the given operational environment (or EUC)
- **Arg 2** - the system **design** satisfies the specification
- **Arg 3** - the **implementation** satisfies the design

Then, by adding two further arguments:

- **Arg 4** - the **transition** from current system to the new (or modified) system will be safe - ie the known risks during this process have been reduced ALARP - and
- **Arg 5** - the **system** will be shown to operate safely **in service**

we have a sufficient, high-level safety argument for developing a new or modified system, bringing it into service and maintaining it throughout its operational life [Fowler et al, 2009]. Since it is the safety argument that determines ultimately what we need to demonstrate, we can use it to drive the whole assurance process as shown in Figure 9.

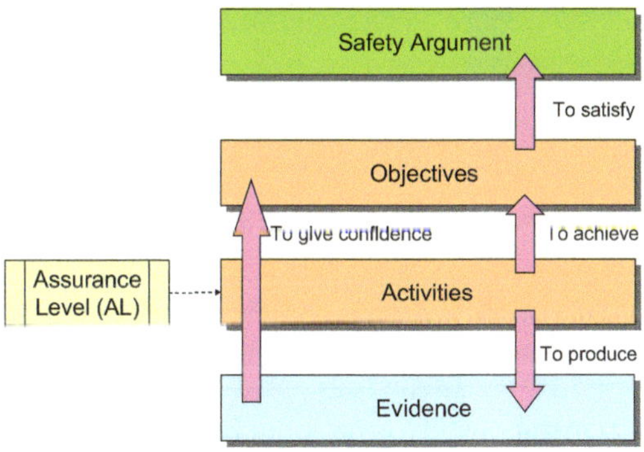

Fig. 9. Safety Assurance within an Argument Framework

The key point about this diagram is that it is the needs of the argument (ie the generation of evidence) that drive the activities - not the other way around - and the lifecycle phases contain only those activities that are necessary to support the argument.

6. Safety assessment in the project lifecycle

The above assurance principles apply to the five phases of a typical project lifecycle, as shown in Figure 10.

In practice, the Safety Plan (produced at the start of a project) should set out a specific safety argument and assurance objectives - with a rationale as to how they were derived to suit the nature and scope of the safety assessment concerned - the lifecycle assurance activities to be carried out, and the tools, techniques etc to be employed. Since the Safety Case (developed during, but finalised at the end, of a project) uses the same argument, it needs only to present the evidence resulting from the activities and provide the rationale as to how that evidence satisfies the argument.

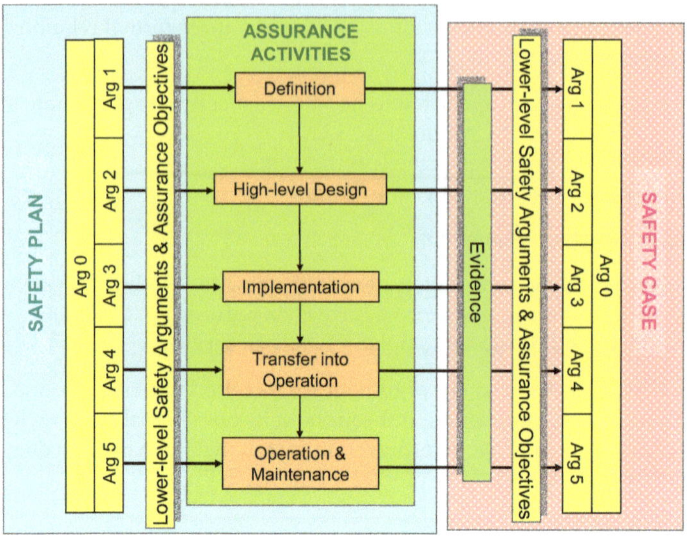

Fig. 10. Overall Safety Lifecycle Process

We will now examine in more detail what is done in the way of safety assurance objectives and activities in each phase of the lifecycle.

6.1 Definition phase

From section 5, we can see that in this phase we need to show that the system has been *specified* to meet the appropriate safety criteria in the given operational environment (or EUC). We will use a new, innovative[16] railway level-crossing[17] control (and possibly protection) system for a planned new two-way suburban road in order to illustrate some of the points in the steps described in sections 6.1.1 to 6.1.4 below - it should be noted that the analysis presented here is not intended to be exhaustive.

6.1.1 Operational environment

For our level-crossing, the properties of the operational environment would need to include:

- users of the crossing - eg passenger and freight trains, road vehicles (of which 80% are cars / light vans and the rest are trucks up to 40 tonnes weight) and occasional pedestrians; exceptionally (once very 1-2 months), large slow-moving vehicles carrying abnormally heavy loads will need to use the crossing
- average length of train - 120 m

---

[16] This is intended to be a hypothetical example; traditional railway standards - eg [RSSB, 2007] - for level crossings would probably not apply.

[17] Where a railway and road intersect at the same level.

- two-way rail traffic levels - 150 passenger trains per day (mainly between 07:00 and 23:00 hours) and 10 freight trains per day (mainly at night)
- road traffic levels - 3000 vehicles per day (mainly between 07:00 and 23:00 hours)
- traffic performance - the current railway speed limit is 140 kph for passenger trains and 60 kph for freight trains. The planned speed limit for the road is 100 kph for cars, 80 kph for vehicles over 2 tonnes and 65 kph for vehicles over 7.5 tonnes
- details of the road / track layout - the geography of the area makes the construction of a bridge / tunnel prohibitively expensive[18]

6.1.2 Pre-existing hazards

The main pre-existing hazard is " **HAZ$_{PE}$#1** - any situation in which, on current intentions, a road user and a train would inadvertently occupy the crossing at the same time". The use of "on current intentions" is crucial since it is describing a hazard not an actual accident. We could use mathematical modelling here to estimate the frequency with which this hazard would occur for a completely uncontrolled crossing and hence estimate the pre-existing risk.

6.1.3 Safety criteria

A suitable quantitative criterion would be that the likelihood of an accident involving multiple fatalities shall not exceed one per 100 years, supported by a second, ALARP criterion. However, given a possible range of outcomes of the hazard in this case, it might be appropriate to make use of a suitable RCS (along the lines of Figure 1 in section 3.2.2 above) in order also to set criteria for outcomes of lesser severity[19].

6.1.4 The specification

We recommend the use of the term Safety Objectives to describe the safety aspects of the specification. The reason is that it helps us remember that, in accordance with the Jackson model of section 4.1 above, what we are seeking to do here is describe, from the users' perspective, what the system must do, not to determine how the system will achieve that in its design.

First of all we need to consider the success case and assess how the pre-existing hazard is mitigated for all *normal* conditions in the operational environment - ie all those conditions that our SRS is likely to encounter on a day-to day basis - constructing various operational scenarios (eg single and multiple trains) as necessary. Two examples of a Safety Objective for this are as follows:

- SO#1 - a road user shall not enter the area defined by the crossing until its exit from the crossing is clear
- SO#2 - road users shall not enter the crossing from [say] 1 minute prior to a single[20] approaching train reaching the crossing until the train has cleared the crossing.

---

[18] This might need to be justified on ALARP grounds.

[19] However, for the purposes of this simple illustration, we will assume that if a train travelling at normal speed collides with a road vehicle there will be some fatalities.

[20] We would need an equivalent rule (ie Safety Objective) for multiple-train situations.

Note that these are genuine objectives (as above) and that the illustrative numbers would need to be verified by some form of dynamic risk modelling based on the environment described in section 6.1.1.

Next we need to assess how well the pre-existing hazard is mitigated for all *abnormal* conditions in the operational environment - ie all those adverse conditions that our SRS might exceptionally encounter - again, constructing various operational scenarios as necessary. An example of a Safety Objective for this is as follows:

- SO#n - in the event that an abnormally large / heavy vehicle (to be defined) is required to use the crossing, all approaching trains shall be prevented from entering the section of track [say] 1 km prior to the crossing until the vehicle has completely cleared the crossing.

This situation is expected to occur infrequently (see section 6.1.1 above) and therefore the system is allowed to operate in a different mode - in this case the train no longer has priority over the road vehicle - from the *normal* case. Again, some form of dynamic risk modelling could be used to determine a suitable exclusion distance for approaching trains.

Finally, we need to consider the potential failure modes of the system, at the service level. At this level, we are not concerned with the causes of failure[21], only with the consequences of failure - for which we would often use Event-tree Analysis for assessing multiple possible outcomes of a particular failure. It is important that the identification of possible failure modes be as exhaustive as possible; a useful starting point is to take each of the success-case Safety Objectives and ask the question what happens if it not satisfied. This will lead to the *system-generated* hazards, an example of which is:

- HAZ_{SG}#1 - road vehicle enters crossing that is in a *closed* state[22] (failure to satisfy SO#2).

Using the operational data from section 6.1.1 we can derive the following Safety Objective to limit the frequency of the hazard such that the appropriate portion of the tolerable-risk criterion (see section 6.1.3) is satisfied for this hazard:

- SO#n+r - the probability of a road vehicle entering the crossing when it is closed shall not exceed $5x10^{-5}$ per operating hour

Note that this illustrative figure takes account of the total number of system-generated hazards (assumed to be four in this illustration), the frequency with which road and rail traffic uses the crossing, and the providential mitigation that even if a vehicle incorrectly enters the crossing there is a significant probability that it would not actually collide with a train. Note also that the hazard occurrence rate is expressed as a frequency even though the SRS is not continuously operating - this was done in accordance with IEC 61508 because the demand rate on the SRS is relatively high (ie up to 150 operations per day).

Thus, at the end of the Definition Phase we should have a set of Safety Objectives which, if they are satisfied in the system design and implementation, would ensure that the pre-existing risk is mitigated, and the system-generated risk is limited, such that the level crossing would satisfy the specified quantitative Safety Criteria.

---

[21] This is done in the failure analysis of the high-level design - see section 6.2 below

[22] *Closed* here is defined by SO#2 (ie from 1 minute before an approaching train reaches the crossing, until the crossing is clear) - it does not necessarily imply a physical closure

6.2 High-level design phase

Having derived what Jackson [Jackson, 1995] refers to as a Specification for the system, the system-development task becomes less safety-specific, and has even more common with general system-engineering principles, except for one key feature - the higher level of confidence (ie assurance) that is required in the results of safety assessment.

Design, as we have seen, is about the internal properties of the system but for this phase we restrict the analysis to a logical design, in which Safety Requirements describe the main human tasks and machine-based functions that constitute the system, and the interactions between them. An illustration, based on our 'futuristic' level-crossing control system (LCCS) of section 6.1, is given in Figure 11.

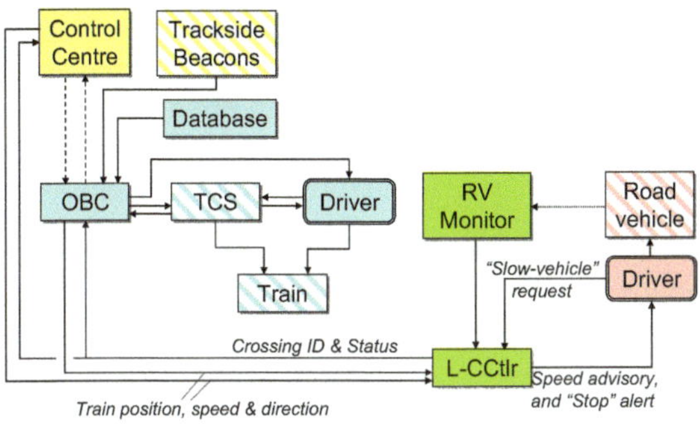

Fig. 11. Example of a Logical Model

A description of this example and the way that the model works is beyond the scope of this chapter - suffice it to say that the new system comprises a fully automated Level-crossing Controller and a Road Vehicle Monitor, which detects the presence of road vehicles within the crossing area. It is to be integrated into a regionally-based "Moving Block" signalling system using Communications Based Train Control - all the train control systems including ATP, but excluding the Onboard Computer, are subsumed into the TCS box[23].

The main points to note are as follows:

- it does not show elements of the physical design[24], such as hardware, software, procedures, training etc - nor does it separately represent human-machine interfaces, these being implicit in every link between a human and machine actor
- since the machine (eg Onboard Computer) and human (eg Driver) elements are shown separately this implies that the degree of automation in the system has been decided, at least provisionally
- the use of colour distinguishes between those elements of the end-to-end system that are in the train (blue), in the road vehicle (pink), in the LCCS (green) and are elements of external, interfacing systems (yellow). All elements that remain unchanged are shown with a striped background.

---

[23] Although level crossing control would normally be integrated with the Control Centre, for the purposes of this illustration we assume a separate subsystem.

[24] As we will see, physical design is taken to be the first stage of Implementation.

Safety Requirements capture what each of those "actors" needs to provide in terms of functionality, performance, reliability and integrity in order to satisfy the specified Safety Objectives. Whereas Figure 11 shows the actors and the way in which they interact quite clearly, the functionality that they provide is contained in the textural Safety Requirements, and the links between those functions are not easily seen. For this reason, on functionally-rich systems we often use a Functional Model, showing an abstract view of the system functions and data, as a bridge between the Specification and the Logical Design, thus increasing the confidence of the completeness and correctness of the latter [Fowler et al, 2009].

It is very important to note that making an argument for a Logical Design is not simply a matter of showing traceability of the individual Safety Requirements, for the Logical Design, back to the Safety Objectives of the Specification. This would ignore three possibilities: that the design as a whole might be in someway internally incoherent; that new failure properties could emerge at the design level that were not apparent at the higher (service) level; or that the Safety Requirements are too demanding of technology and / or human performance. Thus it is necessary to provide assurance that the Logical Design:

- has all of the functionality and performance attributes that are necessary to satisfy the Safety Objectives of the (service-level) Specification
- will deliver this functionality and performance for all *normal* conditions of the operation environment that it is likely to encounter in day-to-day operations
- is robust against (ie work through), or at least resilient to (ie recover easily from), any *abnormal* conditions that it may exceptionally encounter
- has sufficient reliability and integrity to satisfy the Safety Objectives of the Specification
- is realistic in terms of the feasibility of a potential physical system to satisfy the Safety Requirements, and the ability of validation & verification methods to demonstrate, at the appropriate time and to the necessary level of confidence, that the Safety Requirements are eventually satisfied.

By now it will (we hope!) be no surprise that to analyse, verify and validate the design from a safety perspective we use classical systems-engineering techniques, including:

- requirements traceability [Hull et al, 2005]
- Use-case Analysis [ISO/IEC, 2005] and Fast-time / Real-time simulations - for the normal, abnormal and failure scenarios
- Fault-tree Analysis [IEC, 2006b] to assess the causes of failure, "top down" , and
- Failure Modes Effects & Criticality Analysis [IEC, 2006a] to check the FTA, "bottom up".

Furthermore, since the human elements of the system have started to emerge, we can use human factors (HF) techniques such as Cognitive Task Analysis (CTA) and Human Reliability Assessment (HRA) to assess initially whether the task, performance and reliability & integrity demands which the Safety Requirements place on the human operators are at least realistic.

By the end of the High-level Design Phase we should have a set of Safety Requirements - covering the success and failure cases - that are sufficient to ensure that, if they are satisfied in the Implementation, the specified Safety Objectives would be met.

6.3 Implementation phase

We have defined the Implementation Phase such that it comprises development of a Physical Design and the realisation of the Physical Design in the built system. In making an argument for Implementation, we need to show that:

- the Physical Design satisfies the Safety Requirements for the Logical Design
- the causes or effects of any adverse, emergent safety properties (eg common-cause failures) or unwanted functionality have been mitigated in the Physical Design such that they do not jeopardize the satisfaction of the Safety Requirements
- the built system satisfies the Safety Requirements of the Physical Design (ie verification[25])
- the built and integrated system is consistent with the original qualitative Safety Objectives (ie validation).

In the physical design, we take the Safety Requirements from the Logical Design and allocate them to the elements of the Physical System, as follows:

- human tasks map on to, and provide the initial Safety Requirements for, skills, knowledge, procedures and training
- machine-based functions map on to, and provide the initial Safety Requirements for, hardware and software design
- human-machine interactions map on to, and provide the initial Safety Requirements for, human-machine interface (HMI) design.

These in turn lead to further design, Safety Requirements derivation and implementation for each of these elements and then to integration of the complete system - for further reading see [ISO/IEC, 2008b and ISO/IEC, 2008a]. The steps would follow, for example, the classical "V-model" of system development in which the safety engineer must ensure that the physical system as a whole (and its constituent parts) have sufficient reliability and integrity, and complete and correct functionality and performance, to satisfy the higher-level Safety Requirements. These are discussed in turn, as follows.

6.3.1 Building reliable systems – General

The engineering of a safety related system must ensure, as a minimum, that the safety reliability and integrity requirements are met. It is useful to consider first how failures occur and what techniques can be used to reduce failure rates to meet the safety criteria.

Random failures can occur in hardware of any kind due to physical degradation and wear-out mechanisms; the exact time when such a failure will occur is unknown but statistical distributions can be used to predict failure rates and quantification of overall system reliability can be modelled by techniques such as FTA mentioned earlier. Techniques for making hardware elements sufficiently reliable are considered in section 6.3.2 below.

Systematic failures by contrast are caused by design defects - they will occur whenever a system enters a state in which a latent defect is revealed. Software failures are always

[25] It is acknowledged that, in some industries / countries, verification and validation may have the opposite meanings to those used herein.

systematic[26], but hardware designs (especially those such as computer processor chips) can also exhibit systematic failures. Methods of ensuring that software is sufficiently free of defects, such that it can meet its safety function and performance requirements with sufficient reliability and integrity, are discussed in section 6.3.4 below. The concept of a systematic failure may also be applicable to human factors - eg if a procedure is designed incorrectly.

Common-cause failures are ones in which redundant subsystems fail at the same time due to the same external events (eg earthquake or tsunami), internal causes (eg power supply failure) or due to a systematic error affecting multiple systems (known as a common mode failure). This could be a software defect or a human maintenance intervention, for example. Common cause failures in practice often limit the achievable reliability of complex systems. The general approach is to attempt to identify and eliminate sources of common cause failure where possible and also to be conservative with reliability predictions to cater for the possibility of unknown causes.

6.3.2 Hardware safety

The main techniques for ensuring that random hardware failures are sufficiently unlikely are use of high reliability components and redundancy. High-reliability components are expensive so redundancy is used in practice except in special circumstances (such as space applications) where repair is difficult or impossible. Redundancy simply means having two or more subsystems each of which can perform the required safety functions; if one fails then a standby can take over provided that there is some mechanism (automated or manual) to detect the failure. Further gains in reliability can sometimes be achieved by using diversity, where the standby system(s) are not identical to each other. Ideally the diversity should be both conceptual (using different physical processes or measurements) and methodological (different design methods) since this helps to reduce the likelihood of common mode failures (discussed in the next section). Part 2 of IEC 61508 [IEC, 2010], in particular, provides requirements and guidance on hardware safety techniques and measures.

Of course, we must not forget the fundamental principle that (with the possible exception of Information Systems) the functionality and performance properties of hardware is as important to safety as its reliability and integrity is - see the airbag example in section 2.

6.3.3 Human factors safety

HF is a topic that in the past in many industries has had only scant coverage in functional safety [Sandom, 2002]. More recently, things have improved, not the least in European ATM for which EUROCONTROL has developed the "HF Case" [EUROCONTROL, 2007].

In the HF Case, Human Factors are considered on two different levels, the System Level and the Human Performance Level. The underlying philosophy is that the design of tasks, procedures and tools must correspond to the safety requirements on both the level of the overall system as well as on the level of the individual human operator.

---

[26] Although they may be revealed in a quasi-random way.

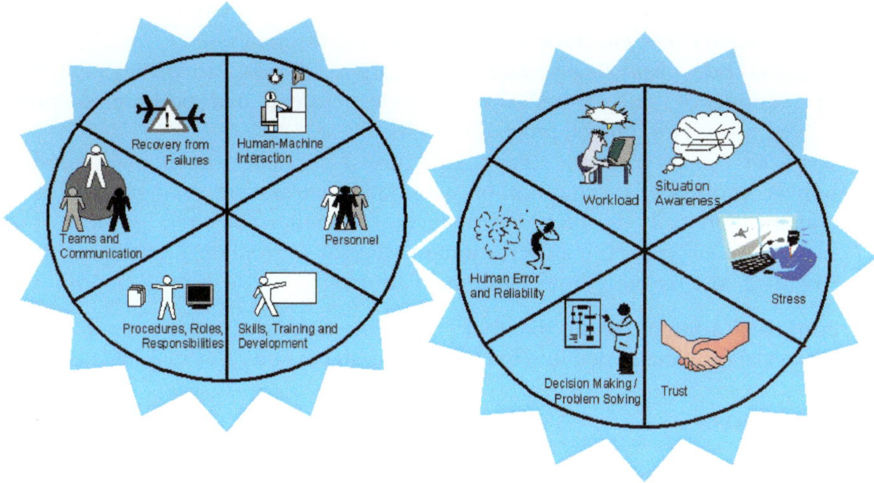

Fig. 12. The HF "Gearbox"

The HF aspects can be classified into six areas at each level, and for each specific task, procedure or tool some of twelve areas may be more important than others; however, the underlying principle is that the HF evaluation should involve both levels. The resulting approach is known as the HF Gearbox, as illustrated in Figure 12.

From a safety perspective, the HF Gearbox, at the human-performance level, addresses:

- *Situation awareness*: the continuous extraction of external information, integration of this information with previous knowledge to form a coherent mental picture, and use of that picture in directing further perception and anticipating future events
- *Workload*: the degree to which the operators' mental resources are consumed by the task at hand.
- *Stress*: a positive or (more often) negative emotional response to a more or less demanding situation.
- *Trust*: the 'right degree of trust' in the equipment permits the operator to exploit the benefits of a equipment whilst being aware of its limitations
- *Human error and reliability*: Human reliability is the operator's capacity to execute a specific task resulting in a performance within specified limits - see section 6.2.
- *Decision making/problem solving*: Decision making is performing a selection between possible options in a specific situation. It can be rational / or emotional.

From a safety perspective, the HF Gearbox, at the system level, addresses:

- *Human-Machine Interaction*: input devices, visual displays, information requirements, alarm handling, HMI usability, fatigue, distraction and concentration, noise, lighting and general comfort as it affects human performance
- *Organization and Staffing*: Staff requirements, manpower availability, operator profile / selection criteria, shift organization
- *Training and Development:* Training needs performance / competence standards, training content, training methods and media, trainer role / responsibilities /

competency, on-the-job training, emergency / abnormal-situation training, testing of training effectiveness, effects on operational task performance

- *Procedures, Roles and Responsibilities*: allocation of task, involvement, workload, trust / confidence, skill degradation, procedure format and positioning, procedure structure, procedure content, procedure realism
- *Teams and Communication*: Team structures / dynamics / relations, coordination, handover processes, communication workload, phraseology, language differences, communication methods, interference effects, information content
- *Recovery from Failures*: Human error potential, error prevention / detection / recovery, detection of, and recovery from, equipment failures.

It is crucial from a safety perspective that HF is not considered to be a separate activity - rather it must be fully integrated into the safety assessment and resulting safety case.

6.3.4 Software safety

6.3.4.1 An IEC 61508 perspective on software safety

Part 3 of IEC 61508 [IEC, 2010] starts at the point where the software requirements specification for a safety related system has been developed, in terms of safety functional behaviour and safety integrity. The system engineering approach described in this chapter should, of course, be used in deriving those software requirements in the first place.

IEC 61508 Part 3 is based on a conventional V lifecycle model for the design and implementation of software. A process-based approach is convenient for software developers who have to follow the standard, but Part 3 stresses that it is not the existence of the process itself but the evidence resulting from the *application* of the process which will demonstrate the achievement of safe software. The development lifecycle stages comprise:

- software architectural design,
- detailed design,
- module design,
- coding.

Verification is required after each stage, to check that the output of the design stage is a correct refinement of the input and has other necessary properties. Verification as a minimum includes software code reviews but can include other forms of analysis ranging from code complexity analysis and rigorous inspection techniques up to formal proof that the software has certain properties. Testing of the software mirrors the design, as follows:

- module or unit testing, to demonstrate that the each software module behaves in accordance with its specification
- integration test at the software design level(s) (which includes hardware/software integration testing), to show that all modules function together as intended
- safety validation testing at the software requirements level, to provide confidence that the safety function and performance requirements are met.

The stress laid by IEC 61508 Part 3 on module testing is justified by experience that software which has been well tested at the module level will usually reveal few errors during later testing, although it is a step often skimped due to time pressures during development.

Detailed techniques and measures are recommended in Annexes A and B of IEC 61508 Part 3 to support each lifecycle stage. Techniques are either Highly Recommended (HR), Recommended (R) or noted without any specific force (-). In a very small number of cases, techniques are Not Recommended. The nature and rigour of the HR techniques and the number to be applied increases with the SIL – this is common to a number of software safety standards and guidelines. However, it is not possible simply to apply all the HR techniques for a given SIL, since some may be contradictory; therefore judgement must be applied in selecting the techniques which will give the greatest benefit. If a relevant HR technique is not used, its omission must be agreed with the independent safety assessor (see below). The standard stresses that it is the combination of testing and analysis that provides the necessary confidence that the software will be safe.

Properties of the design artefact(s) are stated for each lifecycle stage as a guide to selecting and justifying the techniques and measures to be used. Properties include:

- completeness (with respect to the higher level representation)
- correctness
- ease of understanding
- freedom from intrinsic errors.

An "intrinsic" error is one which can be recognised regardless of the functions which the software is to realise – examples at the source code level could include division by zero, numeric overflow, or access via a null pointer. This class of errors typically cause run-time "crashes" and there are analytical tools which can help to eliminate such errors.

Use of pre-existing software elements (such as an operating system or communications software) is allowed, provided that sufficient evidence of reliable operation can be provided. This can include evidence from non-safety applications subject to certain conditions.

Although IEC 61508 Part 3 is based on the V lifecycle, alternative lifecycles can be used. For example, if code is generated automatically from a high-level requirements model (which is possible for some types of control applications) then the software design and coding stages can be omitted, although testing will still be required. The selection of tools to support each lifecycle stage and technique is specifically addressed in Part 3 – a tool which makes a direct contribution to the final software (such as a compiler) must be chosen and justified with greater care than one where the output of the tool is readily amenable to manual inspection and correction.

In common with Parts 1 and 2 of IEC 61508, independent functional safety assessment is required, the degree of independence depending on the SIL to be achieved. For example, a person independent of the designer is sufficient at SIL 1 whereas an independent organisation is required at SIL 4. The assessor should examine both the process itself (the selection of techniques and measures) and the products of the development (for example the test and analysis results and the backing evidence that the testing and analysis have been sufficiently thorough). It is interesting to note that independence between designer/implementer and verifier/tester is not required in IEC 61508, although it is required in other standards and guidelines for safe software.

All evidence resulting from test, analysis and field service experience must be recorded and kept under configuration management along with the design artefacts and software code–

therefore, applying the standard should generate the evidence required to meet SW01 (see subsection 6.3.4.2).

6.3.4.2 A regulatory perspective on software safety

There are two main ways of gaining **assurance** that the safety requirements have been properly and completely implemented in an SRS: assurance which is obtained directly from the attributes of the product itself; and that which is obtained from the characteristics of the processes which gave rise to the product.

So what is an appropriate balance between product and process assurance, and what part should the various standards play in the achievement and demonstration of system safety? The SW01 section of the UK CAA's safety regulation CAP 670 [UK CAA] takes an objective-based approach that provides a sensible answer to these questions[27]. SW01 takes an approach to safety assurance which is deliberately non-prescriptive in terms of the development process; instead, it demands arguments and evidence of the achievement of five safety assurance objectives – ie to show that the:

- software safety requirements correctly state what is necessary and sufficient to achieve tolerable safety, in the system context
- the software satisfies its safety requirements
- each software safety requirement can be traced to the same level of design at which its satisfaction is demonstrated
- software implemented as a result of software safety requirements is not interfered with by other software [that is not safety related]
- all assurance relates to a known executable version of the software, a known range of configuration data and a known set of software products, data and descriptions that have been used in the production of that version.

SW01 defines seven behavioural *attributes* of safety-related software, which the safety assurance must address, with equal rigour (or a valid argument presented as to why a particular attribute has not been addressed), as follows: functionality, timing, accuracy, robustness, reliability, resource usage, and overload tolerance.

In the context of requirements satisfaction, SW01 allows assurance to be offered from three different sources – ie testing, analysis (of design), and field service experience (FSE). For each source of assurance, two forms of evidence are required, for each *attribute*, as follows:

- *Direct* evidence - that which provides actual measures of the product (software) attribute concerned and is the most direct and tangible way of showing that a particular assurance objective has been achieved
- *Backing* evidence would relate to quality of the process by which those measures were obtained and provides information about the quality of the *direct* evidence, particularly the amount of confidence that can be placed in it.

Testing is restricted largely to tests of the final product (executable code) or a very close relation of it. *Direct* evidence is concerned with what tests were carried out and what the

---

[27] Although SW01 covers specifically safety-related software, most of the principles in it can apply equally to the wider aspects of the system

results showed in terms of satisfaction of the safety requirements. *Backing* evidence is concerned with showing that the tests were specified correctly and carried out adequately.

FSE is based on previous operational use of the software. *Direct* evidence is concerned with analysis of data from FSE and what the results of that analysis showed in terms of satisfaction of the safety requirements. *Backing* evidence is concerned with showing that the environment from which the data was obtained is sufficiently similar to that to which the re-used software will be subjected, that an adequate fault-recording process was in place when the software was originally deployed, and that the data-analysis process was adequate and properly carried out.

Analysis covers any proof of requirements satisfaction that is obtained from the design or other representation of the software, including models, prototypes, source code etc. *Direct* evidence is concerned with what the results of the particular *analysis* techniques showed in terms of satisfaction of the safety requirements. *Backing* evidence is concerned with showing that design and other representations of the software were appropriate and adequate and that the *analysis* was adequately specified and properly conducted; where *analysis* was carried out on source code, it is necessary also to show that the object code correctly translates the source code.

In general, the rigour demanded of the evidence increases as the integrity required of the software increases. However, SW01 defined this integrity only in terms of the consequence of failure - ie it does not take account of the probability that such a failure will occur, as is the case, for example, in IEC 61508.

The way in which evidence from the three sources can be used in combination varies according to the attribute for which evidence is offered, and also depends on the integrity required of the software. As the required integrity increases, SW01 allows less dependence on a single source of evidence, and places more emphasis on *analysis* of design as the primary source of evidence.

The advantage of testing over design analysis is that it is carried out on the end product rather than on a representation of that product. On the other hand, the effectiveness of testing can be limited by problems with test coverage and with confidence levels in respect of statistical attributes of the system. For these reasons, assurance from testing usually takes second place to design analysis for the more safety-critical systems, though it is interesting to note that SW01, for example, mandates some degree of testing even where the primary source of assurance is design analysis.

As a source of assurance, design analysis has one further advantage over testing - it is available much earlier in the lifecycle and can therefore make a major contribution to the reduction of programme risk. Iterative development techniques seek to bring forward the availability of test evidence but in so doing bring with them their own problems, including a possible reduction in effectiveness of design assurance unless specific measures are taken to avoid this.

6.4 Transfer phase

The Transfer-into-Operation Phase takes the assurance process up to the point that the system concerned is ready to be brought into operational service. In making an argument for Transfer into Operation, we need to show that:

- everything necessary has been to prepare the new (or modified) system for operational service
- the process of bringing the system into service - ie transitioning from the current system to the full new system - will itself be as safe as reasonably practicable

Preparation for operational service is about showing that all the necessary operational procedures, trained personnel and technical resources are in place to operate and maintain the system and that, under conditions as close a possible to real operational service, the system / resources as a whole will meet the expectations of the service users (ie system validation). Typically, a safety case would be submitted for management approval and regulatory endorsement before transition from the current system to the new system begins. Therefore it is necessary to show beforehand that any risks associated with transition will be managed such that the AFARP criterion will be met throughout this process.

Especially in safety-critical industries that require an uninterrupted operation for 365 days per year, live system testing and the subsequent transition from the old to the new system are often hazardous. Thus a Transition Plan needs to drawn up and executed to ensure that:

- the hazards and risks associated with the transition process have been completely identified and correctly assessed
- measures have been put in place to reduce those risks *ALARP*
- contingency plans are in place to revert to a safe state if the transition is not entirely successful
- the old system components can be removed safely

6.5 Operational phase

The safety focus during the whole of the systems in-service life is on providing assurance of its continuing safe operation. This is vital for a number of reasons, including:

- the *a priori* safety assessment, covered in Arg 1 to 3, might not be complete and correct in every particular
- the system, including the equipment and human elements, might degrade in operational service
- the system will most probably be subject to changes at various times during its operational life
- the operational environment might change during the life of the system.

Thus we need to ensure first of all that the system (comprising equipment, people and procedures) will be supported so as to maintain the required level of safety. The evidence in support of this will be mainly in the form of SMS processes (and related operational and engineering procedures) and how it will be ensured that they will be properly applied - including the use of surveys and audits related to the application of those SMS processes.

Secondly, in order to provide assurance of actual safety achievement we need to show that:

- there is a culture to encourage full and accurate reporting of safety incidents
- the frequency of safety incidents will be measured against pre-defined indicators
- all reported incidents will be properly investigated
- appropriate corrective action will be taken to prevent incident recurrence.

Finally, we need to show that there are procedures in place to manage future changes to the system and / or it operational environment.

7. Conclusions

We started by asserting that safety is not a separate attribute of a system but is a property that depends on other attributes, and on the context in which the system is used. The misconception that adequate reliability and integrity are sufficient to ensure the safety of a system has been prevalent in, for example, the ATM sector [Fowler & Grand-Perret, 2007], but is dispelled in the specific context of software by the results of extensive research [Leveson, 2001] and more generally herein by rationale argument using the example of a car airbag.

After introducing some basics safety concepts and principles, we then showed that safety is as much dependent on correct functionality & performance of the system as it is on system reliability & integrity - the former set of attributes being necessary for the mitigation of *pre-existing* risk (inherent in the operational environment) and the latter for controlling *system-generated* risk (caused by system failure). This led to the view that what was needed was a broader approach (what we called the *success & failure* approach) to system safety assessment, a view that was then shown to be consistent with the principles underlying the generic functional-safety standard IEC 61508 [IEC, 2010] - principles that, however, are not always captured in industry-specific instantiations of this standard.

We then turned to a vital aspect of systems engineering - ie requirements engineering, some important principles of which are advocated in [Jackson, 1995] - and found direct equivalence between the derivation of the required safety properties of a system and the derivation of its non-safety properties.

Finally, we introduced the principles of safety assurance and safety cases and showed how they should drive all the processes of a safety assessment, throughout the lifecycle. Whilst emphasising the importance of ensuring that the level of assurance is appropriate to the safety-criticality of the system, we now leave it to the knowledgeable systems engineer to recognise the common processes of a system development lifecycle in this chapter and to conclude for himself / herself that safety is (with the assurance proviso) actually just a viewpoint (albeit a very important one) on systems engineering!

8. Acknowledgment

The authors would like to record their appreciation for the many very helpful suggestions made by Drs Michael A Jackson and Anthony Hall during the development of this chapter.

9. References

European Commission, 2005, Regulation (EC) No 2096/2005, *Common Requirements for the Provision of Air Navigation Services*, published in the Official Journal of the European Union

Eurocontrol, 2007, *The Human Factors Case: Guidance for Human Factors Integration*

Eurocontrol, 2010, http://www.skybrary.aero/index.php/Risk_Assessment

Fowler D, Grand-Perret S, 2007, *Penetrating the Fog of Safety Assessment - and Vice-versa,* Proceedings of the 2nd IET International Conference on System Safety, London, UK

Fowler D, Perrin E, and Pierce R, 2009, *2020 Foresight - a Systems-engineering Approach to Assessing the Safety of the SESAR Operational Concept,* Proceedings of the 8th USA/Europe Air Traffic Management Research and Development Seminar, Napa, USA

Hull E, Jackson K and Dick J, 2005, *Requirements Engineering,* Springer, ISBN 1852338792

International Electrotechnical Commission, 2006a, IEC 60812, *Analysis techniques for system reliability – Procedure for failure mode and effects analysis (FMEA)*

International Electrotechnical Commission, 2006b, IEC 61025, *Fault Tree Analysis*

International Electrotechnical Commission, 2010, IEC 61508, *Functional Safety of Electrical/ Electronic/ Programmable Electronic Safety Related Systems,* V 2.0

ISO/IEC 12207, 2008a, *Systems and Software Engineering - Software Lifecycle Processes,* V2.0

ISO/IEC 15288, 2008b, *Systems and Software Engineering - System Lifecycle Processes,* V 2.0

ISO/IEC 19501, 2005, *Information technology, Open Distributed Processing – Unified Modelling Language (UML),* V 1.4.2

Jackson M A, 1995, *The World and the Machine,* Proceedings of 17th International Conference on Software Engineering, IEEE, pp283-292

Kelly T and Weaver R, 2004, *The Goal Structuring Notation – a Safety Argument Notation,* http://www-users.cs.york.ac.uk/~tpk/dsn2004.pdf

Ladkin P B, 2008, *An Overview of IEC 61508 on E/E/PE Functional Safety,* http://www.causalis.com/IEC61508FunctionalSafety.pdf

Leveson N G, 2001, *The Role of Software in Recent Aerospace Accidents,* 19th International System Safety Conference, Huntsville AL, USA

Mana P, De Rede J-M and Fowler D, 2007 *Assurance Levels for ATM system elements: Human, Operational Procedure, and Software,* proceedings of the 2nd IET International Conference on System Safety, London, UK

Perrin E and Kirwan B, 2009, *Predicting the Future: The Integrated Risk Picture,* Proceedings of the 4th IET International Conference on System Safety Engineering, London, UK

Pierce, R and Fowler D, 2010, *Applying IEC 61508 to Air Traffic Management,* Proceedings of the Eighteenth Safety Critical Systems Symposium, Bristol, UK

Rail Safety and Standards Board - RSSB (2007). *Engineering Safety Management (The Yellow Book), Volumes 1 and 2 - Fundamentals and Guidance,* Issue 4

Reason J, 2000, http://www.bmj.com/cgi/content/full/320/7237/768

RTCA, 1992, DO-178B, *Software Considerations in Airborne Systems and Equipment Certification,*

Sandom C, 2002, *Human Factors Considerations for System Safety,* in Components of System Safety, Redmill F and Anderson T [Eds.], proceedings of 10th Safety Critical Systems Symposium, 5th-7th February 2002 Southampton, Springer-Verlag, UK

UK CAA, 2011, UK Civil Aviation Authority, CAP670, *Air Traffic Services Safety Requirements*

7

Leveraging Neural Engineering in the Post-Factum Analysis of Complex Systems

Jason Sherwin[1] and Dimitri Mavris[2]
[1]*Columbia University in the City of New York,*
[2]*Georgia Institute of Technology,*
USA

1. Introduction

This chapter is about the pressing problem of, and our proposed response, to data deluge in the analysis of complex systems. We begin by illustrating the problem in certain systems engineering examples, primarily focusing on aerospace-related systems but pointing out the generality of this problem in other data-intensive design problems (Section 2). Having established the need to address this problem, we then propose a solution based on current advances in the intersecting fields of neuroscience and computer engineering, increasingly being called *neural engineering* (Section 3). With a proposed solution in mind, we carry out a case study in which we utilize certain results and algorithms from neural engineering (Section 4). Though this case study gives credible results, we find that we can improve our neural-based models of complex systems data from more direct neuroscience experiments on expertise (Section 5). Finally, we draw conclusions on the current state of the art for leveraging neural engineering results and algorithms on the problem of complex systems post-factum data analysis (Section 6).

2. A problem in systems engineering: Data deluge

The need to engineer within and for both increasingly complex and sophisticated systems is continually growing. In tandem with this problem is the need to analyze ever-increasing amounts of data that describe these systems. In short, the post-factum analysis of an already-built system is a key step in the analysis and, consequently, the design processes.

For instance, within the aerospace community, this problem is not unfamiliar. In that field, the perennial aim has been to balance the various sub-disciplines (e.g., acoustics, aerodynamics, propulsion, structures) to deliver an aircraft or spacecraft that meets a set of pre-defined criteria. But with each added sub-system of an aircraft, there is an added degree of complexity that is contributed to the design process.

This phenomenon is not unique to aerospace systems design. More generally, with the explosion of remote sensing capabilities in recent years, there has been a deluge of data made available about many other complex and intricate systems. But the means to fully analyze this data and to extract a useful comprehension of its content can be a challenge.

Both of these problems - one being a subset of the other - share a common thread: there is a plethora of computation needed to arrive at a design solution. In aircraft design, there is a potential deluge of possible designs as new sub-systems are added to the analysis. Similarly, in the mining of data from complex systems, there is likewise a deluge of possible data interpretations; and no specific interpretation is more 'correct' than any other (via the 'No Free Lunch Theorem', Ho & Pepyne, 2002).

In the midst of this deluge, it is potentially easier to approach the data holistically and to provide a subjective analysis of its content. Not only does this approach allow the data's full scope to be considered, but it also allows comprehension to be communicated rapidly because of its approximate - and therefore, simpler - nature. For instance, many systems engineering techniques have been devised to simplify the potentially overwhelming aspects of a complex system's analysis. Some examples of these are the analytical hierarchy process (Saaty, 2000 and Saaty, 2008), quality function deployment (Chan & Wu, 2002 and Akao, 1990) and other quasi-quantitative methods of subjective evaluation. While these methods have proven to be rapid, their transparency is lacking due to the expert-driven nature of the processing schema. For instance, in quality function deployment (QFD), experts in a particular discipline create subjective mappings from requirements to characteristics for a given product. There is no physics-based model that determines the product's design. Rather, a graded scale is used to map design requirements to characteristics based on a subjective assessment done by an expert. Necessarily, in this and other techniques like it, there is a crucial role for an expert's input to such analysis.

There has also been an opposite response to the data deluge in system analysis: utilize the increasingly available computation power to process the excessive amounts of data. In other words, rather than resign to the need for subjective analysis (e.g., in QFD) due to the problem's complexity, the availability of greater amounts of computing power in recent years has made it possible somewhat to navigate the deluge. For example, this has been the mentality behind the approach of multi-disciplinary optimization (Vanderplaats, 2007), which is used with great success in aircraft design. In multi-disciplinary optimization (MDO), numerical optimization techniques are applied to sets of objective functions whose dependent variables must satisfy various constraints (i.e., inequality, equality and side constraints). The ultimate aim though is not to yield a design that is optimal in any one system, but rather one that is optimal with regard to all systems. Necessarily, such a process is computationally quite costly and as the number of variables grows it becomes infeasible.

Similar examples exist for the analysis of data obtained remotely. For instance, the American military increasingly relies on remote sensing for many of its activities (e.g., the MQ-1 Predator drones, National Commission, 2004). But the exponentially-increasing amounts of data leave the analysts "swimming in sensors and drowning in data" (Drew 2010). In other words, the analytic tools to comprehend the data are well behind the means to gather it.

Such an analysis problem is an inherent precursor to engineering a complex system. For instance, it exists in the case where the system has been human-constructed from many parts (e.g., an aircraft). And it also exists when the system is not human-constructed, i.e., in nature. It is this latter situation that is of the most interest to us now though because, in reality, it is a superset of the first: whether man-made or not, it is a difficult engineering analysis problem to figure out how complex systems work. In particular, although the human-constructed parts may behave in predictable ways in many situations, there are always new interactions arising between component sub-systems that reveal a previously unknown system-level behavior. Therefore, while reductionist approaches to system construction can be successful in most cases, we have still not obtained the hoped for deterministic prediction of behavior.

Instead, it seems that a degree of uncertainty is inherent in the analysis and consequently the engineering of all systems. This does not mean that we throw the previously successful reductionist approaches to the wind though. But for the analysis and engineering of those systems for which such approaches are impossible (e.g., not accurately quantifiable, not modelled within computational constraints), the only mechanism for design choices thus far has been the aforementioned systems engineering techniques (QFD, AHP, MDO, etc.). A new approach is needed.

3. A new path for handling data deluge in analysis: Neural engineering

As a result of this problem, we suggest here to consider the breadth of results and techniques emerging from neural engineering to bolster systems analysis for engineering purposes. In particular, instead of relying on an inconsistent mapping made by human experts to design analysis (e.g., as in QFD), why not understand some cognitive elements to expertise and, in turn, apply that comprehension to both systems analysis and manipulation? Of course, these are both monumental tasks to perform, considering not only the breadth of cognitive abilities that comprise expertise but also determining how to implement them in real engineering contexts.

Despite the seemingly daunting nature of these endeavors, certain elements of expert decision-making, human situational awareness and cortical biology can inform some of the details as to how we can understand and, in turn fine tune, the ways by which we as engineers collect observations and integrate them into a cohesive analysis; such an analysis is then the foundation of ensuing engineering choices. Nowhere is this need as great as it is in the analysis of complex and large-scale systems. Therefore, a true test as to the utility of neural engineering for systems purposes would be to implement these ideas within a complex or large-scale analysis and engineering task. In this chapter, we will demonstrate a simulated use of such an application.

As a demonstration, we discuss the application of neural engineering to the analysis of Iraq's stability during 2003-2008. This application was never used in a real context, however we frame the problem within the context of its utility to a decision-maker whose actions influence the outcome of such a system. In other words, he/she must analyze and then manipulate this system. Our assumption is that the decision-maker only has access to a stream of data that measures certain conditions related to Iraq's stability. More importantly, we assume that there is no possibility of developing an analytic model to describe the time-evolution of Iraq during these years. Rather, we cast aside that futile aim and attempt to glean useful patterns directly from the data. As part of this demonstration paraphrase (seeing as the full-blown analysis comprises a Ph.D. thesis and several papers), we emphasize the importance of learning algorithms to do so. Specifically, we consider algorithms based off of some anatomical and behavioral features of the human cortex. Built-in to the rationale behind using these algorithms is the assumption that many of the cognitive faculties comprising the development and use of expertise reside in the cortex.

Building off of the hope (and shortcomings) provided by the Iraq example, we then review some of the latest developments in neural engineering that have possible applications in the analysis of other large-scale systems. As we will see, it is important to maintain an awareness of how the biological hardware (e.g., a neuronal network) "computes" its analysis of complex, time-evolving, often self-conflicting and/or occluded data. We will offer the results of an experiment from audio cognition in which expertise of subjects is tracked directly from neural data. Finally, we will then consider how these insights would then translate back to the actual solid-state computations done in modern computers to drive systems engineering analysis.

4. Case study of neural engineering-assisted analysis: Iraq war, 2003-2008

The focus of this case study is to create a computational situational awareness (SA) usable by the Department of Defense to gauge Iraq's stability during 2003-2008. Situational awareness is a term from psychology used to describe elemental steps of an expert's mental processes (Endsley, 1995). In other words, situational awareness is the description of the cognitive processes involved in an expert's analysis of a situation. So if we can design an appropriate computational SA for the Iraq context then it is equivalent to developing a means to analyze that context for driving it to a desired state - in other words, to engineer it.

As a theoretical construct, SA was developed to analyze the decision-making processes of aircraft pilots, yet its general usage has extended into many other areas in which expertise are employed by a human controller. In general, an SA can be particular to a given scenario or context. For example, pilots have SA for flying airplanes, pianists have SA for playing music, etc. In our case study, the SA of interest applies to the stability of Iraq during the war from 2003-2008, which henceforth will be called the 'Iraq context'.

To develop this SA computationally, we implement a method that is summarized by the information flow of Fig. 1.

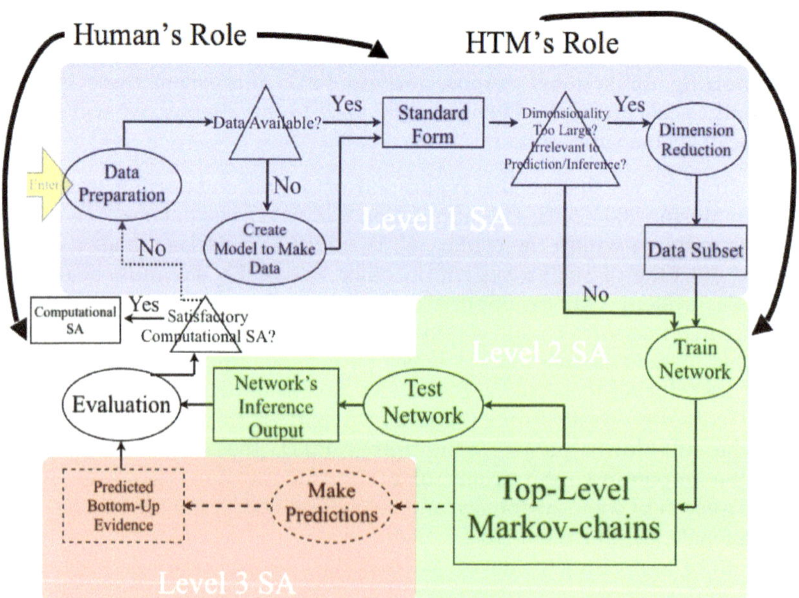

Fig. 1. Method for building/maintaining computational SA with HTM

This method maps the neurally-inspired machine learning algorithms to be used here (Hierarchical Temporal Memory, or HTM, see Section 3.2.1 for details, George & Hawkins, 2009 and George, 2008) to the three levels of SA first hypothesized by Endsley (Endsley, 1995). These three levels are Level 1 (perception of relevant elements), Level 2 (comprehension of those elements) and Level 3 (prediction). Here, we focus on Levels 1 and 2, since they are a necessary antecedent to Level 3. In particular, we present here a way to implement Levels 1 and 2 for this problem via data preprocessing and HTM training/testing.

4.1 Why use such a high-level representation of mental processes?

While this approach enhances awareness of trends in the data of the Iraq context, it also mimics the basic tenets of what constitutes SA in actual decision-makers. In particular, Endsley and others have shown that the selection of a goal is crucial to SA formation. In other words, the collection of data, its analysis and the engineering goal are all inextricable in forming SA. Here, we assume the criteria for success established by the U.S. Department of Defense: to bring Iraq to political, economic and social stability between 2003-2008 (United States House of Representatives, 2005). Consequently, we rely on data related to these aspects of the Iraq context, so that not only do we enhance a decision-maker's own SA of the Iraq context but we also create one – albeit, a rudimentary one – with a computer. By starting from such a high-level representation of expert mental processes, we can then specialize the computational tools used to find problem-relevant patterns in the data.

4.2 Deploying these processes computationally

Once the collection of data is focused onto problem-relevant elements, the analysis of that data becomes a learning problem. By conceiving of the expert's assessment of data as a learning problem (due to Endsley), we are in a position to mimic some of these processes computationally.

However, there is a question to consider before doing so: What is the right answer? In particular, no matter what machine learning approach is used, it is difficult to validate the SA learned about this context, since we do not know the right answer a priori. In other words, we do not know if our assessment of Iraq's stability is 'correct'. Although this is possible in other learning problems, such as invariant visual pattern recognition (i.e., the pattern is either object A or object B), we cannot do this here.

So to verify the accuracy of the computational SA formed in this context, another method will be introduced that has influence from system dynamics: we call it extreme-case bounding. This method has assumptions built into it that creates fictitious extreme cases of stability, either extremely unstable or extremely stable. With these fictitious bounds used for HTM network training/testing (e.g., extreme dystopia or utopia based on the data), some insight into the actual progression of events in Iraq during 2003-2008 can be obtained. Needless to say, this method is not perfect and it is somewhat arbitrary because we arbitrarily select a peg against which to measure Iraq's stability. Nevertheless, it provides an intriguing foothold in an avenue of computational SA that has thus far been difficult to probe concretely.

4.2.1 The final computational piece: A neurally-inspired machine learning algorithm

Thus far, Hierarchical Temporal Memory (HTM) algorithms have been mentioned as a way to execute the various stages of SA accounted by Endsley and have been adapted into Fig. 1. However, we have not yet discussed why these algorithms in particular are of interest. In what follows, we will argue that the hierarchical storage of spatio-temporal data and the way by which temporally adjacent data points are related to each other lend well to steps of SA laid out in Fig. 1. To make these points clear, Fig. 1 includes learning and inference steps involved in HTM training and testing as well.

An HTM attempts to mimic two crucial aspects of cortical function and anatomy. These are particularly of use for determining better ways to handle highly-varied data, so their potential utility in forming SA are apparent. First, these algorithms rely on the temporal adjacency of observed events when storing spatial patterns. The anatomical inspiration for this procedure comes from observations of cortical function. In particular, there are numerous cell groups and types in the cortex that have been identified as 'sequence detectors' (e.g., PPA, Broca's Area, FFA). Secondly, and in relation to the first aspect, the algorithms store these sequences in a hierarchical arrangement across both space and time. The result of this division of spacetime is that local regions' spatio-temporal patterns are first encoded from which more global regions' patterns are then encoded, etc. The anatomical inspiration for this compartmentalization of information comes directly from the different hierarchical functional areas observed in the human cortex (e.g., visual cortex, audio cortex, etc.).

Since an HTM is implemented on a computer, it is perhaps useful to consider a mathematical description of HTM. For starters, a trained HTM is in fact a Bayesian inference network. In particular, it is a network of nodes, each of which solving the same problem: learning spatio-temporal sequences. On a network-level, the goal behind the algorithms is to learn a schema (S) that describes data related to a given problem. That problem exists locally for each node in a vector space (v_i) and, upon grouping all nodes, exists on a global level that concerns all data related to the problem (V). Considering the local version of the problem, each vector (x_k) in v_i is an observation of an aspect of a given complex phenomenon at the kth time step. The HTM node's goal then is to create q Markov-chains to which any one of the vectors in v_i can be assigned. For compression purposes, it is highly desirable that $q < k$. By collecting sets of observations in this way, each node's Markov-chain (m_q) corresponds to some spatio-temporal high-level feature of the local complex phenomenon. Consequently, the set of all M Markov-chains constitutes a schema (S) of the global phenomenon. In other words, S is a reduction of the phenomenon witnessed in each node's vector space, v_i. In particular, by using HTM, the aim is to use learning mechanisms akin to certain aspects of neural coding to develop a schema, i.e., a situational awareness of the data (V).

4.3 Implementation for the Iraq context

For the Iraq context, the phenomenon is the Iraq War during 2003-2008. The goal of the HTM then is to create a schema (S_{Iraq}) that is a suitable description of the Iraq context. This schema is based on what kind of data is used to describe the Iraq War (V_{Iraq}). Recalling that the analysis and collection of data are inextricably linked in forming SA, and due to the DoD goal of achieving political, economic and security stability, we have chosen metrics of these aspects of Iraq stability to track during 2003-2008. In the following sub-sections, we show what specific data is used (part of forming Level 1 SA) and how that data is ‚comprehended' using a trained HTM for inference (Level 2 SA).

4.3.1 Level 1 SA: Data preparation

Entering the information flow of Fig. 1, the first task (represented by an oval) is to prepare the data. Before doing so, we must address the Boolean question (represented by a triangle) about whether data is available. For the Iraq context, we actually have data. But some effort is needed to prepare this data into standard form.

There are four issues we must confront in doing so. First, the primary source (United States Department of Defense, 2005, 2006, 2007, 2008, and O'Hanlon & Campbell, 2008) from which the data is extracted contains many blanks in the data, depending on how many metrics are used. So a set of metrics must be selected from the actual data that exhibits a minimal number of blanks.

Second, the primary source has not prepared the data in a temporally structured format suitable for HTM learning. Dileep George pioneered work on HTM algorithms and he gives guidelines for generalizing their usage in other domains. In particular, George writes, "if there is no temporal structure in the data, application of an HTM to that data need not give any generalization advantage." So the data must be arranged in this fashion if HTM is to be of use. Specifically, observations at specific time intervals should follow one another.

Third, the relative magnitudes of the chosen metrics will be necessary to consider. Consequently, a transformation of the data may be necessary before training/testing.

Fourth and finally, one of the metrics we use to describe the Iraq context is only known within given bounds at each time step. Consequently, we must select a technique to get only one value at each time step, rather than a range.

Considering all of these points, it is possible to pick a subset of metrics from the primary source that we can use to describe the Iraq context in a data-driven fashion related to our goal of tracking stability. These selected metrics are shown in Table 1 with identifying numbers next to each of them.

Metric	Units	Metric #
Iraqi_Civilian_Fatalities	persons	1
Multiple_Fatality_Bombings	persons	2
US_Troop_Fatalities	persons	3
Improvised_Explosive_Device_US_Troop_Deaths	persons	4
Car_Bomb_US_Troop_Deaths	persons	5
Mortar_and_Rocket_US_Troop_Deaths	persons	6
Rocket_Propelled_Grenade_US_Troop_Deaths	persons	7
Helicopter_Loss_US_Troop_Deaths	persons	8
Other_Hostile_Fire_US_Troop_Deaths	persons	9
Total_US_Troop_Deaths	persons	10
Attacks_on_Iraqi_Infrastructure_and_Personnel	incidents	11
Coalition_Troop_Strength	persons	12
Crude_Oil_Production	millions of barrels per day	13
Crude_Oil_Export	millions of barrels per day	14
Nationwide_Electricity	Megawatts	15
Nationwide_Unemployment_Rate	%	16

Table 1. Sixteen metrics to describe Iraq context

While it is possible to select fewer metrics, a drop off in performance was seen when this was done. We have shown this in other works (Sherwin & Mavris, 2011 and Sherwin, 2010). We believe that this occurs because the degree of stability in an operational theater already lacks a clear definition amongst stakeholders. Consequently, the more metrics that are incorporated into the analysis, the more complete the description of stability will be. Inversely, the fewer metrics that are incorporated, the more narrow the description will be. Here, we stopped at sixteen because this number approaches the upper limit of what was publically available, although more metrics may make the analysis that much more rich.

Finally, to give the HTM a baseline for stability and instability, artificial data generated from a rudimentary system dynamics was created based on the selected metrics. For instance, in this model (for stability, for instance), the number of troop deaths due to car bombs fell off to zero over time (roughly the same 60 months of time for which actual data exists). Alternatively, in this model, (for instability, e.g.), the nationwide electricity would flatten out to zero. In general, metrics associated with stable or unstable situations would monotonically be driven to an extreme maximum or minimum over the course of 60 months, starting from a real data point. In other words, we use extreme-cases to bound the reality observed in actuality - and this reality is more of a complex mix of certain features of instability and/or stability along different avenues (such as politically, socially, or economically).

4.3.2 Level 2 SA: HTM-aided comprehension of the data

With the ability to generate data for both progressively stable and unstable situations, as well as the actual time series of data on the Iraq context, it is possible to attempt HTM as an unsupervised machine learning mechanism. Recall, an HTM is a network trained is built to find a schema, S, that describes the Iraq context. This is based on each vector observed in each node's local vector space, v_i, all of which considered together constitute V. To aid the spatio-temporal grouping, these vectors are first grouped with K-means clustering before temporal adjacency is learned and grouped into the network's Marko-chains, M. These Markov-chains are then used to perform evidence-based Bayesian inference on novel data.

With HTM, the aim now is to fuse the data and to extract possibly implicit meaning from it pertinent to the Iraq context. We emphasize the unsupervised nature of the learning here because our goal is to extract implicit meaning and not to impose our possibly biased judgments. Furthermore, we attempt to extract this meaning from a system that is not ergodic (i.e., there is no end-state), not separable into components (hence, model-able) and not completely observable (i.e., uncertain data describes the system).

It has been found to be more effective to first train the HTM on the extreme-case data and then to test its inference capabilities on the actual data (Sherwin & Mavris, 2011). Therefore, we implement this approach so that the HTM can learn from the extreme-case boundaries and then use them to classify the reality in between.[1]

The evaluation of this computational SA is not entirely straightforward and so additional techniques were employed to probe the SA formed about the Iraq context. These studies will

[1] The former method has been tried and has proven unsuccessful (Sherwin, 2010).

not be reviewed in too much depth now, but a summary of them is important for our purposes.

For starters, we do not know if too many or too few metrics are being used here to describe the Iraq context. Therefore, studies were done to see what effects there are from reducing the number of metrics used to train and infer with the network. It was found that fewer metrics reduce the semantic richness of the data, thereby causing certain volatile metrics to dominate the learning and subsequent inference.

Also, we examined the degree to which information is hierarchically stored in intermediate levels of the networks. We found that, true to the promise of HTM, this was the case. Finally, we considered alternative ways of feeding the data into the networks to see what effects - if any - there are on the simulated SA.

Why would we use these techniques in particular though? For instance, what purpose could there be in probing the hierarchical storage of information in an HTM? It is necessary to recall that our purpose in using HTM for computational SA has been its declared ability to condense information into hierarchies of both space and time. For the Iraq context, we test hierarchical storage directly because it is not clear what the top-level node's output should be, as it might be for simpler recognition tasks (e.g., invariant visual pattern recognition, George & Hawkins, 2009, and George, 2008). One possible outcome of this analysis is that it might in turn help us to identify what aspects of the Iraq context are not well observed. This would then provide the beginnings of a feedback mechanism with Level 1 SA to search for more data.

In fact, in the course of this research, it was one possible feedback mechanism between Levels 1 & 2 SA that informed us to improve our extreme-case model. This resulted in the monotonic extreme-case models used below. Necessarily, this is not the only possible feedback mechanism, but as we will see, it helps to strengthen the credibility of the Level 2 SA formed here computationally. If we use data for training that becomes monotonically extreme from a realistic starting condition then we would expect an HTM network to learn to recognize clear progressions towards stability/instability.

We employ an evolutionary approach to network design here and modify a network used in an HTM demonstration example (see Numenta, 2008).[2] In order to exploit the HTM network's temporal learning algorithms, we modify the network parameters to accommodate how the metrics' values change in time. The complete network parameters employed for this computational Level 2 SA can be seen in another work (see appendix C.9 in Sherwin, 2010).[3]

From training the network, we can survey the resulting schema created by the network. As for all HTM nodes, this is described in terms of coincidence patterns (i.e., distinct spatial patterns) that form the schema's Markov-chains. Here, we find that there are sixty-one

---

[2] All analysis has been done with Vitamin D Toolkit 1.3.0 as a graphical user interface to NuPIC 1.6.1, which is run on Python 2.5.2. It should be noted that slightly different results are obtained if networks are created, trained and tested directly in NuPIC. See appendix D in Sherwin, 2010 for more information on this topic.

[3] Even though this work says that these parameters are for a progressively trained network, they are the same ones used for the monotonically trained one.

coincidence patterns ($C_{3,1}$) and fifty-nine Markov-chains ($G_{3,1}$) in the top-level node.[4] The coincidence patterns are the result of the K-means clustering in this top-level node, while the Markov-chains are the result of first-order transition probabilities between these coincidence patterns. This is a standard calculation for each of the nodes in an HTM network (see George & Hawkins, 2009).

After proper training, an HTM network is most valuable as an inference tool, so now we evaluate its performance. We will start plaintively by looking at inference on the training data, moving onto novel data later.

When we perform inference on the monotonic training data, we see a clear progression of Markov-chains as instability increases, but stability is still not clear. We can see this by following the probability over Markov-chains $\lambda_t(g_r)$ of the top-level node, given the bottom-up evidence ($-e_t$) at each t. In particular, we examine the maximum of this distribution ($\max\left[\lambda_t(g_r)\right]$) to see what stability state is most likely. What we find is that the progression indicated by $\max\left[\lambda_t(g_r)\right]$ is $g0$, $g1$, $g2$, ..., $g58$. Since we know the data is monotonic towards instability, we can reasonably claim that the Markov-chain labels are monotonic towards instability as well. For example, the bottom-up evidence when g45 is most likely in the top level indicates a situation that is less stable than when g5 is most likely.

Having trained the network to recognize progressions in instability, it would be useful now to test this ability on novel data. In particular, we feed into the network real data of the Iraq context. When we look for instability gradations in the actual data, we see some interesting results (Fig. 2). In Fig. 2, as in similar figures to follow, each row is a time point. The first, second, third, etc. columns indicate the groups (i.e., Markov-chains) that are most likely, second most likely, third most likely, etc., given the bottom-up evidence. The significance of this ordering of group numbers at each time point is that we can quantitatively say how unstable Iraq is at each of them. Note throughout that time points $t \in [0, 60]$ correspond to each month from May 2003 to April 2008. In particular, at $t = 11, 12$, the entire probability distribution over top-level Markov-chains shifts towards higher number Markov-chains. At $t = 11$, $g25$, $g24$, $g23$ are in the top three (see Fig. 2).

At $t = 18$, the probability distribution shifts as well, indicating $g12$, $g13$, $g14$ in the top three. In light of our results from inference on the monotonic-extreme-case instability data, it would seem that the Iraq context is increasingly unstable during these months. Furthermore, the actual data during these months indicates this in comparison to those months that come before them.

Let us expand our purview to those time points leading up to and coming out of $t \in [41,49]$, another region of heightened instability according to the network. If we consider the top seven Markov-chains of the top-level for $t \in [36,60]$ then we see something quite interesting. For $t \in [36,41]$, the distribution shifts increasingly towards $g12$, $g13$, $g14$, $g15$, $g16$, $g17$, $g18$. Also, we can see the demotion of $g0$ over these time steps (Fig. 3), indicating increasing instability.

[4] Here and throughout the remainder of the paper, we follow George's notation for HTM theory (George & Hawkins, 2009 and George, 2008).

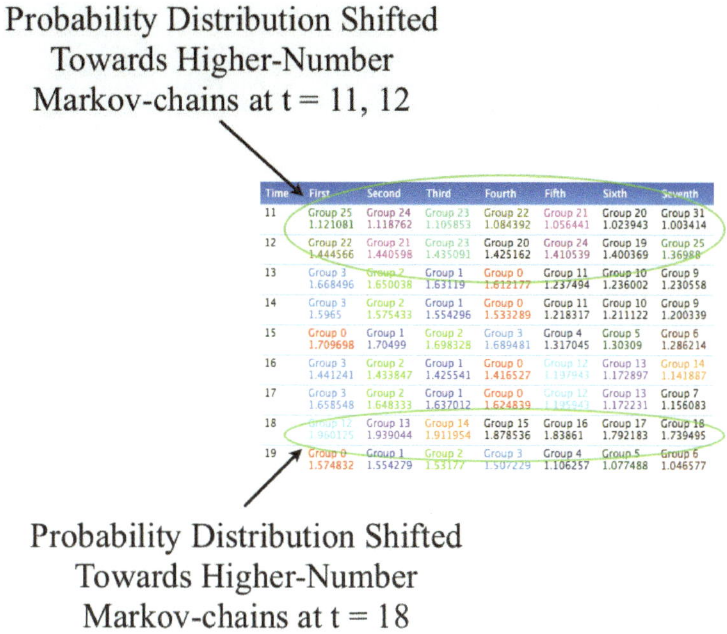

Time	First	Second	Third	Fourth	Fifth	Sixth	Seventh
11	Group 25 1.121081	Group 24 1.118762	Group 23 1.105853	Group 22 1.084392	Group 21 1.056441	Group 20 1.023943	Group 31 1.003414
12	Group 22 1.444566	Group 21 1.440598	Group 23 1.435091	Group 20 1.425162	Group 24 1.410539	Group 19 1.400369	Group 25 1.36988
13	Group 3 1.668496	Group 2 1.650038	Group 1 1.63119	Group 0 1.612177	Group 11 1.237494	Group 10 1.236002	Group 9 1.230558
14	Group 3 1.5965	Group 2 1.575433	Group 1 1.554296	Group 0 1.533289	Group 11 1.218317	Group 10 1.211122	Group 9 1.200339
15	Group 0 1.709698	Group 1 1.70499	Group 2 1.698328	Group 3 1.689481	Group 4 1.317045	Group 5 1.30309	Group 6 1.286214
16	Group 3 1.441241	Group 2 1.433847	Group 1 1.425541	Group 0 1.416527	Group 12 1.197943	Group 13 1.172897	Group 14 1.141887
17	Group 3 1.658548	Group 2 1.648333	Group 1 1.637012	Group 0 1.624839	Group 12 1.195943	Group 13 1.172231	Group 7 1.156083
18	Group 12 1.960125	Group 13 1.939044	Group 14 1.911954	Group 15 1.878536	Group 16 1.83861	Group 17 1.792183	Group 18 1.739495
19	Group 0 1.574832	Group 1 1.554279	Group 2 1.53177	Group 3 1.507229	Group 4 1.106257	Group 5 1.077488	Group 6 1.046577

Fig. 2. Instability recognition of real data

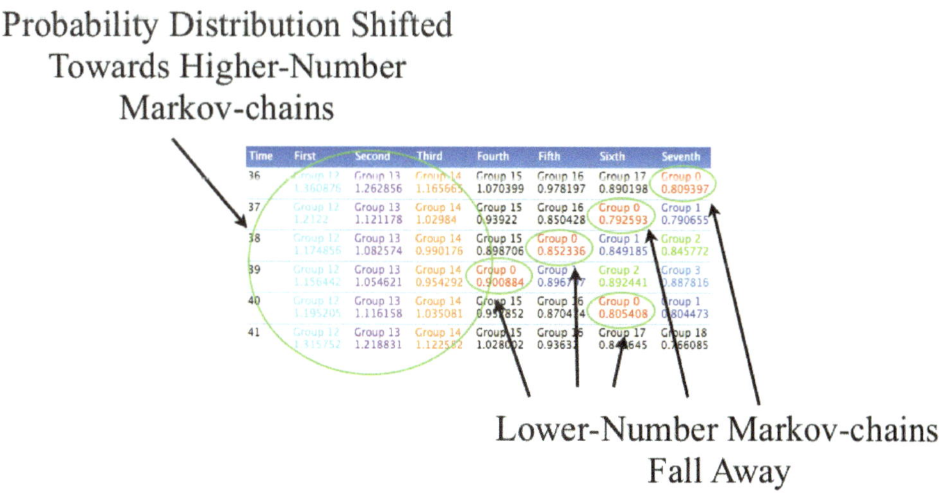

Time	First	Second	Third	Fourth	Fifth	Sixth	Seventh
36	Group 12 1.360876	Group 13 1.262856	Group 14 1.165665	Group 15 1.070399	Group 16 0.978197	Group 17 0.890198	Group 0 0.809397
37	Group 12 1.2122	Group 13 1.121178	Group 14 1.02984	Group 15 0.93922	Group 16 0.850428	Group 0 0.792593	Group 1 0.790655
38	Group 12 1.174856	Group 13 1.082574	Group 14 0.990176	Group 15 0.898706	Group 0 0.852336	Group 1 0.849185	Group 2 0.845772
39	Group 12 1.156442	Group 13 1.054621	Group 14 0.954292	Group 0 0.900884	Group 1 0.896[illegible]7	Group 2 0.892441	Group 3 0.887816
40	Group 12 1.195205	Group 13 1.116158	Group 14 1.035081	Group 15 0.95[illegible]852	Group 16 0.8704[illegible]4	Group 0 0.805408	Group 1 0.804473
41	Group 12 1.315752	Group 13 1.218831	Group 14 1.122582	Group 15 1.028002	Group 16 0.9363[illegible]	Group 17 0.8[illegible]645	Group 18 0.7[illegible]6085

Fig. 3. Probability Distribution Shifts Towards Higher-Number Markov-chains

Then for $t \in [41,49]$, these seven Markov-chains are the most likely, given the bottom-up evidence (Fig. 4).

Time	First	Second	Third	Fourth	Fifth	Sixth	Seventh
41	Group 12 1.315752	Group 13 1.218831	Group 14 1.122552	Group 15 1.028002	Group 16 0.93632	Group 17 0.848645	Group 18 0.766085
42	Group 12 1.378272	Group 13 1.276451	Group 14 1.176123	Group 15 1.07841	Group 16 0.984456	Group 17 0.895383	Group 18 0.812243
43	Group 12 1.639405	Group 13 1.555941	Group 14 1.471043	Group 15 1.385653	Group 16 1.300823	Group 17 1.217681	Group 18 1.137402
44	Group 12 1.878105	Group 13 1.80413	Group 14 1.727491	Group 15 1.648984	Group 16 1.56955	Group 17 1.490251	Group 18 1.412246
45	Group 12 1.517151	Group 13 1.444452	Group 14 1.368938	Group 15 1.291384	Group 16 1.212711	Group 17 1.133971	Group 18 1.056317
46	Group 12 1.330323	Group 13 1.251093	Group 14 1.169857	Group 15 1.087495	Group 16 1.005014	Group 17 0.923518	Group 18 0.844182
47	Group 12 1.548336	Group 13 1.463013	Group 14 1.37652	Group 15 1.289824	Group 16 1.203992	Group 17 1.120162	Group 18 1.039504
48	Group 12 1.650163	Group 13 1.548342	Group 14 1.448014	Group 15 1.350301	Group 16 1.256347	Group 17 1.167274	Group 18 1.084134
49	Group 12 1.44882	Group 13 1.369168	Group 14 1.287562	Group 15 1.204894	Group 16 1.122174	Group 17 1.040509	Group 18 0.961074

Probability Distribution
Completely Shifted Towards
Higher-Number Markov-chains

Fig. 4. Complete Shift Towards Markov-chains Indicating Instability

As we know from the actual data, there were peaks in violence and other attacks, sagging economic metrics, etc. during this time. But there were also metric trends that favored stability. Consequently, this conflicting data makes it difficult to characterize the stability level during this time period. But here we see the probability distribution shift for the entire time period towards these mid-grade instable states.

The situation in the Iraq context changes though with time. For $t \in [50,51]$, the probability distribution begins to shift back. Finally, for $t \in [52,60]$, $g0$, $g1$, $g2$, $g3$ are the top four most likely Markov-chains (see Fig. 5).

Even though this does not indicate stability, it does indicate a dramatic drop in instability, according to how we trained the network. So we see here how the monotonic training data has provided a peg against which to categorize evidence that trends towards instability. But what about stability recognition?

As mentioned earlier, direct stability recognition is less clear, even with the monotonic training data. Rather, we can only infer stability recognition with the network. Why is this so? If we consider the types of metrics used here then we notice that only four of them increase with stability. So, as a more stable situation is reached, the remaining twelve metrics drop close to zero. Consequently, the bottom-up evidence does not provide enough magnitude to propagate through the network. All the change comes from the four metrics that increase with stability. In the current permutation of the data, one of them is in the receptive field of the third node in level one ($N_{1,3}$) and the other four are in the field of $N_{1,4}$. The entire left receptive field (covered by $N_{1,1}$ and $N_{1,2}$) therefore produces blank recognition. This is because there is simply not enough bottom-up evidence coming up through this side of the network. So we are not able to determine gradations of stability because the utility function of these metrics can be assumed to be inversely proportional to stability. Consequently, as stability is reached, the magnitude of $-e_t$ goes to zero. In future implementations, it might be possible to alleviate this problem by transforming the data by an inverse or offsetting the zero. We have not done this though because we have devised an approach to recognize degrees of instability in real data, as judged against the extreme-case baseline. Furthermore, these results imply stability recognition due to the monotonic utility function of stability/instability.

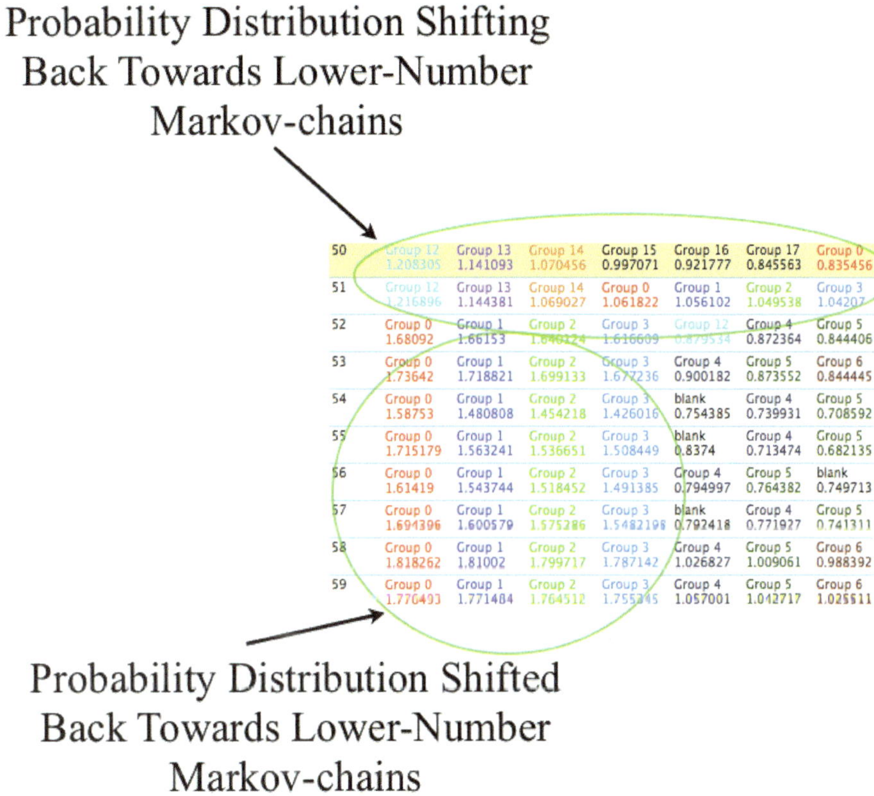

50	Group 12 1.208305	Group 13 1.141093	Group 14 1.070456	Group 15 0.997071	Group 16 0.921777	Group 17 0.845563	Group 0 0.835456
51	Group 12 1.216896	Group 13 1.144381	Group 14 1.069027	Group 0 1.061822	Group 1 1.056102	Group 2 1.049538	Group 3 1.04207
52	Group 0 1.68092	Group 1 1.66153	Group 2 1.640124	Group 3 1.616609	Group 12 0.879534	Group 4 0.872364	Group 5 0.844406
53	Group 0 1.73642	Group 1 1.718821	Group 2 1.699133	Group 3 1.677236	Group 4 0.900182	Group 5 0.873552	Group 6 0.844445
54	Group 0 1.58753	Group 1 1.480808	Group 2 1.454218	Group 3 1.426016	blank 0.754385	Group 4 0.739931	Group 5 0.708592
55	Group 0 1.715179	Group 1 1.563241	Group 2 1.536651	Group 3 1.508449	blank 0.8374	Group 4 0.713474	Group 5 0.682135
56	Group 0 1.61419	Group 1 1.543744	Group 2 1.518452	Group 3 1.491385	Group 4 0.794997	Group 5 0.764382	blank 0.749713
57	Group 0 1.694396	Group 1 1.600579	Group 2 1.575286	Group 3 1.548219	blank 0.792418	Group 4 0.771927	Group 5 0.741311
58	Group 0 1.818262	Group 1 1.81002	Group 2 1.799717	Group 3 1.787142	Group 4 1.026827	Group 5 1.009061	Group 6 0.988392
59	Group 0 1.770493	Group 1 1.771484	Group 2 1.764512	Group 3 1.755345	Group 4 1.057001	Group 5 1.042717	Group 6 1.025511

Fig. 5. Complete Shift Back Towards Markov-chains Indicating Less Instability

4.4 Consequences of the Iraq context implementation

We should be very clear at the outset: the schema formed with the HTM-based computational SA created here is not how a human brain of an expert decision-maker would function. Rather, it is an alternative analysis of the data at hand that can be used by a decision-maker in such a scenario. The key element to its utility though is the fact that the computational SA functions in analogous ways to some neuro-anatomical processes, such as coincidence detection and spatio-temporal pattern condensation into a flexible schema. In particular, the HTM-based SA learns from its experiences to infer about uncertain and novel situations. To be implemented on a computer, it does so with the aforementioned hierarchical and temporal breakdown of its 'experiences' (in the form of spatio-temporal vectors).

Furthermore, this is not a perfect implementation of neural engineering to analyze complex systems. But as the preceding sections demonstrate, it provides a quantifiable analysis of a system that is otherwise left in the hands of subjective analyses whose justifications are missing or obfuscating. Perhaps with better insight into human expert's mental processes the results would have stronger impact in systems engineering analysis.

5. Recent neuroscientific results on expertise

It should be clear from the preceding implementation of computational SA that the following is true:

1. Situational awareness (SA) is a nebulous term used to define an equally nebulous ability of humans
2. The machine learning-based approximation of this process with HTM is imperfect
3. More details on neural markers of expertise would inform any future computerizations of human mental processes

Considering these points in succession, it is clear that a refined perspective on human-borne expertise can add tremendous value to our first attempt of forming SA computationally. In particular, we aim to highlight some recent advances in the neuroscience of expertise that can ultimately be of use in how we analyze and design complex systems in engineering.

5.1 What happens when you poke an expert's brain?

The most insightful way to examine an expert's brain is to subject him/her to stimuli that violate their expertise-borne predictions. This experimental paradigm has seen tremendous success in tracking the neural markers of unexpected stimuli (most notably in analysis of the P300, a positivity that emerges around 300ms after a repeatedly unexpected stimulus is observed). By tracking neural signatures like the P300 and others, it is possible to see how a human brain - experienced in a certain stimulus domain - responds to errant stimuli.

Although research in many modalities have been done (e.g., functional magnetic resonance imaging (fMRI) and magnetoencephalography (MEG)), we focus here on electroencephalography (EEG) measurements of neural data. We do this for two reasons: 1) single-trial classification of neural data from EEG is generally more robust than it is from fMRI (and not very developed for MEG), 2) EEG neural data has a much higher temporal resolution than fMRI (and slightly higher than MEG), making it an ideal candidate for more immediate integration into systems engineering problems.

5.1.2 A simple experiment in error-detection

To illustrate the kinds of neural processes observable with EEG systems, we will summarize some experimental work on expectation violation. While this experiment may seem removed in the specific sense from analyzing expertise for the purposes of systems engineering analysis, the abstract concept at the heart of this experiment could not be more on target. In particular, subjects are asked to listen to an audio stimulus with which they are quite familiar. In this case, American popular songs, such as „Eye of the Tiger" or „Sweet Home Alabama," are used because the subjects all have strong prior expectations for the course of these songs once they start listening to them. In other words, they are experts at how these songs unfold. In this experiment, we analyze the subject's expertise about these audio stimuli.

The aim of this experiment is to alter the song in such a way that the following occurrences are true: 1) a subject with normal hearing should be able to discern the alteration, 2) the subject should be able to reorient his/her expectations after the alteration has taken place. The latter condition is a requirement if multiple alterations are to be performed within one hearing of the song. To balance these two requirements, it was chosen that the song's key should be altered either up or down by a semi-tone at various points in the recording.

After analyzing the neural data with learning algorithms based on logistic regression classification (Parra et al., 2002 and Parra et al., 2005), it is found that we can distinguish from neural data alone when the subject perceived an alteration and when he/she did not, regardless of where they were in the song. In other words, we are able to distinguish times in the experiment when the subject's expertise (and associated predictions) were at a conflict with the data (i.e., the audio stimuli) in front of him/her. An example of this phenomenon can be seen in Fig. 6.

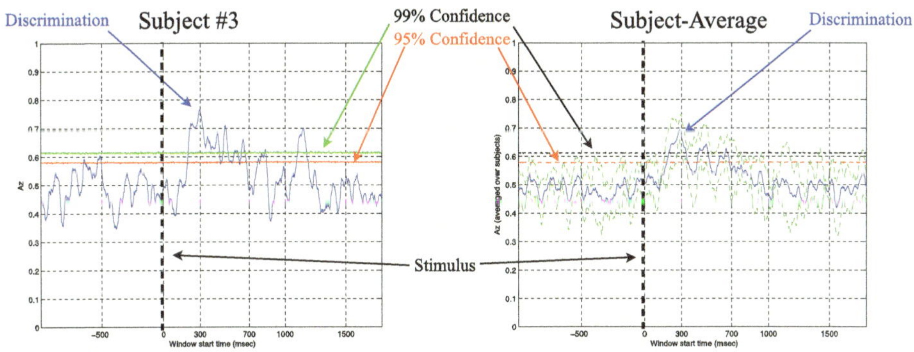

Fig. 6. Individual subject and subject-average discriminations of expertise violation

What this plot shows is the classification performance in blue (measured by Az, or the area under the receiver-operator characteristic curve, see Green & Swets, 1966) vs. the time when this classification is performed relative to when the alteration happened (in the case of the alteration) or did not happen (in the case of the corresponding condition in a control listening). The alteration is shown in the dotted-black line. The green and red solid lines in the left figure indicate the 99% and 95% confidence lines. In the right figure, the subject-averaged significance lines are differentiated from the individual lines (because they are computed differently) by being dashed. The red-dashed line is the 95% and the black-dashed line is the 99% significance line. Finally, the dashed-green line is the 5%-standard deviation from the averaging. As the average plot shows, similar plots exist for other subjects in the experiment than the one shown on the left.

This experiment shows that there is a measurable neural process happening when an expert witnesses something unexpected. In particular, just as with HTM, the key component to deciphering the incoming phenomenon (here, it is audio, V_{Audio}) is the subject's schema (S_{Audio}) and how it predicts the evolution of V_{Audio} in time. In other words, we are playing with the subjects' Level 3 SA (i.e., prediction). If this result is to be used in the analysis of complex systems then it must involve a dynamic system. Luckily, this challenge in complex system analysis is actually a boon for an expert's assessment of a phenomenon, V.

5.2 What does this mean for systems engineering?

As described earlier, systems engineering is a complex process that is largely driven by expert insight into a range of problems. But we see both from the computational SA demonstration and one result (of many related others) from neuroscience that a new way to think about expertise is developing. Furthermore, both results depend on a temporal evolution of data as a means by which SA is created, either computationally or by a human.

It may be possible to integrate this understanding technologically with the needs of systems engineering today. For instance, why not handle the data deluge with an approach that couples the speedy perception of cortical circuitry to the vast computational power of modern computers? This is already being done in remote sensing in the form of cortically-coupled computer vision (Sajda, 2010). Similar versions of this problem exist in aircraft analysis and design, as described earlier. As we become more familiar with the insights on expertise provided by neuroscience, it could only be a matter of time before neural engineering is a needed integrative enabler of analyzing and manipulating increasingly complex and dynamic systems.

6. Conclusion

The systems engineering task is only getting more difficult. As systems become more complex and demands for reliability increase, there is a growing need – already being met in certain ways – to build appropriate analytic tools to engineer within this context. But systems engineers are increasingly required to analyze an already existing system before the design process begins. In other words, post-factum analysis is a key aspect to the systems engineering process. Consequently, subject matter expertise becomes a key enabler for the design process. However, the elusive nature of expertise remains an intractable aspect to that process. Here, we have shown a computational approach built on insights into expertise and how it was used in a decision-making context. In other words, we showed how the analysis of the provided data enabled justifiable conclusions on an otherwise unpredictable and already established complex system (e.g., the Iraq War, 2003-2008). However, we noticed some shortcomings in this approach and so turned our focus to more basic neuroscience questions about what neural processes occur as expertise is used. In particular, we discussed one recent experiment as an example of current work going in the neural markers of expertise. Since such markers are stimulus driven, in particular, the expert reacts to data being either in line with expectations or not, we forecasted a potential role for neural engineering in the analysis, and consequent design, of complex systems. Not only does this need exist but such an approach would also complement the current techniques used in this endeavor.

7. Acknowledgment

We want to thank the Aerospace Systems Design Laboratory at Georgia Institute of Technology for supporting this research.

8. References

Akao, Y. (1990). *Quality Function Deployment: Integrating Customer Requirements Into Product Design,* Productivity Press, ISBN 1-56327-313-6, New York, USA

Chan, L. & Wu, M. (2002). Quality function deployment: A literature review. *European Journal of Operational Research,* Vol.143, No.3, (December 2002), pp. 463-497

Drew, C. (2010). Military is awash in data from drones, In: *New York Times,* 10.01.2010, Available from http://www.nytimes.com/2010/01/11/business/11drone.html?pagewanted=all

Endsley, M. (1995). Toward a theory of situation awareness in dynamic systems. *Human Factors,* Vol.37, No.1, (March 1995), pp. 32-64

George, D. & Hawkins, J. (2009). Towards a mathematical theory of cortical micro-circuits. *PLoS Computational Biology,* Vol.5, No.10, (January 2009), pp. 1-26.

George, D. (2008). *How the Brain Might Work: A Hierarchical and Temporal Model for Learning and Recognition,* Ph.D. Dissertation, Stanford University, Palo Alto, USA

Green, D. & Swets, J. (1966). *Signal Detection Theory and Psychophysics,* Wiley & Sons, Inc., ISBN 0-932146-23-6, New York, USA

Ho, Y. & Pepyne, D. L. (2002). Simple explanation of the no-free-lunch-theorem and its implications. *Journal of Optimization Theory,* Vol.115, No.3, (December 2002), pp. 549-570

National Commission on Terrorist Attacks upon the United States (2004). *The 9/11 Commission Report,* Norton and Co., ISBN 0-393-32671-3, New York, USA

Numenta, Inc. (2008). Getting Started with NuPIC, Numenta, Inc., Available from http://www.numenta.com/archives/education/nupic_gettingstarted.pdf, pp. 34-35

O'Hanlon, M. E., & Campbell, J. H. (2008). *Iraq Index: Tracking Variable of Reconstruction & Security in Post-Saddam Iraq,* The Brookings Institution, Washington, DC, USA

Parra, L. et al. (2002). Linear Spatial Integration for Single Trial Detection in Encephalography. *NeuroImage,* Vol.17, (December 2002), pp. 223-230

Parra, L. et al. (2005). Recipes for the linear analysis of EEG. *NeuroImage,* Vol.28, No.2, (May 2005), pp. 326-341

Saaty, T. L. (2000). *Fundamentals of Decision Making and Priority Theory with the Analytic Hierarchy Process,* RWS Publications, ISBN 0-9620317-6-3, Pittsburgh, USA

Saaty, T. L. (2008). Decision making with the analytic hierarchy process. *International Journal of Services Sciences,* Vol.1, No.1, (2002), pp. 83-98

Sajda, P. et al. (2010). In a blink of an eye and a switch of a transistor: Cortically-coupled computer vision, *Proceedings of the IEEE,* vol. 98, no. 3, March, 2010

Sherwin, J. S. & Mavris, D. N. (2011). A computational approach to situational awareness: A follow-up on the details. *Journal of Battlefield Technology,* Vol.11, No.1, (March 2011), pp. 1-11

Sherwin, J. S. (2010). *A Computational Approach to Achieve Situational Awareness from Limited Observations of a Complex System,* Ph.D. Dissertation, Georgia Institute of Technology, Atlanta, USA

United States Department of Defense (2005). Report to Congress: Measuring Stability and Security in Iraq, July 2005

United States Department of Defense (2005). Report to Congress: Measuring Stability and Security in Iraq, October 2005

United States Department of Defense (2005). Report to Congress: Measuring Stability and Security in Iraq, May 2006

United States Department of Defense (2006). Report to Congress: Measuring Stability and Security in Iraq, February 2006

United States Department of Defense (2006). Report to Congress: Measuring Stability and Security in Iraq, August 2006

United States Department of Defense (2006). Report to Congress: Measuring Stability and Security in Iraq, November 2006

United States Department of Defense (2007). Report to Congress: Measuring Stability and Security in Iraq, March 2007

United States Department of Defense (2007). Report to Congress: Measuring Stability and Security in Iraq, June 2007

United States Department of Defense (2007). Report to Congress: Measuring Stability and Security in Iraq, September 2007

United States Department of Defense (2007). Report to Congress: Measuring Stability and Security in Iraq, December 2007

United States Department of Defense (2008). Report to Congress: Measuring Stability and Security in Iraq, March 2008

United States Department of Defense (2008). Report to Congress: Measuring Stability and Security in Iraq, June 2008

United States House of Representatives (2005). Making Emergency Supplemental Appropriations for the Fiscal Year Ending September 30, 2005, and for Other Purposes, *Conference Report 109-72*, May 2005

Vanderplaats, G. N. (2007). *Multidiscipline Design Optimization,* Vanderplaats R&D, Inc., ISBN 0-944956-04-1, New York, USA

8

System Engineering Method for System Design

Guillaume Auriol, Claude Baron,
Vikas Shukla and Jean-Yves Fourniols
CNRS, LAAS,
Université de Toulouse, UPS, INSA, INP, ISA, UT1, UTM, LAAS
France

1. Introduction

The purpose of this chapter is to present some educational materials, the process and the outcomes to teach an engineering approach applied to a practical development case. The starting point is the requirements of an application of remote supervision of a room with several parameters: light, temperature and movement (intrusion into the room or movement of an object within the room). This application is based on wireless terminal nodes composed of a sensor, a microcontroller and a telecommunication module. Several rooms can be interconnected, so it must be possible to use the sensors of each room of a given site simultaneously. Various issues can be raised during teaching on wireless sensor networks (Kotzl & Essien, 2005): electronic design, risks to humans (Brownsell et al., 1999), energy management, telecommunication technologies, etc.

During the course, students have to learn and apply a 'systems engineering' (Ullrich K.T. and Eppinger S.D, 2003), (Terry A. Bahill and Clarck Briggs, 2001) approach based on standards in the field (Martin, 1998), (ISO15288, 2008), (IEEE1220, 2005) to solve a problem with numerous design options. Several off-the-shelf software and hardware components are at the students' disposal: a panel of telecommunication modules, different communication and signalling protocols, etc. They start by studying the requirements to extract an exhaustive list of needs. They must then propose and evaluate functional and architectural solutions, and finally implement the chosen solution in order to validate their 'systems engineering' approach.

Section 2 gives an overview of the method to follow to design a telecom system. Section 3 depicts the application through stakeholder's needs. Sections 4 to 6 detail the four steps of the method with (4) definition of stakeholders' needs and definition of technical requirements, (5) design of functional and (6) physical architectures. Section 7 presents the component realization, the component integration and the system validation. Finally, in the conclusion highlights the educational benefits to use a system engineering method.

2. Overview of the method

This section presents the main steps of the methodology based on UML diagrams (Bock, 2009), (Weilkiens, 2008) that the students have to follow. In a "V" development cycle, engineering processes cover the usual activities of a top-down process: (1) definition of stakeholders' needs, (2) definition of technical requirements, (3) design of functional and (4) physical architectures (see figure 1).

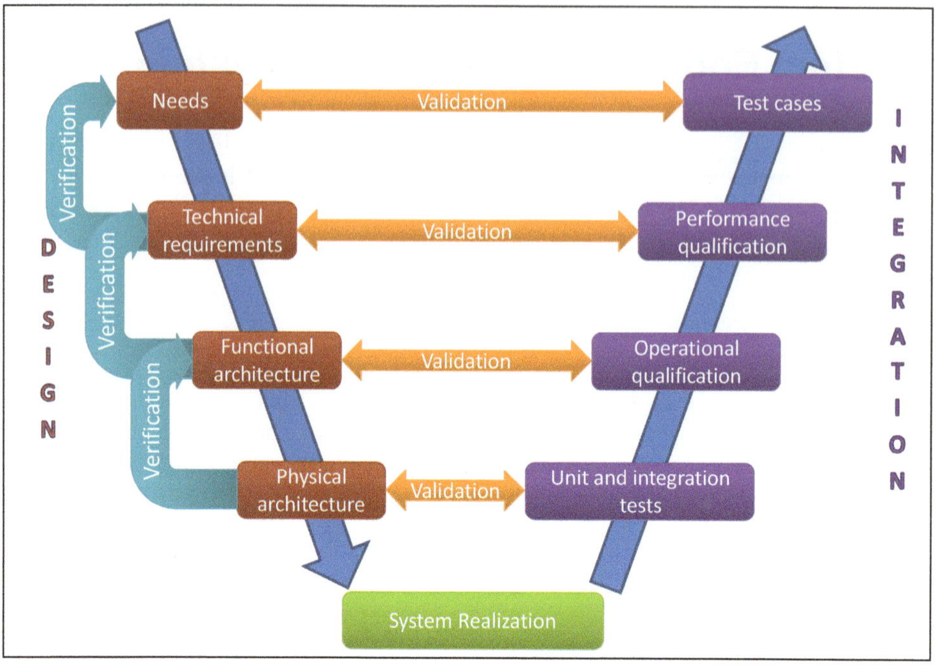

Fig. 1. Typical "V" cycle development

The *definition of stakeholders' needs* process first consists in identifying what kind of information the customer has given, generally a set of systems specifications which may not be either very well structured or even complete. The problem is then to understand the system in its specific context: define its purpose, mission, objectives and stakeholders, with the help of several operational scenarios. This process produces a document which lists and classifies stakeholders' needs and system constraints in every aspect of the system's use and lifecycle.

The goal of the *definition of technical requirements* process is to translate each need into an expression allowing the design of a feasible solution. It proceeds by refining the system mission and by breaking down the operational modes and scenarios into activities, in order to obtain corresponding technical requirements. This process also leads to complete and precise initial statements. The result is a document containing technical requirements that are coherent, formalized and verifiable (technical or physical features) and that will be useful for the designer.

After this essential step, it remains to build high-level *functional architectures*. The aim of this process is to establish and evaluate several functional architectures that could be candidates and retain one.

The *physical architectures* for the system, describing the chosen solution, as well as its physical interfaces, are given during the physical design processes.

At each step, a *verification process* is invoked, in order to justify the expression of needs, technical requirements, design choices, and to ensure traceability right through the development process.

Finally, a *validation process* is performed to compare technical requirements to performances obtained during in situ tests.

3. Description of the application

The students have to develop an application for the remote monitoring of several parameters of a room (luminosity, temperature) and of movements (intrusion detection). This application is based on nodes made up of a sensor, a microcontroller and a telecommunication module, in addition to the power module. Several rooms can be inter-connected while the sensors inside each room must interact, as illustrated in figure 2. Three categories of nodes are used according to the nature of the data they have to transmit. These data are different by their:

- nature: some are binary (detection of a threshold), others are analog,
- criticality,
- periodicity: some transmissions are periodic, while others are event-triggered.

As far as their transfer is concerned, these data have different needs concerning the quality of service. These needs must also be taken into account for the choice of a specific telecommunication technology and during the development of appropriated communication protocols.

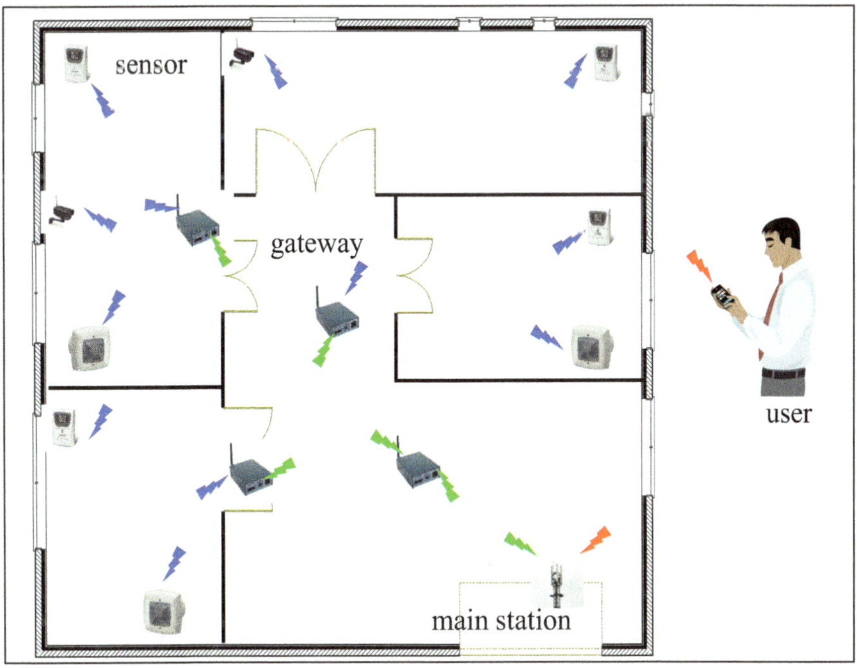

Fig. 2. Application of monitoring

4. Definition of stakeholders' needs and technical requirements

4.1 Needs

This step consists in enumerating the different elements of the context with which the system interacts when in use (physical and functional boundaries). The relationships between the system and these external elements are clearly illustrated in two distinct diagrams: a use case diagram, and an initial class diagram. The use case diagram is obtained by imagining global services showing the main interactions between the elements of the context and the system of interest. For example, in our surveillance application, the system collects energy readings from its environment and reports the collected temperatures to an operator; it is configured and repaired by a maintenance operator. On the basis of the use case diagram, we can draw up an initial class diagram containing the elements of the context and their physical links with the system of interest.

Students have to apply a system engineering (SE) method to design a sensor network. They use this network to validate their solution: choice of a telecommunication transceiver, communication and signalling protocols... suiting to a targeted application. This training includes 7 supervised sessions of practical works (3 hours each); free sessions are also scheduled so that the students can have access to the technical equipment. The starting point is the application specifications. Various items are available: software development tools, sensors (this teaching is synchronized with another one which objective is the development by the students of all the electronic part of the sensors), microcontroller evaluation boards and several kinds of transceiver with a detailed technical documentation. The interface boards between evaluation board and three transceivers are also given from the start.

At the end of this need identification process, they obtained two schemas with the main services provided or required by the system (figure 3) as well as the main components interacting with environment (figure 4).

4.2 Definition of technical requirements

Next step is to define what are the high-level stages of the system life and, in each one, what are the systems states (also called 'modes'). We usually find three cycles: upstream, utilization, downstream cycles. The upstream cycle includes four classical modes: design, realization, validation and installation. The utilisation cycle depends of the system of interest; for example, in our case, we distinguish maintenance, waking and monitoring states. In the downstream cycle, we usually find the retrieval mode.

Students are essentially involved in upstream and utilization modes. They obtain the general operational modes depicted in figure 5.

The technical requirements express the needs in the language of the project manager, or prime contractor, whereas the needs were previously expressed in the users' language. It is now necessary to complete and refine the information supplied by the users so that they lead to potential solutions. This is the goal of the technical requirements definition process.

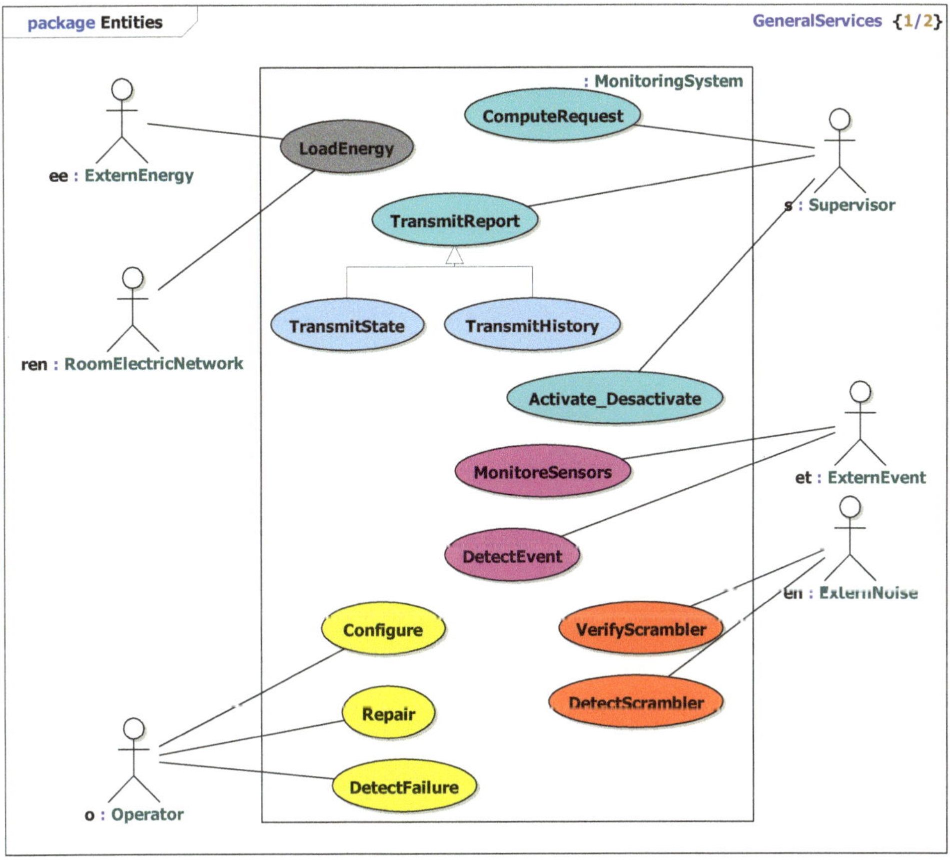

Fig. 3. General service diagram

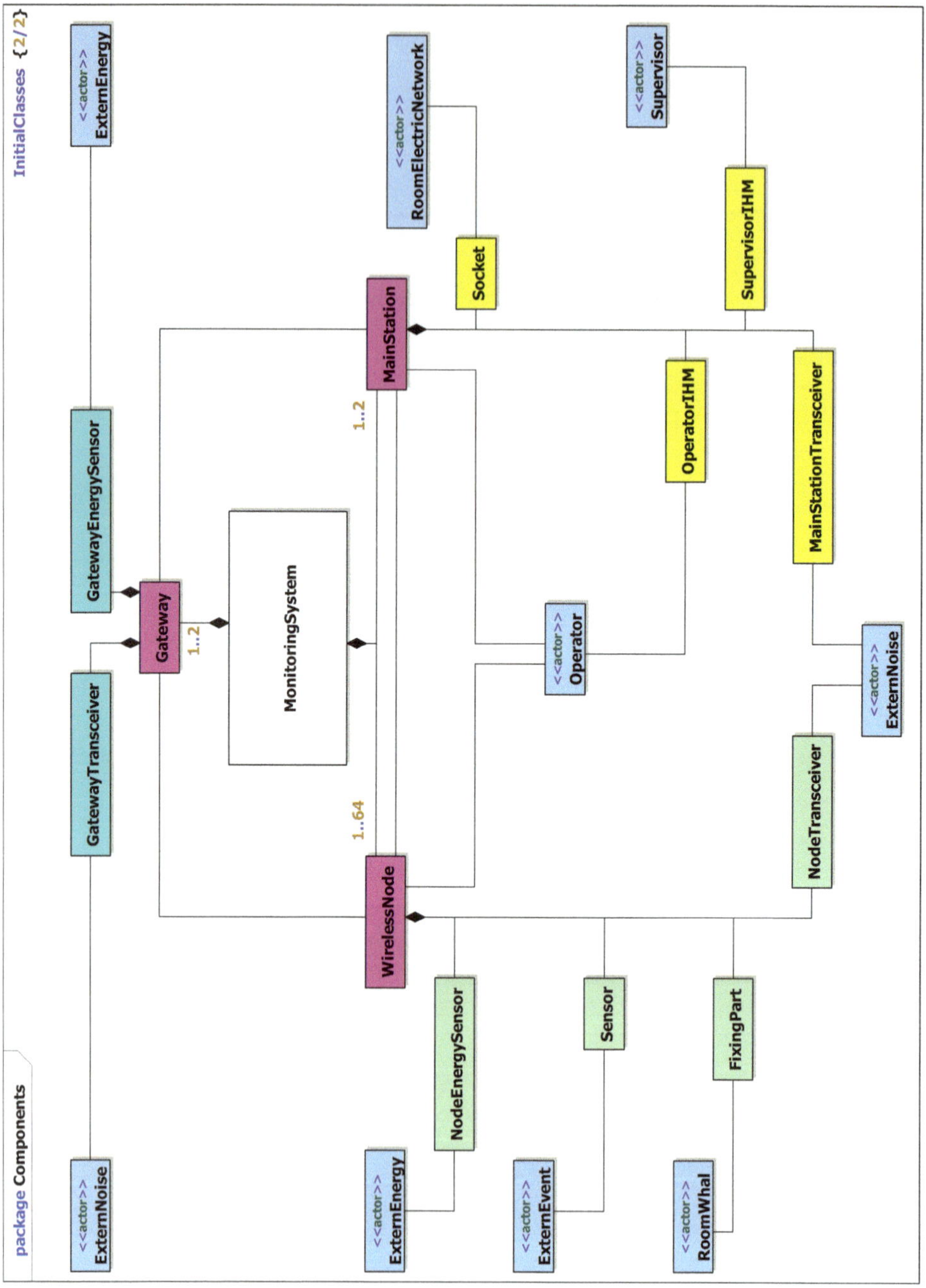

Fig. 4. Initial class diagram

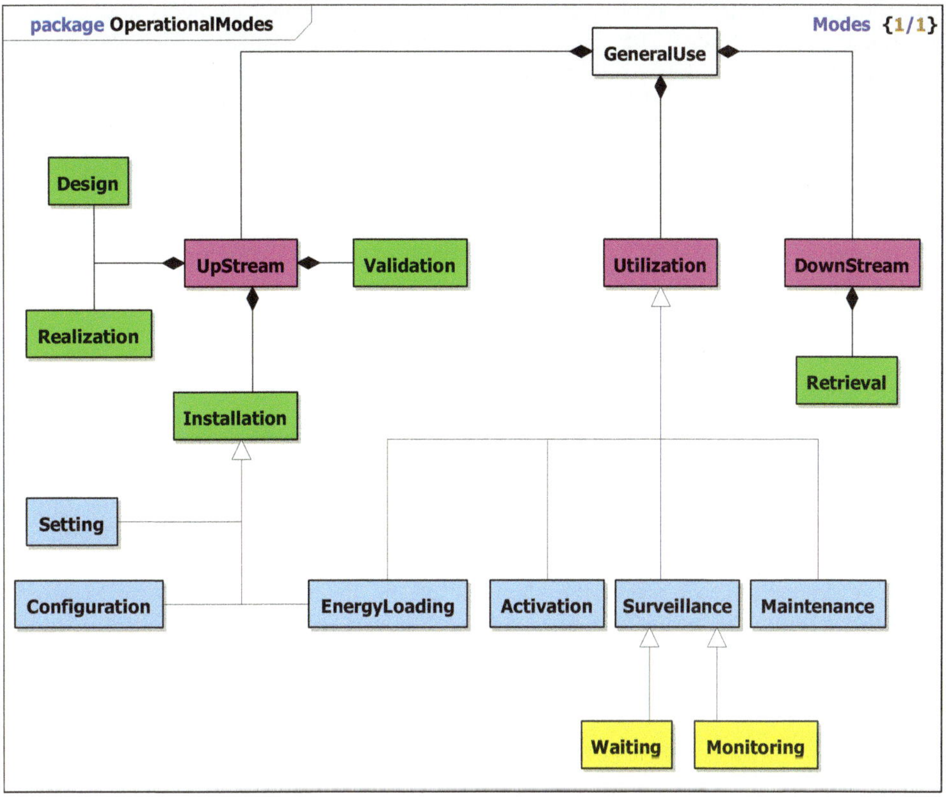

Fig. 5. Operational modes

An example of technical requirements found by students and obtained by translating need into expressions allowing the design of a practicable/realizable solution is resumed in Table 1. Some previous works (Auriol *et al.*, 2008) explain a way to introduce students to requirement engineering.

Id	Needs	Technical Requirements
Operational Need 9	A node represents a position on the site	Define a unique ID for a node Define a process for associating a node with a position Define a specific grid for authentication Define a mechanism to upload node routing table

Table 1. Example of a need and definition of corresponding technical requirements

5. Design of functional architecture

The functional design process consists in identifying functional elements and designing functional architectures. The goal is to establish and evaluate several functional architectures that could be candidates and retain one. The identification of functional elements is directly obtained by an analysis of technical requirements: functional, interface and operational requirements, operational scenarios, and a breakdown of expected services. Performance requirements must then be allocated to functions. Once several functional architectures have been obtained, we need to identify and solve any conflicts between the elements of each functional solution (optimization process) and verify that each functional architecture correctly and fully satisfies the technical requirements. An evaluation of the various alternative functional architectures compared according to several parameters (quality, costs, times, performances, risks, etc.) leads to the best trade-off.

For example, when students deal with the services found in the first step, they obtain the Activity Diagram depicted in figure 6.

6. Design of physical architecture

Once a functional architecture for the system has been defined, the goal of the physical design process is to design various physical architectures to support these functions. The effort in this step is focused on identifying classes of components, establishing parameters and choosing criteria to assign the elements of the functional architecture to physical components, and the evaluation of several solutions. The physical architecture design process takes as its starting point the result of the functional design step, and refines it. Indeed, for each architecture, the first task is to decide whether the functional breakdown is sufficient to identify physical components and/or technologies capable of supporting the execution of the end functions of the functional architecture. The objective is then to consider various possible physical architectures and to estimate their feasibility. Once various possible physical architectures have been obtained, it only remains to choose a final architecture. Once this choice has been made, the final task is to fully specify the solution, to validate and justify it.

Students extract the components of the system from the initial class diagram (figure 7) and progressively complete the physical architecture diagram with components according to the functions found during the precedent step (figures 8&9).

At this level of breakdown, students add the available solutions. In this chapter, we only give some details about transceivers and communication protocols. For example, they can choose transceivers among:

- a half-duplex FM transmitter, manufactured by Telital, using FSK modulation at 433MHz
- a EEE 802.15.4 [9] transmitter with a *ZigBee* [10] stack manufactured by Microchip
- a GSM / GPRS modem manufactured by Sagem

The choice of these technologies is driven by the diversity of the services that they offer. To simplify, four technical parameters are retained: cost, range, consumption and access BUS and the embedded Medium Access Control (MAC) layer. During the integration step, students have to refine and to extend these performances

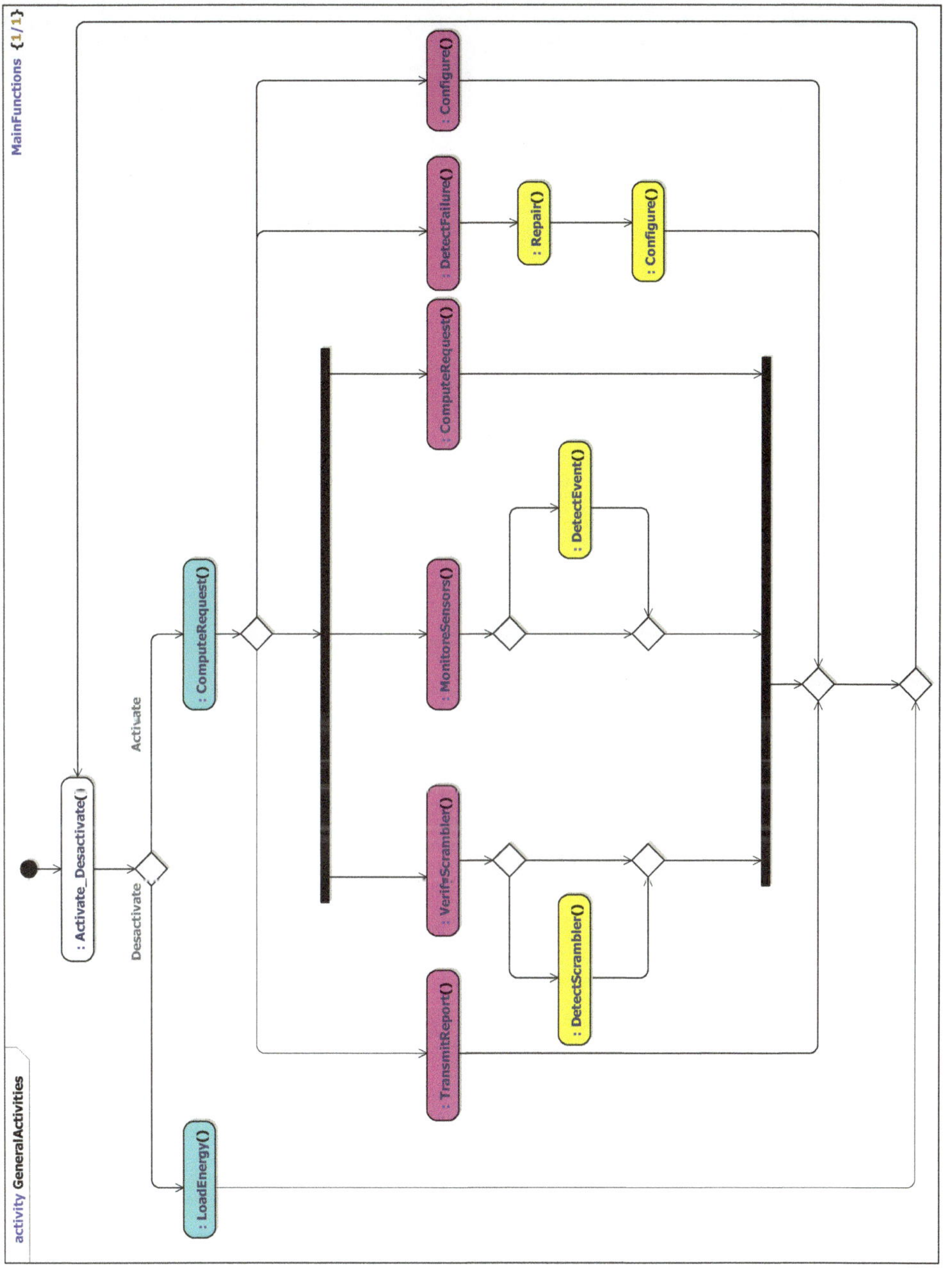

Fig. 6. General activity diagram

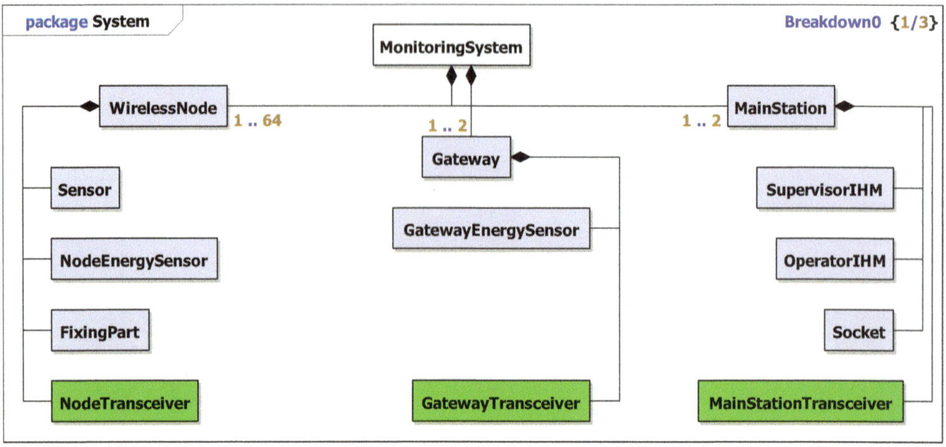

Fig. 7. Main components

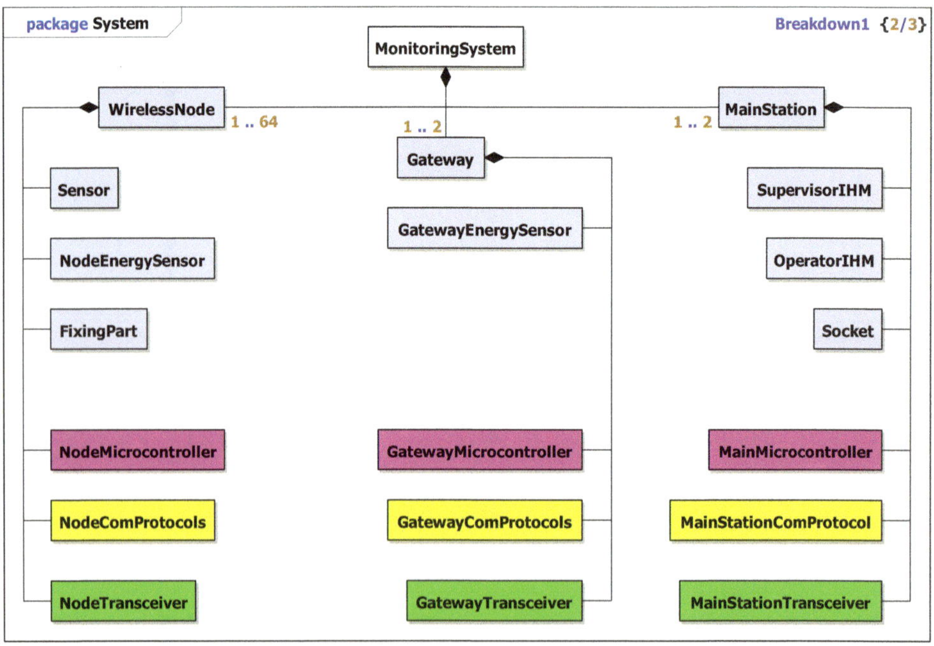

Fig. 8. Microcontrollers and protocols components

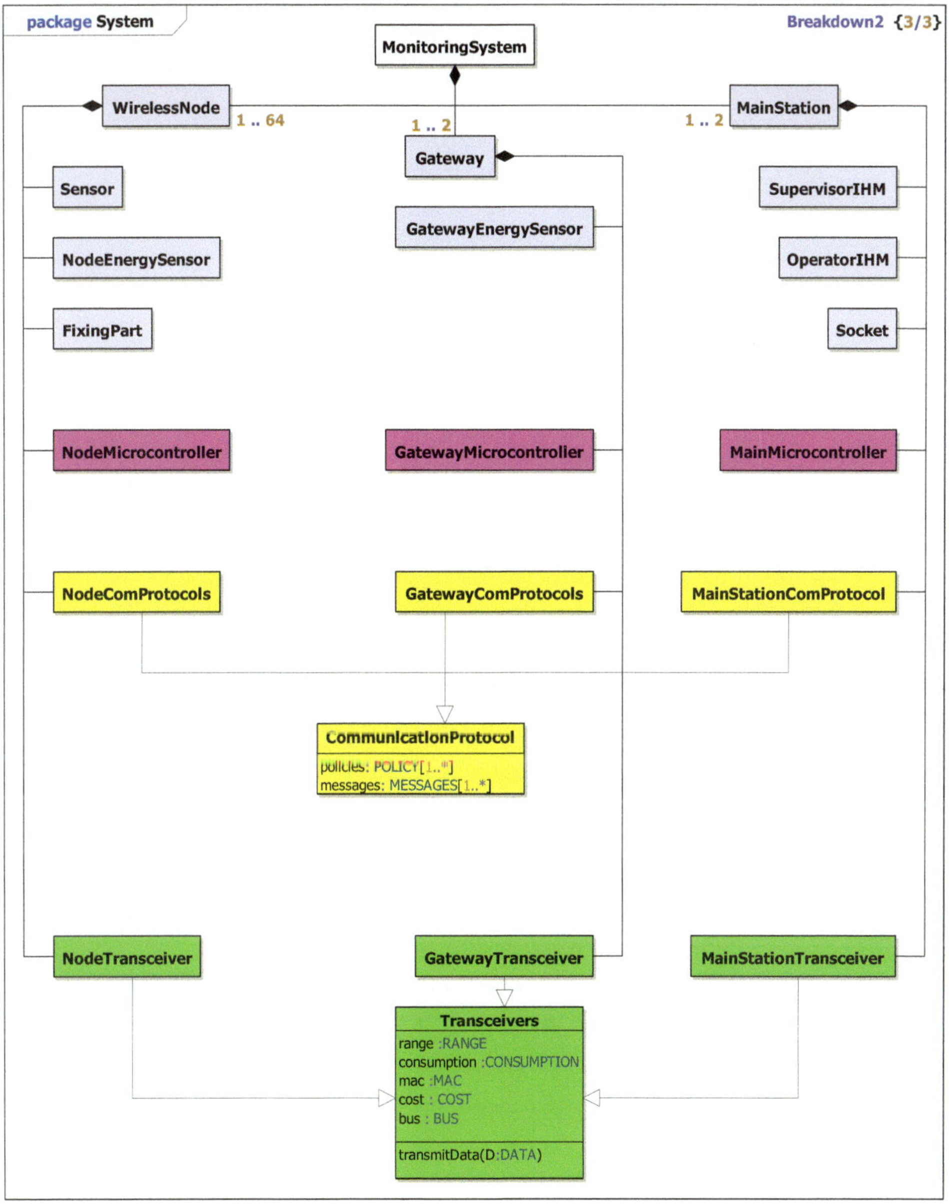

Fig. 9. Final class diagram

Then, students obtain the schema in figure 10.

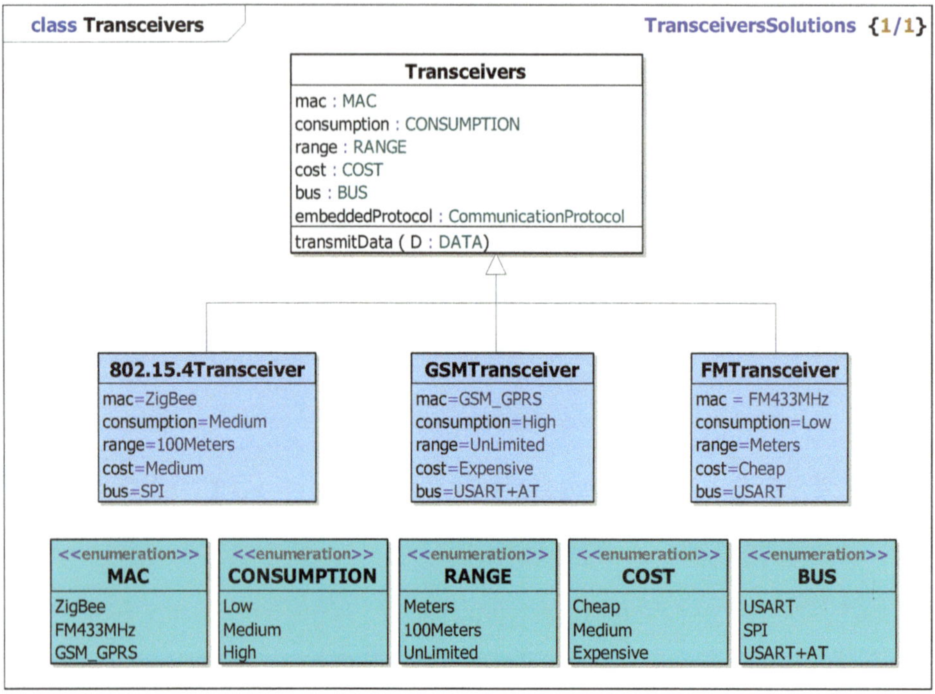

Fig. 10. Details of transceiver diagram

Each transceiver studied has specific data transfer and signaling protocol which can be extremely basic, or may include complex embedded applications. The implementation of these services also differs a lot from one device to another. To develop the application of remote monitoring, students must initially understand the need for these services, and then extend them if necessary. Mainly, the retained parameters to characterize the communication protocols concern:

- services as: connection, loss detection, carrier detection, acknowledges,
- policies of deployment,
- reliability to transmit several kinds of messages which could be analog or binary, critic or not, periodic or not.

Students obtain the schema of figure 11.

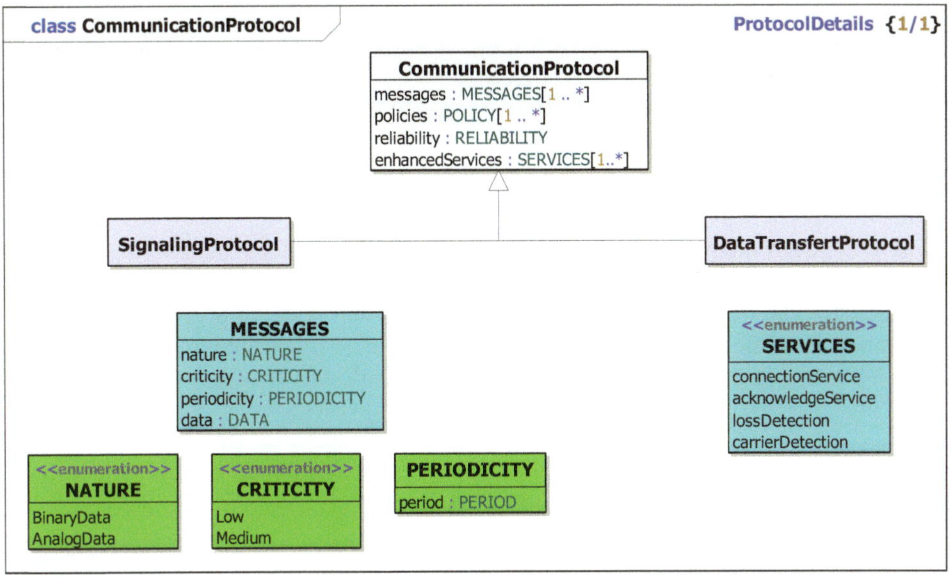

Fig. 11. Details of protocol diagram

7. Integration and validation

During this step of integration, students connect component via customized boards to the evaluation board of a microcontroller (MCBSTM32 [7] of Keil, processor ARM / ST). This configuration represents a "less embedded" solution than a dedicated electronic board which would have specifically been developed and offers a high flexibility of use and of debug to the students by supplying a multi line LCD screen for display, free zones for oscilloscope measures, series connections for connections to a terminal PC, diodes... The data transfer and signalling protocols are implemented on the microcontroller by means of the environment µVision3 of Keil [8] which offers functions of simulation and transfer towards a microcontroller.

Students have to discover and understand the manipulation of each device taken separately before starting to completely implement the platform to validate their choice of components.

The students follow the evolution detailed below:

Step 1. discovery of devices. For that purpose, they test every device by establishing a direct and basic communication (send of an *ASCII* characters) between a single transmitter and a receiver. This step is common to every device and requires no elaborated configuration nor to add services to those already offered by modules.

Step 2. several transmitters on the network. By testing services offered by the first device (FM technology), students understand that it is necessary to add a field "address" in any emission of information. They also note a risk of collision and loss of frames which grows with the transmitter number if they do not use a device offering suitable services or if they do not extend those basic ones proposed by the device.

Step 3. the network spreads out. The initial range of the first two modules (FM and *ZigBee*) is not sufficient to cover communication needs on more important distances. It is then essential to set up several modules with a suitable signaling service on intermediate devices.

Gradually, the students thus take into account a more and more complex topology until they consider the complete platform of test compatible with the application of remote monitoring. The objective for the students is to validate their choice of components by the mean of a platform whose technological solution is represented in figure 12. Indeed, this solution appears as the best compromise if all the parameters are taken into account.

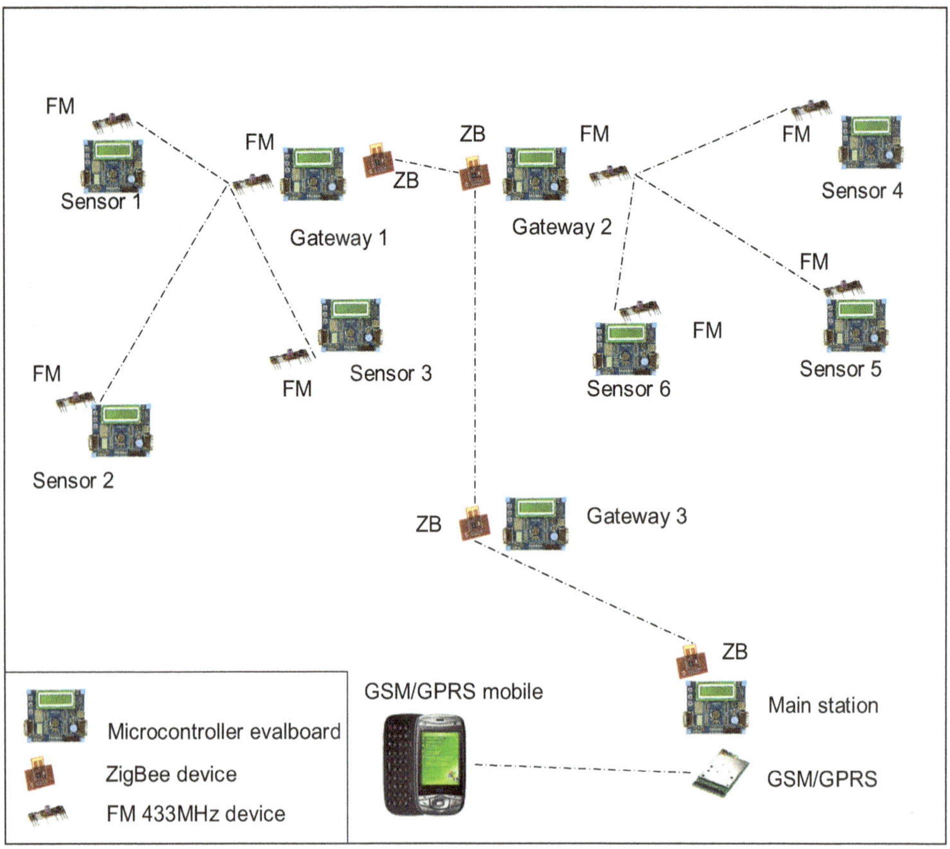

Fig. 12. Technological solution

8. Conclusion

This chapter describes a teaching experiment during which the students apply a systems engineering approach to design a solution for a complex system when numerous design and implementation options are available. The chosen application is the remote surveillance of several rooms simultaneously taking into account three parameters: light, temperature and movement. This application is based on wireless terminal nodes composed of a sensor, a microcontroller and a telecommunication module. They dispose of a set of off-the-shelf software and hardware components from which they must design the best functional and architectural solutions, by drafting a technical requirements dossier to satisfy the users' needs.

For that, they are guided to progress following the steps of the V cycle. They start by studying the requirements to extract an exhaustive list of needs. Then they propose and evaluate functional and architectural solutions. They finally implement the chosen solution by integrating the different modules of the physical architecture in order to validate their 'systems engineering' approach.

This experiment was positive in that it taught students that even if they had no previous specific knowledge of the field of wireless Personal Area Networks, a formalised systems engineering approach allowed them to develop a solution.

9. Acknowledgement

A part of the research leading to these results has received funding from the European Community's Seventh Framework Programme (FP7/2007-2013) (www.crescendo-fp7.eu) under grant agreement n°234344.

10. References

ANSI/IEEE 1220, 2005, Standard for Application and Management of the Systems Engineering Process, www.ansi.org

Auriol G., Baron C., Fourniols J-Y., 2008, Teaching requirements skills within the context of a physical engineering project, *The Third International Workshop on Requirements Engineering Education and Training (REET'08)*, September 9th, 2008, Barcelona, Spain

Bock C., 2006, SysML and UML Support for Activity Modeling, *Systems Engineering*, Vol. 9, Number 2, Summer 2006.

Brownsell S. J., Williams G., Bradley D. A., Bragg R., Catlin P., Carlier J, 1999, Future systems for remote health care, *Jour. Telemedicine & Telecare*, vol 5, pp 141-152

Honour E. C. , 2004. Understanding the value of systems engineering, *INCOSE International Symposium*, Toulouse, France.

ISO-IEC 15288, 2008, System Life Cycle Processes, www.iso.org

Kotzl D. and Essien K., 2005, Analysis of a Campus-Wide Wireless Network, *Wireless Networks*, Vol. 11, Numbers 1-2, pp 115-133.

Martin J. N., 1998, Overview of the EIA 632 Standard - Processes for Engineering a System, *17th Digital Avionics Systems Conference (DASC)*, 31 Oct-7 Nov 1998, Bellevue, WA, USA

Terry A. Bahill and Clarck Briggs. "The systems engineering started in the middle process : A consensus of systems engineers and project managers", Systems Engineering, 4(2) pages 156–167, 2001.

Ullrich K.T. and Eppinger S.D., "Product design and development", McGraw Hill International Edition, 2003

Weilkiens T., 2008, *Systems Engineering with SysML/UML: Modeling, Analysis, Design*, Morgan Kaufmann Publishers In, 28 March 2008, The MK/OMG Press, ISBN 0123742749, 320p

Assessing the Capacity for Engineering Systems Thinking (CEST) and Other Competencies of Systems Engineers

Moti Frank[1] and Joseph Kasser[2]
[1]*HIT-Holon Institute of Technology,*
[2]*NUS-National University of Singapore,*
[1]*Israel*
[2]*Singapore*

1. Introduction

This chapter introduces a tool for assessing engineers' interest in what is required from successful systems engineers, or in other words, assessing the extent of engineers' systems thinking. What is required from successful systems engineers (the characteristics of successful systems engineers) is commonly called 'competencies of successful systems engineers' and much activity to develop systems engineering competency models has been done in recent years. A summary of several systems engineering competency models is presented in the chapter. The competency model that has been used as the underpinning basis for the developing of the assessment tool presented in this chapter is the CEST (Capacity for Engineering Systems Thinking) model. The main reason for choosing this model is presented in the chapter and then the model itself and several principles for assessing engineers' systems thinking are presented. Finally, the assessment tool is presented as well as the main methods that have been used for validating the tool.

2. Systems thinking and CEST

Systems thinking is what makes systems engineering different from other kinds of engineering and is the underpinning skill required to do systems engineering" (Beasley & Partridge, 2011). *Systems thinking*, according to Senge (1994), is a discipline for seeing the whole. *Engineering Systems Thinking* is hypothesized as a major high-order thinking skill that enables individuals to successfully perform systems engineering tasks (Frank, 2000; 2002). Systems engineers need a systems view or a high capacity for engineering systems thinking (CEST) to successfully perform systems engineering tasks. Research found that this ability is a consistent personality trait and that it can be used to distinguish between individual engineers (Frank, 2006). CEST may be developed through experience, education and training (Davidz & Nightingale, 2008; Kasser, 2011) and can be assessed (Frank, 2010). Moreover, well designed and taught systems engineering courses may accelerate systems thinking development.

The chapter introduces a tool for assessing engineers' CEST. Since there is no known way for directly 'measuring' thinking skills of individuals, an indirect way is needed, for example, IQ tests are pen-and-paper indirect tests for 'measuring' the intelligence of individuals.

One of the main assumptions made by Frank (2010) is that in order to be a successful systems engineer, one must have both a will and an interest to be a systems engineer.

In addition, as mentioned, successful systems engineers possess a high level of engineering systems thinking (CEST). Thus, the three components discussed here - success in a systems engineering position, an interest in systems engineering positions and CEST- are all interconnected and interrelated. The will and interest to be a systems engineer basically means the desire and interest to be involved with job positions that require CEST. In other words, we may hypothesize that there is a high positive correlation between the engineering systems thinking extent (CEST) of an individual and his/her interest in what is required from successful systems engineers. Figure 1 is a simple concept map that depicts the relationships between these three components:

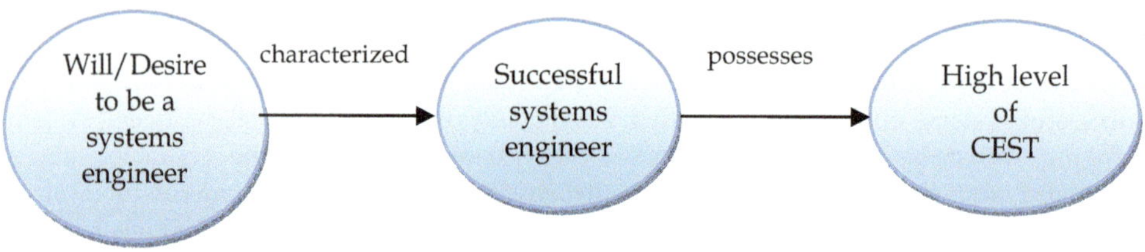

Fig. 1. the relationships between the desire, successful SE and CEST

If this hypothesis is supported, then it enables developing a method for assessing the extent of CEST of individuals. This is because interests may be assessed by an interest inventory which is a very common and frequently used to help people choose a profession, and as a selection tool (to determine whether a certain individual is suitable for a certain role) in the recruiting process (Anastazi, 1988). This chapter introduces a tool for assessing engineers' interest in what is required from successful systems engineers, or in other words, assessing the extent of the engineering systems thinking.

3. Systems engineering competency models

What is required from successful systems engineers (the characteristics of successful systems engineers) is commonly called 'competencies of successful systems engineers' and much activity to develop systems engineering competency models has been done in recent years. A summary of the following models is presented below:

- INCOSE UK SE Competencies Framework
- MITRE Systems Engineering Competency Model
- Systems Thinking Enablers
- Advancing the Practice of Systems Engineering at JPL
- Characteristics of the Ideal Systems Engineer

3.1 INCOSE UK SE competencies framework

According to the systems engineering competencies framework of the United Kingdom chapter of the International Council on Systems Engineering (INCOSE UK, 2010), systems engineering ability comprises four key elements: competencies, basic skills and behaviours, supporting techniques and domain knowledge. The competencies are grouped into three categories: systems thinking, holistic lifecycle view and systems engineering management. The full document presents the following information for each competency: a description, why it matters and effective indicators of knowledge and experience in four levels - awareness, supervised practitioner, practitioner and expert. Examples of basic skills and behaviours are:

- abstract thinking - ability to see multiple perspectives, ability to see the big picture, knowing when to ask for advice, engaging an expert, peer review, requesting training;
- knowing when to stop - the Pareto principle, the 80:20 rule, decision making skills;
- creativity - lateral thinking (six thinking hats), brainstorming, TRIZ;
- objectivity - reference of policy, baselining, viewpoint analysis;
- problem solving - TQM tools (cause/effect, force field, Pareto, etc.), SWOT analysis, PESTEL analysis, decision trees, logical reasoning;
- developing others - coaching, mentoring, training;
- two way communicating - listening skills, verbal and non-verbal communication, body language, writing skills, presentation skills;
- negotiating - win-win, bartering, diplomacy, cultural awareness, stakeholder management, management of expectations;
- team working - Belbin team roles, Meyers-Briggs type indicator, TQM tools;
- decision making - risk/benefit analysis, Pareto analysis, pair-wise comparison, decision trees, force field analysis, six thinking hats.

3.2 MITRE systems engineering competency model

The MITRE competency model (Metzger & Bender, 2007) consists of 36 competencies organized into five sections: enterprise perspectives, systems engineering life cycle, systems engineering planning and management, systems engineering technical specialties, collaboration and individual characteristics. For example, the section 'enterprise perspectives' consists of three competencies - comprehensive viewpoints, innovative approaches and foster stakeholder relationships and the section 'collaboration and individual characteristics' consists of nine competencies - building trust, building a successful team, communicating with impact, persuasiveness and influence, facilitating, managing and championing change, high quality standards, results orientation, adaptability and integrity.

3.3 Systems thinking enablers

According to Davidz and Nightingale (2008), the primary mechanisms that enable systems thinking development include: experiential learning, a supporting environment and certain individual characteristics, such as thinking broadly, curiosity, questioning, being open-minded, communication, tolerance for uncertainty, strong interpersonal skills and 'thinking out of the box'.

3.4 Advancing the practice of systems engineering at JPL

The JPL (Jet Propulsion Laboratory) competency model presented by Jansma and Jones (2006) refers to personal behaviours and processes. The personal behaviours are presented in five groups:

- Leadership Skills - has the ability to influence; has the ability to work with a team; has the ability to trust others; communicates vision and technical steps needed to reach implementation; mentors and coaches less experienced systems engineers.
- Attitudes and Attributes - has intellectual self-confidence; has intellectual curiosity; has ability to manage change; remains objective and maintains a healthy scepticism.
- Communication - advances ideas and fosters open two-way discussions; communicates through storytelling and analogies; listens and translates information.
- Problem Solving and Systems Thinking - manages risk; thinks critically and penetrates a topic in a methodical manner.
- Technical Acumen - successfully expresses a technical grasp of system engineering at all levels; is a generalist in nature; with proven technical depth in one or two disciplines; has proven knowledge of systems engineering practices.

3.5 Characteristics of the ideal systems engineer

Burk (2008) found that the characteristics of the ideal systems engineer are: systems outlook, customer/user/consumer orientation, inquisitiveness, intuition, discipline, communication and cooperation (but not capitulation).

4. The maturity model framework

The maturity model for the competency of systems engineers is based on an assessment of an individual's skill against ability in each of three broad dimensions - knowledge (systems engineering and domain), cognitive characteristics (systems thinking and critical thinking) and individual traits. The maturity model is designed in such a manner so as to be a generic maturity model for assessing competency in many practitioner professions simply by changing the knowledge requirements (Kasser & Frank, 2010).

The maturity model is a two-dimensional model. The vertical dimension covers the following three broad areas:

- **Knowledge** of systems engineering and the application domain in which the systems engineering is being applied.
- **Cognitive characteristics**, namely the ability to think, identify and tackle problems by solving, resolving, dissolving or absolving the problems in both the conceptual and physical domains.
- **Individual traits**, namely the ability to communicate with, work with, lead and influence other people.

The horizontal dimension is based on Kasser, Hitchins and Huynh (2009) who argue that anecdotal evidence exists for five types of systems engineers:

- Type I. This type is an "apprentice" who can be told "how" to implement the solution and can then implement it.

- Type II. This type is the most common type of systems engineer. Type IIs have the ability to follow a systems engineering process to implement a physical solution once told what to do.
- Type III. Once given a statement of the problem, this type has the necessary know-how to conceptualize the solution and to plan the implementation of the solution, namely create the process to realize the solution.
- Type IV. This type has the ability to examine the situation and define the problem.
- Type V. This type is rare and combines the abilities of the Types III and IV, namely has the ability to examine the situation, define the problem, conceptualize the solution and plan the implementation of the physical solution.

The two-dimensional maturity model framework shows the assessment of the competency in increasing levels of competency (Type I to V) as presented in the following Table. Declarative knowledge is knowledge that can be declared in some manner. It is "knowing that" something is the case. Describing a process is declarative knowledge. Procedural knowledge is about knowing how to do something. It must be demonstrated; performing the process demonstrates procedural knowledge. Conditional knowledge is about knowing when and why to apply the declarative and procedural knowledge (Woolfolk, 2011). This usually comes from experience. In the Table, where knowledge is required at the conditional level, it includes procedural and declarative. Similarly, where knowledge is required at the procedural level, it includes declarative knowledge.

	Type I	Type II	Type III	Type IV	Type V
Knowledge					
Systems engineering	Declarative	Procedural	Conditional	Conditional	Conditional
Domain (problem solution)	Declarative	Declarative	Conditional	Conditional	Conditional
Cognitive characteristics					
System Thinking					
Operational	Declarative	Procedural	Conditional	Conditional	Conditional
Functional	Declarative	Procedural	Conditional	Conditional	Conditional
Big picture	Declarative	Procedural	Conditional	Conditional	Conditional
Structural	Declarative	Procedural	Conditional	Conditional	Conditional
Generic	Declarative	Procedural	Conditional	Conditional	Conditional
Continuum	Declarative	Procedural	Conditional	Conditional	Conditional
Temporal	Declarative	Procedural	Conditional	Conditional	Conditional
Quantitative	Declarative	Procedural	Conditional	Conditional	Conditional
Scientific	No	No	Procedural	No	Conditional
Critical Thinking	Confused fact finder	Perpetual analyser	Pragmatic performer	Pragmatic performer	Strategic revisioner
Individual traits (sample)					
Communications	Yes	Yes	Yes	Yes	Yes
Management	No	Yes	Yes	Yes	Yes
Leadership	No	No	Yes	Yes	Yes

Table 1. The two-dimensional maturity model

The maturity model may serve both as a competency model and a framework for assessing/comparing other competency models (Kasser et al., 2011).

Other systems engineering competencies models found in the literature include:

- NASA Systems Engineering Competencies (NASA, 2009).
- Systems Engineering Competency Taxonomy (Squires et al., 2011).
- Generic Competency Model (Armstrong et al., 2011).

5. The CEST competency model

However, the competency model that has been used as the underpinning basis for the developing of the assessment tool presented in this chapter is the CEST model (Frank, 2002; 2006). The main reason for choosing this model is that in order to assess systems thinking in engineers, it is necessary, first, to elaborate this thinking skill to elements that can be assessed. The CEST Competency Model presents a list of cognitive competencies that are all related to systems thinking and each one of them can be assessed separately.

Eighty-three competencies of successful systems engineers have been found in the studies and these findings were used to create the CEST Competency Model. These 83 competencies were then aggregated into 35 competencies - 16 cognitive competencies, nine skills/abilities (all also related to cognitive competencies), seven behavioural competencies and three related to knowledge and experience.

The 16 cognitive competencies are as follows for successful systems engineers:

1. understand the whole system and see the big picture;
2. understand interconnections; closed-loop thinking;
3. understand system synergy (emergent properties);
4. understand the system from multiple perspectives;
5. think creatively;
6. understand systems without getting stuck on details; tolerance for ambiguity and uncertainty;
7. understand the implications of proposed change;
8. understand a new system/concept immediately upon presentation;
9. understand analogies and parallelism between systems;
10. understand limits to growth;
11. ask good (the right) questions;
12. (are) innovators, originators, promoters, initiators, curious;
13. are able to define boundaries;
14. are able to take into consideration non-engineering factors;
15. are able to "see" the future;
16. are able to optimize.

The nine skills/abilities that are all related to cognitive competencies of successful systems engineers are the ability to:

1. analyze and develop the needs and mission statement, and the goals and objectives of the system;
2. understand the operational environment and develop the concept of operation (CONOPS);

3. analyze the requirements (requirements analysis) including capturing requirements, defining requirements, formulating requirements, avoiding suboptimizing, generating System Requirements Documents (SRD), "translating" the concept of operations and the requirements into technical terms and preparing system specifications, validating the requirements, tracing therequirements, ensuring that all needs, goals and external interfaces (context diagram) are covered by the requirements, and allocating the system requirements into lower levels;
4. conceptualize the solution;
5. generate the logical solution - functional analysis;
6. generate the physical solution - architecture synthesis;
7. use simulations and SE tools;
8. manage systems processes including interface management, configuration management, risk management, knowledge/data management, resource management, integration, testing, verification and validation;
9. conduct trade studies, provide several options and rate them according to their cost-effectiveness.

The seven behavioural competencies of successful systems engineers are as follows:

1. be a team leader;
2. be able to build, control and monitor the project (technical management);
3. possess additional management skills (negotiators, resolving conflicts. etc.);
4. be characterized by good communication and interpersonal skills; be able to collaborate; be strong team players; establish trusting relations with stakeholders;
5. be capable of autonomous and independent self-learning;
6. characterized by having a strong desire/will to deal with systems projects;
7. characterized by seeing failures not as "the end of the road" and by having tolerance for failure.

The three competencies related to knowledge and experience for successful systems engineers are as follows:

1. expert in at least one science or engineering discipline (core disciplines such as physics, electrical engineering, mechanical engineering, aeronautical engineering and industrial engineering);
2. possesses technical general knowledge in additional science/engineering disciplines (interdisciplinary and multidisciplinary knowledge);
3. experience of several years in working as a domain and as a junior systems engineer in several systems projects.

In organizations and projects there are many different kinds of job positions that may be included in the systems engineering category. Different positions require different competencies, for example, a systems engineer who works in marketing needs different knowledge, skills and behavioural competences from those of a systems engineer who deals with integration or a systems engineer who deals with verification and validation. In addition, it is unlikely that a successful systems engineer would possess all of these 35 competencies. It is more likely that a certain systems engineer possesses part of the listed competencies and is employed in a position that requires these specific characteristics. Thus, it is not enough to assess CEST by the final score of the assessment tool presented below. Analyzing the answers to each question is important as well.

However, it appears that a set of core competencies do in fact exist, necessary to all systems engineers, independent of their specific position. It is a matter of hierarchy. Every job level requires competencies suitable for the said level. The higher the systems engineering position in the organization/project hierarchy, the higher the level of required cognitive competencies, skill/ability and behavioural competencies, and broader knowledge needed.

6. Assessing CEST

The battery for assessing CEST in its final stages will comprise:

- *Paper-and-pencil tests.* These tests will include three inventories:
  - *An interest inventory* - will be discussed in detail in Section 7 below.
  - *A knowledge and skills test.* The present paper does not discuss the knowledge and skills test. Much work in this field has already been done by the International Council on Systems Engineering (INCOSE), the INCOSE Certification of Systems Engineers Exam working group. This exam is based on the INCOSE SE Handbook (INCOSE, 2006).
  - *An aptitude test.* Please see several sample questions in Frank (2007).
- *Field tests.* In the field test the examinee will be asked to develop and present a functional flow block diagram that describes the functional (logical) and physical architecture of a system that meets a given specification.
- *Lab test.* In the future, the possibility of adding a lab test will be considered. In this lab test the capability for global processing by the right hemisphere will be tested (Evert & Kmen, 2003). The field test and the lab test are not in the purview of this chapter.

7. The interest inventory for assessing CEST

As said earlier, the will/desire and the interest to be a systems engineer (to be involved in systems projects) mainly means the will and interest to deal with situations that require systems thinking. In addition, one of the seven behavioural competencies of a successful systems engineer is a will/desire to be a systems engineer (to be involved in systems projects) - see competency number 6 in the list of the seven behavioural competencies aforementioned in the CEST competency model section. These two findings lead to the conclusion that the will/desire to be involved in positions that require engineering systems thinking predicts success in systems engineering positions. This will/desire can be assessed by an interest inventory. As mentioned above, an interest inventory is a very common tool which is frequently used to help people choose a profession and as a selection tool in the recruiting process (Anastazi, 1988).

Usually, the items in interest inventories deal with preferences, specifically likes and dislikes regarding a diverse group of activities, jobs, professions or personality types. Likewise, the items included in the tool discussed in this chapter refer to ranges of likes and dislikes regarding systems engineering activities, various disciplines and knowledge required from systems engineers, systems engineering activities and types of people involved in projects.

In its present version the tool consists of 40 pairs of statements. For each pair, the examinee has to choose between the two statements according to his/her preference. The examinee checks answer "A" if he/she prefers the first statement or answer "B" if he/she prefers the second statement. In order to improve the questionnaire's reliability, questionnaire items were reorganized, so in some cases "A" represented the systems thinking answer and in other cases "B" represented the systems thinking answer. Each "A" answer receives 2.5 points, while each "B" answer receives no point. Thus, the range of the scores is 0-100.

Several examples of the questions in the tool are presented below. The following three sample questions are based on the finding that successful systems engineers understand the whole system and see the big picture - see competency number 1 in the list of the cognitive competencies of successful systems engineers aforementioned in the CEST competency model section.

Sample question No. 2

- A. When I take care of a product, it is important for me to see how it functions as a part of the system.
- B. When I take care of a product, it is important for me to concentrate on this product, assuming that other engineers will take care of the other parts of the system.

Sample question No. 3

- A. It is important for me to identify the benefits derived from embedding several products/sub-systems/systems.
- B. I prefer not to deal with combinations of products/sub-systems/systems, but rather to concentrate on the product for which I am responsible.

Sample question No. 4

- A. It is important for me to know what other employees in my department/project do.
- B. It is important for me to do my best and not interfere to the work of other employees in my department/project.

The following sample question is based on the finding that successful systems engineers understand systems without getting stuck on details - see competency number 6 in the list of the cognitive competencies of successful systems engineers aforementioned in the CEST competency model section.

Sample question No. 6

- A. I don't like to be involved with details; I prefer to deal with the system's aspects.
- B. In areas in which I'm involved, I like to understand all the details.

The following sample question is based on the finding that successful systems engineers understand interconnections - see competency number 2 in the list of the cognitive competencies of successful systems engineers aforementioned in the CEST competency model section.

Sample question No. 11

- A. When I deal with a product, I always look at the interconnections and mutual influences between the main product and the peripheral products.
- B. I prefer to thoroughly take care of the part for which I am responsible and leave the issue of interconnections between a system's parts to the integration engineers.

The following sample question is based on the finding that successful systems engineers possess interdisciplinary and multidisciplinary knowledge - see competency number 2 in the list of the competencies related to knowledge and experience aforementioned in the CEST competency model section.

Sample question No. 17

- A. I think that every employee should gain interdisciplinary knowledge and general knowledge in several fields.
- B. I think that every employee should become an expert in his/her field. Learning more fields may lead to sciolism (to know a little about many subjects).

The following sample question is based on the finding that successful systems engineers are able to analyze and develop the needs and mission statement, and the goals and objectives of the system - see competency number 1 in the list of the abilities and skills of successful systems engineers aforementioned in the CEST competency model section.

Sample question No. 22

- A. I like to discuss the needs with the customer.
- B. I prefer to leave the contact with the customer to marketing experts.

The following sample question is based on the finding that successful systems engineers are innovators, originators, promoters, initiators and curious - see competency number 12 in the list of the cognitive competencies of successful systems engineers aforementioned in the CEST competency model section.

Sample question No. 39

- A. It is important for me to continuously think what else can be improved.
- B. It is important for me to determine the finish line and to finish my jobs in time.

The following sample question is based on the finding that successful systems engineers possess management skills - see competency number 3 in the list of the behavioural competencies of successful systems engineers aforementioned in the CEST competency model section.

Sample question No. 30

- A. I like to integrate and to lead interdisciplinary teams.
- B. I'm a professional; I prefer not to be involved with managerial issues.

8. Validity of the interest inventory

Four types of validity have been checked in a series of studies (Frank, 2010) - content validity, contrasted groups validity, concurrent validity and construct validity.

Content Validity

The proposed tool was developed and the content validity was achieved by basing the items of the interest inventory discussed here on a literature review including the INCOSE SE Handbook Version 3 (INCOSE, 2006), laws of the fifth discipline and systems archetypes (Senge, 1994), systems thinking principles (Kim, 1994; Waring, 1996; O'Connor and

McDermott, 1997; Sage, 1992), some principles of systems dynamics (Sweeney and Sterman, 2000; Ossimitz, 2002), the seven 'thinking skills' of systems thinking (Richmond, 2000) and on the findings of a Ph.D. study presented in Frank (2002).

Contrasted Groups Validity

This type of validity is determined by comparing the grades of two contrasted groups. In one study it was found that systems engineers achieved significantly higher scores, as compared to other engineers. In another study the contrasted group validity was checked by comparing the tool's CEST scores of four groups - senior Electrical Engineering students, senior Technology Management students, systems engineers and other engineers. Statistical analyses revealed that: (1) the systems engineers achieved significantly higher scores than the other engineers, (2) the systems engineers achieved significantly higher scores than the Technology Management students and the Electrical Engineering students, while (3) the senior Technology/Engineering Management students achieved significantly higher scores as compared to the senior Electrical Engineering students. This result is not surprising because Technology/Engineering Management students are trained to look at problems holistically.

Concurrent Validity

This type of validity is the correlation between the scores obtained by two assessment tools. In one study, the concurrent validity was checked by calculating the correlation between the participants' scores using the proposed tool and the appraisal of their supervisor. It was found that the Pearson Correlation Coefficient was close to 0.4 ($p<0.05$). This result is very similar to the predictive validity of other selection tools. In another study the concurrent validity was checked by calculating the correlation between systems engineers' scores using the tool and the appraisal of their supervisor. The supervisor had been familiar with the participants' systems thinking capabilities for many years. The subjective assessments were all made by the same senior supervisor to decrease bias. It was found that the Pearson Correlation Coefficient between the participants' scores and the supervisor assessments was 0.496 ($p<0.05$).

Construct Validity

Construct validity indicates the extent to which the tool measures a theoretical construct or characteristic (Anastasi, 1988). The construct validity was checked by factor analysis. The analysis revealed five factors that may be labelled as follows: seeing the big picture, implementing managerial considerations, using interdisciplinary knowledge for conceptualizing the solution, analyzing the needs/requirements and being a systems thinker. These results are compatible with the factors found in an earlier study (Frank, 2006).

9. Some possible implementations of the assessment tool

Every enterprise strives to fill positions in the organization with employees who have the best chance to succeed. Employees are also interested in entering positions that fulfil their aspirations. Selection and screening processes can help match the interests of both parties, thus contributing both to the organization and the individual.

Many studies show that individuals do not behave and function in the same way in every organizational environment. The meeting point between the characteristics of an individual and the specific environment of his/her workplace often determines the quality of the functioning of the individual. Hence, the goal of the selection process is to help find the optimal meeting point and match the right employee to the right job within an organization.

The selection process for systems engineering positions should reliably predict those employees who can succeed and reject those who are likely to fail. Out of the employees who can succeed as systems engineers, it is necessary to choose those who have the highest chance of succeeding. Since no selection process is perfect, two types of errors are possible - choosing candidates that fail after they have been placed and rejecting candidates who might have succeeded. These errors have an influence on both the organization and the individual.

From the organization's point of view, rejection of candidates who might have succeeded in systems engineering positions can be critical, especially under conditions of an ever-increasing shortage of systems engineers. Likewise, placing engineers who later fail in systems engineering positions is also an expensive error, taking into consideration the necessary training which will be invested and the subsequent damage which might be caused to the projects in which they are involved. The tool presented in this chapter may be used for selection, filtering, screening of candidates for systems engineering job positions and for placing the 'right person to the right job'.

10. References

Beasley, R., & Partridge, R. (2011). The three T's of systems engineering - trading, tailoring and thinking. Paper presented at the 21st Annual Symposium of the International Council on Systems Engineering (INCOSE). Denver, CO, USA. June 20-23, 2011.

Davidz, H.L., & Nightingale, D.J. (2008). Enabling systems thinking to accelerate the development of senior systems engineers. *INCOSE Journal of Systems Engineering,* vol. 11, no. 1, pp. 1-14.

Evert, D.L., & Kmen, M. (2003). Hemispheric asymmetries for global and local processing as a function of stimulus exposure duration. *Brain Cognition,* vol. 51, no. 1, pp. 42-115.

Frank, M. (2000). Engineering systems thinking and systems thinking. *INCOSE Journal of Systems Engineering,* vol. 3, no. 3, pp. 163-168.

Frank, M. (2002). Characteristics of engineering systems thinking - A 3-D approach for curriculum content. *IEEE Transaction on System, Man, and Cybernetics,* vol. 32, no. 3, Part C, pp. 203-214.

Frank, M. (2006). Knowledge, abilities, cognitive characteristics and behavioral competences of engineers with high Capacity for Engineering Systems Thinking (CEST). *INCOSE Journal of Systems Engineering,* vol. 9, no. 2, pp. 91-103.

Frank, M. (2007). Toward a quantitative tool for assessing the capacity for engineering systems thinking. *International Journal of Human Resources Development and Management,* vol. 7, no. 3/4, pp. 240-253.

Frank, M. (2010). Assessing the interest for systems engineering positions and other engineering positions' required capacity for engineering systems thinking (CEST). *INCOSE Journal of Systems Engineering,* vol. 13, no. 2, pp. 161-174.

INCOSE (2006). The International Council on Systems Engineering's Systems Engineering Handbook, Version 3. Seattle WA: INCOSE.

INCOSE UK (2010). INCOSE UK Systems Engineering Competencies Framework. Retrieved June 24, 2011 from http://www.incose.org/members/index.aspx

Jansma, P.A., & Jones, R.M. (2006). Systems Engineering Advancement (SEA) Project. Retrieved June 24, 2011 from http://trs-new.jpl.nasa.gov/dspace/bitstream/2014/38979/1/05-3271.pdf

Kasser, J. (2011). Systems engineering a 21st century introductory course on systems engineering. *INCOSE Journal of Systems Engineering* (in press).

Kasser, J.E., Hitchins, D., & Huynh, T.V. (2009). Reengineering Systems Engineering. Paper presented at the 3rd Annual Asia-Pacific Conference on Systems Engineering (APCOSE), Singapore, 2009.

Kasser, J.E., & Frank, M. (2010). A Maturity Model for the Competency of Systems Engineers. Paper presented at the 20th Anniversary INCOSE (International Council on Systems Engineering) Symposium (INCOSE 2010), Chicago, USA, 12-15 July 2010.

Kasser, J.E., Hitchins, D., Frank, M., & Yang Yang Zhao (2011). A framework for competency models of systems engineers. *INCOSE Journal of Systems Engineering* (in press).

Kim, D.H. (1995). *Systems Thinking Tools.* Cambridge, MA: Pegasus

Metzger, L.S., & Bender, L.R. (2007). MITRE Systems Engineering Competency Model. Retrieved June 24, 2011 from http://www.mitre.org/work/systems_engineering/guide/10_0678_presentation.pdf

NASA (2009). Systems engineering competencies. Retrieved June 24, 2011 from: http://www.nasa.gov/pdf/303747main_Systems_Engineering_Competencies.pdf

O'Connor, J., & Mcdermott, I. (1997). *The art of systems thinking.* San Francisco, CA: Thorsons.

Ossimitz, G. (2002). Stock-flow-thinking and reading stock-flow-related graphs: an empirical investigation in dynamic thinking abilities. Paper presented in the *System Dynamics Conference* (Palermo, Italy). Albany, NY: System Dynamics Society.

Richmond, R. (2000). *The "Thinking" in Systems Thinking,* Waltham MA: Pegasus.

Sage, A.P. (1992) *Systems Engineering.* Wiley, NY.

Senge, P. (1994). *The fifth discipline: the art and practice of the learning organization.* New York, NY: Doubleday.

Squires, A., Wade, J., Dominick, P., & Gelosh, D. (2011) Building a competency taxonomy to guide experience acceleration of lead program systems engineers. Paper presented at CSER 2011, University of Southern California, April 15-16.

Sweeney, L.B., & Sterman, J.D. (2000). Bathtub dynamics: initial results of a systems thinking inventory. *System Dynamics Review,* vol. 16, no. 4, pp. 249-286.

Waring, A. (1996). *Practical systems thinking*. Boston, MA: Thomson Business Press.
Woolfolk, A. (2011). *Educational psychology* (3rd ed.). Boston, MA: Allyn & Bacon.

10

An Abstracted and Effective Capabilities Portfolio Management Methodology Using Enterprise or System of Systems Level Architecture

Joongyoon Lee and Youngwon Park
Ajou University/SE Technology Ltd.
Republic of Korea

1. Introduction

The purpose of this chapter is to provide an abstracted methodology for executing Capabilities Portfolio Management (hereafter, CPM) effectively based on the Department of Defense Architecture Framework version 2.0 (hereafter, DoDAF V2.0)[1]. A methodology is the specification of the process to follow together with the work products to be used and generated, plus the consideration of the people and tools involved, during a development effort. Based on the definition of methodology of ISO 24744 (ISO, 2007), this chapter provides process, product and modeling related technology as considerations of people and tools involved in CPM. From DoDAF V2.0, the purpose of developing an architecture is for beneficial use of it. A good set of architectural artifacts facilitates the manipulation and use of them in meeting its purposes of use. Systems engineering methodologies evolve to accommodate or to deal with problems in enterprise level, system of systems (hereafter, SoS) level and family of systems (hereafter, FoS) level. And the CPM of the United States Department of Defense (hereafter DoD) is a good example which demonstrates enterprise or SoS level problems. However, the complexity of the metamodel of DoDAF makes it difficult to develop and use the architecture models and their associated artifacts. DoDAF states that it was established to guide the development of architectures and to satisfy the demands for a structured and, repeatable method in evaluating alternatives which add value to decisions and management practices. One of the objectives of DoDAF V2.0 is to define concepts and models usable in DoD's six core processes. DoDAF V2.0 provides a particular methodology in the architecture development process. However, DoDAF as well as other guidelines states requirements for CPM which is one of DoD's six core processes, rather than how to perform CPM. This chapter provides an abstracted methodology for CPM which includes the process, abstrated products and tailored meta-models based on DoDAF Meta Model (hereafter, DM2).

---

[1]The Department of Defense Architecture Framework (DoDAF) is an architecture framework for the United States Department of Defense, that provides structure for a specific stakeholder concern through viewpoints organized by various views. (quoted from http://en.wikipedia.org/wiki/DODAF)

2. Current issues on system of systems problems

The definition of system of DoDAF V2.0 (DoD, Aug. 2010) has been changed from that of DoDAF V1.5. A system is not just computer hardware and software. A system is now defined in the general sense of an assemblage of components (machine or, human)- that perform activities (since they are subtypes of Performer) and interact or become interdependent. The Federal Enterprise Architecture Practice Guidance (Federal Government, 2007) has defined three types of architecture: enterprise architecture, segment architecture, and solution architecture. "Enterprise architecture" is fundamentally concerned with identifying common or shared assets - whether they are strategies, business processes, investments, data, systems, or technologies. By contrast, "segment architecture" defines a simple roadmap for a core mission area, business service, or enterprise service. "Solution architecture" defines agency IT assets such as applications or components used to automate and improve individual agency business functions. The scope of solution architecture is typically limited to a single project and is used to implement all or part of a system or business solution. From the viewpoint of a system hierarchy, the solution architecture addresses system level problems whereas enterprise architecture and segment architecture address SoS/FoS problems respectively. Systems engineering methodologies have evolved to deal with enterprise or SoS level problems.

The purpose of DoDAF V2.0 is to define concepts and models usable in DoD's six core processes:

1. Capabilities Integration and Development (JCIDS)
2. Planning, Programming, Budgeting, and Execution (PPBE)
3. Acquisition System (DAS)
4. Systems Engineering (SE)
5. Operations Planning
6. Capabilities Portfolio Management (CPM)

The DoD's six core processes are good examples of addressing SoS level problems. However, DoDAF V2.0 and other guidelines state requirements rather than how to perform these processes. This chapter provides a methodology for CPM which contains detailed processes, methods, artifacts and tailored meta-model of DM2.

3. Capability Portfolio Management methodology development guide

ISO/IEC 24744 (ISO, 2007) defines that a methodology specifies the process to be executed, usually as a set of related activities, tasks and/or techniques, together with what work products must be manipulated (created, used or changed) at each occasion possibly including models, documents and other inputs and outputs. So a methodology is the specification of the process to follow together with the work products to be used and generated, plus techniques which are the consideration of people and tools involved, during a development effort.

3.1 Methodology development requirements

3.1.1 Process, methods, tools, and environment concept of methodology element

According to Martin (Martin, 1997), it is important to have a proper balance among process, methods, tools, and environment (PMTE) when performing systems engineering tasks. He defines that a process is a logical sequence of tasks performed to achieve a particular objective, a method consists of techniques for performing a task, and a tool is an instrument when applied to a particular method. While, in ISO/IEC 24744, a method is used as a synonym of methodology, this chapter adopts Martin's PMTE paradigm. So this chapter provides the CPM methodology which has its own process, method (technique), and tool (model or artifacts).

ISO/IEC 24744 (ISO, 2007) also states that a methodology element is a simple component of a methodology. Usually, methodology elements include the specification of what tasks, activities, techniques, models, documents, languages and/or notations can or must be used when applying the methodology. Methodology elements are related to each other, comprising a network of abstract concepts. Typical methodology elements are Capture Requirements, Write Code for Methods (a kind of tasks), Requirements Engineering, High-Level Modelling (kinds of activities), Pseudo-code, Dependency Graphs (notations), Class, Attribute (kinds of model building blocks), Class Model, Class Diagram, Requirements Specification (kind of work products), etc. From this concept, the elements for CPM methodology of this chapter are Capture Requirements (top level CPM requirements), High-Level Model of CPM process (kinds of activities), metamodel (Class Diagram), and Attribute.

3.1.2 Metamodel development requirements

A metamodel is the specification of the concepts, relations and rules that are used to define a methodology. This metamodel should be simple and consistent with the analysis methodology. And the metamodel is a schema for semantic data and a language that supports a particular process, method (technique), and tool (model or artifacts).

Probability and set theory have axioms of mutually exclusive and collectively exhaustive (hereafter, MECE) concepts, and decomposition concepts. This means no overlap, no omission of concept and complete decomposition of a concept also. Axiomatic design theory (Suh, 1990) states that the design axiom No.1 is the independence axiom, "Maintain the independence of functions (not affecting other functions)" and the design axiom No. 2 is the information axiom, "Minimize the information content of the design (functionally uncoupled design)." These are the same MECE principle concept of different viewpoints, one is a set viewpoint and the other is a functional design viewpoint. A past study (Lee and Park, 2009) adopted this concept to the metamodel design. The study pointed out that if the metamodel design satisfies the MECE principle, the classes within the metamodel is distinguished from each other clearly, the model composes a complete set of semantic, and relates to each other clearly. The metamodel requirements of this study are summarized in Table 1.

No.	Metamodel requirements
1	Each class of the metamodel should be mutually exclusive and collectively exhaustive concepts as defined in the axiomatic design theory.
2	Metamodel should be consistent, integrated and balanced among processes and methods to achieve the greatest benefits from the disciplined systems engineering practice.
3	The requirement space and the solution space shall be separated strictly as the systems engineering teaches to ensure the solution space is open for multiple candidates.
4	The attributes of the metamodel should result in effective benefits from the viewpoint of SoS architecting and its usage (e.g. CPM).

Table 1. Metamodel requirements

And the study (Lee & Park, 2009) also proposed five rules for developing metamodel and those metamodel development requirements are presented in Table 2.

No.	Metamodel development requirements
1	Create the minimum number of data groups
2	Do not overlap concept across data groups
3	Make the relation names among groups clear and meaningful.
4	Make the relations among the groups to represent systems engineering methodology.
5	Include the operational viewpoint and system viewpoint category while creating groups.

Table 2. Metamodel development requirements

Current proposed DM2 shows many similar type of classes which violates Lee & Park 's metamodel requirement No.1.

As mentioned before, the metamodel must be consistent, integrated and balanced between process and methods to achieve the greatest benefits from the good systems engineering practice. The systems engineering method teaches that the requirement space and the solution space shall be divided strictly. These attributes of the metamodel resulted in effective benefits from the viewpoint of building architecture (e.g. SoS architecting) and the usage (e.g. CPM).

3.2 Capability Portfolio Management methodology requirements

3.2.1 Capability Portfolio Management requirements

DoDD 7045.20 (DoD, Sep. 2008) defines that capability portfolio management (CPM) is the process of integrating, synchronizing, and coordinating Department of Defense capabilities needs with current and planned DOTMLPF[2] investments within a capability portfolio to better inform decision making and optimize defense resources and capability portfolio is a collection of grouped capabilities as defined by JCAs[3] and the associated DOTMLPF programs, initiatives, and activities. The top level requirement of CPM is that CPMs shall provide recommendations regarding capability requirements to capability investments. And other requirements for recommending capability requirement to the Heads of the DoD Components, and to the Deputy's Advisory Working Group (DAWG) are that the CPM should evaluate capability demand against resource constraints, identify and assess risks, and suggest capability trade-offs within their capability portfolio. DoDD 7045.20 (DoD, Sep. 2008) provides CPM requirements and responsibilities but does not provide process and method.

---

[2] DOTMLPF is an acronym used by the United States Department of Defense. DOTMLPF is defined in the The Joint Capabilities Integration Development System (JCIDS) Process. The JCIDS process provides a solution space that considers solutions involving any combination of doctrine, organization, training, materiel, leadership and education, personnel and facilities (DOTMLPF).

[3] Joint Capability Area (JCA) - Collections of like DOD capabilities functionally grouped to support capability analysis, strategy development, investment decision making, capability portfolio management, and capabilities-based force development and operational planning. (http://www.dtic.mil/futurejointwarfare/cap_areas.htm).

3.2.2 Current status of Capability Portfolio Management methodology

DM2 of DoDAF V2.0 provides Conceptual Data Model (hereafter, CDM), Logical Data Model (LDM), and Physical Exchange Specification (PES). LDM provides data groups (classes) and their usage in DoD's six core processes including CPM. DoDAF V2.0 provides metamodel which support method but does not provide process and methods itself for CPM. Table 3 shows DM2 CDM core concepts which represent the relation among DM2 Data Groups and DoD's six core processes. Table 3 also shows twenty five data groups that are used to develop architectures across DoD's six core processes including CPM.

No	DM2 Data Groups	JCIDS Cap. Mgmt.	JCIDS Iteop. & Supp.	DAS	PPBE	CPM	SE / SOSE	Ops. Planning	Class usage level
1	Activity	6	2	3	3	5	3	5	27
2	Agreement	2	3	3	0	0	3	3	14
3	Capability	6	3	4	3	6	3	3	28
4	Condition	4	2	1	2	2	2	4	17
5	Data	3	4	2	1	3	3	2	18
6	DesiredEffect	6	1	3	0	4	2	6	22
7	Guidance	1	3	3	2	2	3	4	18
8	Information	4	4	2	1	3	3	4	21
9	Location	2	2	1	2	2	1	5	15
10	Materiel	3	1	3	2	2	4	4	19
11	Measure	6	4	4	2	4	3	2	25
12	MeasureType	6	4	4	2	4	3	2	25
13	Organization	2	1	1	2	2	1	5	14
14	Performer	4	4	4	4	4	5	4	29
15	PersonType	3	2	2	1	2	2	4	16
16	Project	0	0	4	2	3	2	3	14
17	Resource	3	3	2	1	3	3	4	19
18	Rule	2	4	3	1	2	4	2	18
19	Service	2	3	3	2	3	5	0	18
20	Skill	3	2	2	1	2	2	4	16
21	Standard	2	6	3	1	2	4	2	20
22	System	2	5	5	5	5	6	3	31
23	Vision	2	0	2	1	3	0	1	9
24	ArchitecturalDescription	4	5	3	2	4	5	3	26
25	Constraint	2	4	3	1	2	4	2	18
	Class usage level	80	72	70	44	74	76	81	-

Legend 6: Critical role, 5: Substantial role, 4: Significant role, 3: Moderate role
2: Supporting role, 1: Minor role, 0: No role

Table 3. Relation among DM2 Data Groups and DoD's six core processes

The study (Lee & Park, 2009) points out that DoDAF metamodel is too complex to use and proposed more simplified metamodel to enhance usability. Fig. 1 shows many classes used in DM2. There are still many classes which generate complexity when architecting.

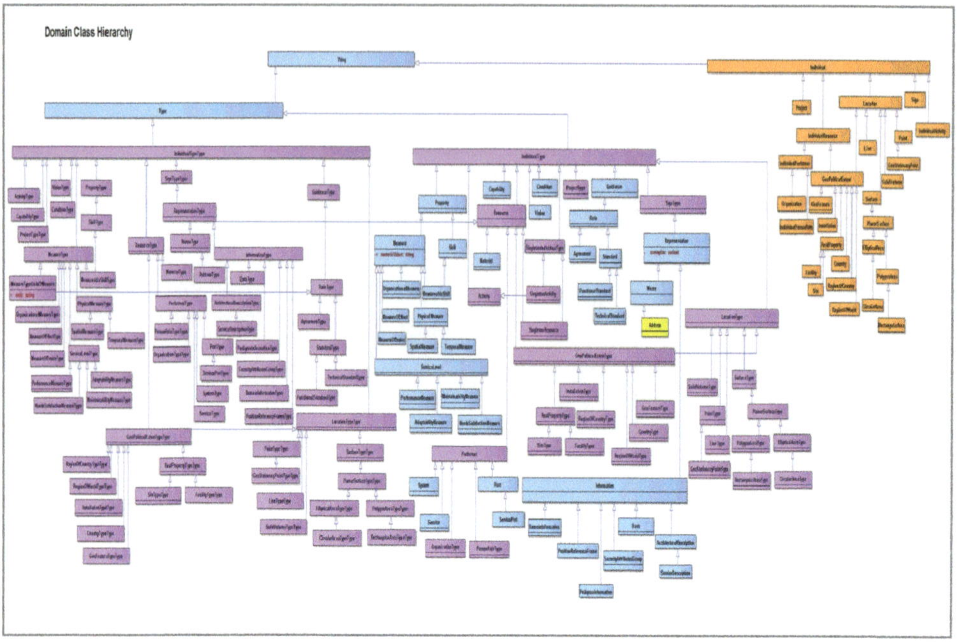

Fig. 1. Notional class hierarchy of DM2

In order to overcome complexity and enhance usability of metamodel, Lee & Park proposed another metamodel based on DoDAF 2.0 JCIDS overlay protocol. Fig. 2 shows Lee & Park's proposal for CDM. The study articulate that the proposed metamodel is the product of an integrating effort that combines the MECE principles, systems engineering principles. The study also demonstrates that it is a simple and effective process to develope and use the artifacts of an architecture.

The CDM of current DM2 of DoDAF V2.0 is similar to the proposed Lee & Park's metamodel. Fig. 3 shows DM2 CDM overlay with the Lee & Park's proposed metamodel. Table 4 shows the relation between classes of DM2 CDM and Lee & Park proposed metamodel. From the contents viewpoint the total of eighteen classes of DM2 CDM are matched with classes of Lee & Park's proposed metamodel. Unmatched classes of DM2 CDM with Lee & Park's are seven classes as follows: Data, Information, Agreement, Location, Skill, MeasureType, and PersonType. To maintain consistency with DM2 CDM, Lee & Park's metamodel complemented with these 7 classes. Three classes of Data, Information, and Location are added, two classes of Agreement and Skill are excluded for the reason of not directly related to the CPM and the other two classes of MeasureType and PersonType go for attribute of Measure and Person. Based on these analysis results the metamodel of CPM methodology could maintain consistency conceptually with DM2 CDM.

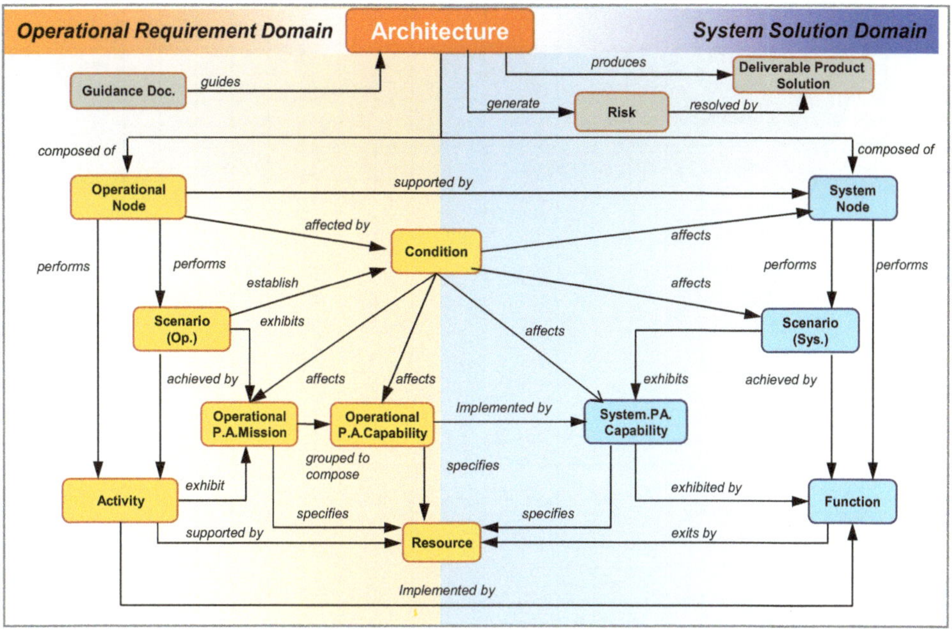

Fig. 2. Lee & Park proposed metamodel for capability based assessment (CBA) methodology

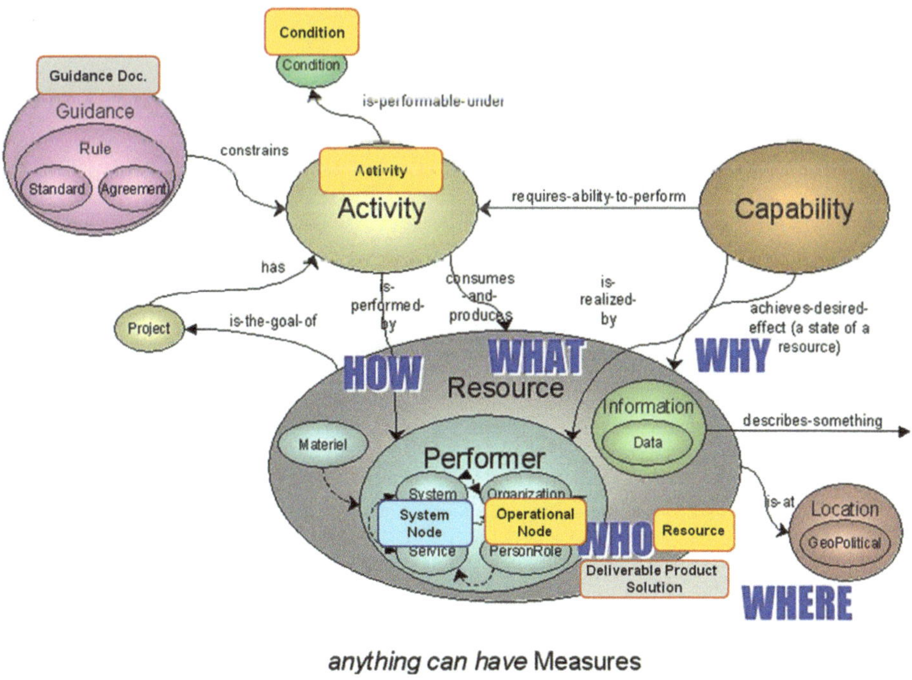

Fig. 3. DM2 CDM overlay with Lee & Park proposed metamodel

DM2 CDM No.	Lee & Park proposed classes / Classes of DM2 CDM core concepts	Operational Node	Scenario	Activity	Op. Perf. Attribute (Oriented)	Op. Perf. Attribute (Capability Oriented)	System Node	Scenario (Sys)	Function	System Perf. Attribute (Capability Oriented)	Architecture	Guidance Document	Risk	Deliverable Product Solution	Condition	Resource	Executables	Ref
13	Organization	○														●		
14	Performer	●																
19	Service		●	○				○	○									
1	Activity			●														
6	DesiredEffect				●													
3	Capability					●				○								
11	Measure				○	○				○								
23	Vision										○							
24	Architectural Description										●							
7	Guidance											●						
21	Standard											●						
10	Materiel													●		●		
22	System													●		●		
17	Resource	○					○							●		●		
16	Project													○				
4	Condition														●			
18	Rule														●			
25	Constraint														●			
2	Agreement																	Class
5	Data																	Class
8	Information																	Class
9	Location																	Class
12	MeasureType																	Attribute
15	PersonType																	Attribute
20	Skill																	Class

Table 4. Relation between classes of DM2 CDM and Lee & Park proposed metamodel

And process viewpoint of methodology status, CBA guides (DoD, Dec. 2006) have relatively detailed information about CBA process and methods. The CBA process and related information could be used to perform CPM but that is not sufficient for CPM method. The following part provides CPM process, product and method which manipulate information of the product and supporting metamodel.

4. Proposal of Capability Portfolio Management methodology

As mentioned before, a methodology specifies the process to be executed, usually as a set of related activities, tasks and/or techniques, together with work products possibly including models, documents. CPM methodology has its own process, method (technique), and product (model or artifacts) as the tool category of Martin's PMTE. According to these requirements, the CPM methodology of this chapter shows CPM process, product and model related technique. The CPM process consists of a set of activities/tasks. Each step of activity has corresponding output product and model related technique which is used to build model and/or generate the output products.

In order to facilitate further discussions, key terms quoted from DoDD 7045.20 capability portfolio management (DoD, Sep. 2008) are defined as follows. Capability portfolio is a collection of grouped capabilities as defined by JCAs and the associated DOTMLPF programs, initiatives, and activities. And CPM is the process of integrating, synchronizing, and coordinating capability requirements with current and planned DOTMLPF investments within a capability portfolio to better inform decision making and optimize defense resources. From this definition, CPM can make a balanced capability requirements to maximize mission effects within limited resources and the capability requirements are originated from a group of capabilities defined by JCAs.

4.1 Capability Portfolio Management process

CPM requirement is to provide recommendations regarding capability requirements to capability investments. So CPM process has to generate balanced capability requirements. The capability requirements should be generated with DOTMLPF investments within a capability portfolio (a collection of grouped capabilities as defined by JCAs).

To achieve these CPM requirements a proposed process is composed of following 5 activities: (1) Define top level missions and develop scenarios (2) Build trace relation among elements of JCA, universal joint task list (hereafter, UJTL) and activity and identify mission essential task list (hereafter, METL) of DoD (3) Develop capabilities and the related conditions and resources (4) Analyze mission effectiveness and derive (transform) capability requirements (5) Derive integrated & balanced capability requirements. And more detailed tasks are listed in Table 5.

4.2 Capability Portfolio Management method and product

In order to provide CPM method and product which could be a model or artifact. This part provides descriptions, products and model related techniques for each task of CPM process.

Activities of CPM process		Tasks of CPM Process	
A.1	Define top level missions and develop scenarios	T.1	Define top level missions
		T.2	Define states & modes for each missions
		T.3	Develop mission threads for each states & modes
		T.4	Design operational scenarios for each missions
A.2	Build trace relation among JCA, UJTL and activity and identify METL	T.5	Trace each activity to UJTL
		T.6	Check alignment JCA, UJTL and allocated activity
		T.7	Identify METLs for each mission scenario
A.3	Develop capabilities and related conditions and resources	T.8	Develop capability instance which aligned to activity (attributed in METLs)
		T.9	Develop condition instances for each activity (attributed in METLs)
		T.10	Develop resource instances for each activity (attributed in METLs)
A.4	Analyze mission effectiveness and derive(transform) capability requirements	T.11	Analyze operational effectiveness (MOEs) for each operational missions (e.g. Joint Operating Concepts, hereafter, JOC)
		T.12	Analyze operational effectiveness (MOEs) for functional missions (e.g. Joint Functional Concepts, hereafter, JFC)
A.5	Derive integrated & balanced capability requirements	T.13	Allocate operational element to supporting systems element
		T.14	Synthesize operational performances to system performances
		T.15	Optimize resources to maximize MOEs for a capability
		T.16	Define integrated capability requirements

Table 5. Activities and tasks of proposed CPM process

4.2.1 Define top level missions

- Description: Defining top level mission is a process to define top level missions of an enterprise to provide the point of reference or directions which CPM aims to attain.
- Product: Top level mission statement of an enterprise
- Model related technique: Mission is a kind of activity and the mission activity is the top level activities of an activity hierarchy. And, the level attribute of the mission activity should be set as 'Mission level'.

4.2.2 Define states & modes for each mission

- Description: Defining states & modes for each mission is a process to define states and modes in which top level missions of an enterprise are implemented, and this process provides categories (e.g. war-time & piece-time) of thinking for developing threads which encompass tasks to be analyzed.
- Product: States & modes definition
- Model related technique: States & modes are kind of activity and the abstraction level of this activity is below the mission activities of an activity hierarchy. Thus the level attribute of the states & modes activity should be set as 'States & modes'.

4.2.3 Develop mission threads for each states & modes

- Description: From CJCSI 6212.01E (DoD, Dec. 2008) definition, a mission thread could be defined as an operational and technical description of the end to end set of activities and systems that accomplish the execution of a mission. Developing mission threads means producing a list of threads which needed for each states & modes to accomplish a mission. Each thread composed of a series of required tasks.
- Product: Mission threads
- Model related technique: Mission threads are kind of activity and this activity is the below the states & modes activities of an activity hierarchy. And so, the level attribute of the thread activity should be set as 'Thread'.

4.2.4 Design operational scenarios for each mission

- Description: Designing operational scenarios for each mission is a process to define a series of activities in each thread. Through this process, the end to end sets of activities of operational nodes are designed, and states of mission accomplishments are designed by integrating those threads
- Product: Operational scenario template
- Model related technique: The activities within an operational scenario are kind of activity and these activities are the leaf-node activities of an activity hierarchy. Thus the level attribute of the leaf-node activity should be set as 'Leaf-node'. The leaf-node activity could directly allocate to supporting entity e.g. operational node and system node.

4.2.5 Trace each activity to Universal Joint Task List

- Description: Tracing each activity to UJTL is a process allocating each activity to the related tasks listed in UJTL which contains information that identifies conditions and standards to analyze Leaf-node-level activities
- Product: Activity to UJTL traceability table
- Model related technique: UJTL class is required and the elements of UJTL class (tasks) are allocated by leaf-node activities of scenario.

4.2.6 Check alignment Joint Capability Area, Universal Joint Task List and allocated activity

- Description: The Joint Capability Area Management System (JCAMS) of DoD provides JCA linkages to Universal Joint Tasks. The allocated activities to UJTL should be checked by the alignment with JCA from the viewpoint of semantics. From the viewpoint of semantics, tracing relation between activity-UJTL-JCA should be meaningful.

- Product: Traceability table of Activity-UJTL-JCA
- Model related technique: JCA class is required and the elements (contents) of JCA could be traceable to UJTL. Then the traceability from leaf-node activity via UJTL to JCA is established.

4.2.7 Identify Mission Essential Task Lists for each mission scenario

- Description: METLs are decided through a process to identify key tasks, which directly contribute to achieve mission effectiveness, among leaf-node level activities of a mission scenario. The designated activities as METL have a role to develop capability requirements. The activities designated as METL are base activities for following analysis of CPM methodology.
- Product: METL List
- Model related technique: The activity class needs the importance attribute. And so, the activity importance attribute of the METL activity should be set as 'METL'.

4.2.8 Develop capability instance which aligned to activity

- Description: The activities which are identified as METLs should be carried out CBA separately and develop appropriate capabilities in the light of JCAs. The developed capability is an instance of capability class which are traced to activity instances. The developed capabilities will be traced to the functions of systems or other requirements of DOTMLPF.
- Product: Traceability table of Activity - Capability
- Model related technique: Capability class is required aside from JCA class and the capability class should have relation with JCA and activity class.

4.2.9 Develop condition instances for each activity

- Description: For the purpose of carrying out CBA, proper conditions for missions are developed and allocated to activities which are identified as METLs. The developed conditions are instances of the conditions of UJTL.
- Product: Traceability table of Activity – Condition
- Model related technique: Condition class is required aside from UJTL class and the UJTL class has 'provide relation' with Condition class.

4.2.10 Develop resource instances for each activity

- Description: Required resources (DOTMLPF) are defined to fulfill relevant activities. Those resources realize capabilities to support related activities.
- Product: Traceability table of Activity – Resource – Capability
- Model related technique: Resource class is separately required with other performer type classes e.g. organization and system. The resource class has relation with activity and capability class. Resource class has resource type of DOTMLPF. Especially the Resource class typed with organization is equivalent to organization class and resource class typed with materiel is equivalent to system class.

4.2.11 Analyze operational effectiveness for each operational mission

- Description: Performance levels for each activity are analyzed to estimate the performance level of activities to produce best mission effectiveness within constrained resources under the given condition for activities to accomplish operational missions. It is required to determine performance indicators or standards of each activity to achieve mission effectiveness. Especially this analysis works are performed in aspect of operational missions in this step. And so, the performance levels of activities for each operational mission (e.g. JOC) are measurements of operational effectiveness (MOEs).
- Product: Operational mission effectiveness and performance of activity table
- Model related technique: Within activity class, mission level attributed activity element could have sub attribute of operational mission type and functional mission type. And to display and analyse the performance of scenario, a performance measure class is required. And according to the type of activity of mission scenario, the attribute of a measure could be operational effectiveness (reflect JOC), functional performance (reflect JFC) or system performance.

4.2.12 Analyze operational effectiveness for functional missions

- Description: Performance levels for each activity are analyzed to estimate optimized performance level of activities which produce best operational mission effectiveness within 'constrained resources' which support activities to accomplish functional missions. From the viewpoint of opposite direction, the performance level of activities performing a functional mission should be optimized to enhance the total performance of the activities performing various operational missions. This opposite directional task will be explained at 4.2.15 again. It is required to determine performance indicators or standards of each activity to achieve mission effectiveness. Especially in this step, the analysis works are performed from the viewpoint of functional missions relative to the operational missions. And so, the performance levels of activities for each functional mission (e.g. JFC) are measurements of operational effectiveness (MOEs).
- Product: Functional mission effectiveness and performance of activity table
- Model related technique: Within Activity class, mission level attributed Activity element could have sub attribute of operational mission type and functional mission type. And to display and analyse the performance of scenario, a Performance Measure class is required. And according to the type of activity of mission scenario, the attribute of a measure could be operational effectiveness (reflect JOC), functional performance (reflect JFC) or system performance.

4.2.13 Allocate operational element to supporting systems element

- Description: This phase changes operational viewpoint to system viewpoint. And this phase allocates defined organization, operational nodes, activities and input/output information to systems, system nodes, system functions and input/output data. Lessons learned from systems engineering imply that system elements are not considered before this step, and instead, requirements are defined in operational viewpoint, then operational requirement are converted into system viewpoint in order to support operational requirements.

- Product: Organization vs. system relation table, Operational node vs. system node relation table, operational activity vs. system function relation table, Operational information vs. system data relation table.
- Model related technique: To reflect the principle of systems engineering which divide requirement space and the solution space strictly, the following relations are built. Organization class is supported by system class. Operational node class is supported by system node class. Activity class is supported by function class. Operational information class is supported by system data class.

4.2.14 Synthesize operational performances to system performances

- Description: This step aims that operational performances, which are derived from operational activities, are changed to system performance, which are derived from system functions. A system function is employed to support several operational activities. Those operational performances are synthesized into an optimized system performance from the view point of cost-effectiveness.
- Product: Operational performances vs. system performances matrix
- Model related technique: Measure class of operational performances type is traced to measure class of system performances type

4.2.15 Optimize resources to maximize operational effectiveness for a capability

- Description: This is the most peculiar phase of CPM process. Perform cost-effectiveness analysis repeatedly to achieve maximum effectiveness under the condition of limited resources. And define the capability requirements, which are the requirements for all resources to encompass DOTMLPF, and those resources are traced to one capability under certain items of JCA. The resulted performances of resources are synthetically maximize return on invest (ROI) for the relevant capability.
- Product: Capability vs. Resources matrix
- Model related technique: Capability class has been realized by relation with DOTMLPF type of resources.

4.2.16 Define integrated capability requirements

- Description: According to the definition of capability, the capability elements e.g. desired effects of various missions, a set of tasks and combination of means & ways are defined for a capability using performance measures of activity, function and resources. The elements contributing critically to the resulted capability should be marked.
- Product: Capability recommendation document
- Model related technique: 'Capability decisive element' attribute required for classes of resource, activity and function.

4.3 Metamodel for Capability Portfolio Management

From the proposed CPM process, and based on DM2 CDM and Lee & Park's metamodel, the additionally required classes (Entity type), attributes of classes and relations for each task are identified. The additionally identified classes, relations, and attribute are used to complement metamodel for CPM methodology. Table 6 shows the additionally required classes, relations and attributes.

No.	Tasks of CPM Process	Required class	Required attribute & relation
T.1	Define Top Level Missions	Activity	Activity_type-attribute: Mission
T.2	Define States & Modes for each missions	Activity	Activity_type-attribute: State&Mode
T.3	Develop Mission Threads for each States & Modes	Activity	Activity_type-attribute: Thread
T.4	Design Operation Scenarios for each missions	Activity	Activity_type-attribute: Scenario
T.5	Trace each Activity to UJTL	UJTL	UJTL traces_to Activity relation
T.6	Check alignment JCA, UJTL and allocated Activity	JCA	JCA categorize UJTL relation
T.7	Identify METLs for each mission scenario	UJTL	Activity_type-attribute: METL
T.8	Develop Capability instance which aligned to Activity (attributed in METLs)	Capability	Capability requires_ability_to_perform Activity relation
T.9	Develop Condition instances for each Activity (attributed in METLs)	Condition	Activity performable_under Condition relation
T.10	Develop Resource instances for each Activity (attributed in METLs)	Resource	Resource _type-attribute: DOTMLPF, Capability realized_by Resource relation
T.11	Analyze Operational Effectiveness (MOEs) for each operational missions (e.g. JOC)	Measure	Measure_type-attribute: MOE for JOC
T.12	Analyze Operational Effectiveness (MOEs) for functional missions (e.g. JFC)	Measure	Measure_type-attribute: MOE for JFC
T.13	Allocate operational element to supporting systems element	Data	Organization supported_by System relation, OperationalNode supported_by SystemNode relation, Activity supported_by Function relation, Information supported_by Data relation
T.14	Synthesize operational performances to system performances	Measure	Measure_type-attribute: Operational performance Measure_type-attribute: System performance Measure traces_to Measure relation
T.15	Optimize resources to maximize MOEs for a capability	Capability	Capability realized_by Resource relation
T.16	Define integrated capability requirements	Document	Document documents AllClass

Table 6. Proposed CPM Process and required classes and attributes for CPM Process

Like the study (Lee & Park, 2009) proposed metamodel, CPM metamodel should be developed in accordance with the metamodel requirement and metamodel development requirements of Table 1 and 2. And also CPM metamodel should be aligned with DM2 CDM for interoperability with DoDAF V2.0. Table 7 shows the proposed CDM classes for CPM which is aligned with classes of DM2 CDM and additionally added classes originated from Lee & Park' metamodel and the CPM task analysis. The additional JCA and UJTL classes comes from CPM task analysis of Table 6 and System Node and Function classes reflect the systems engineering concept of strict separation of requirement space and solution space.

DM 2 No.	Classes of DM2 CDM core concepts	CPM usage level	relation with proposed classes	Proposed CDM classes for CPM
3	Capability	6	correspond to	Capability
1	Activity	5	correspond to	Activity
22	System	5	correspond to	System
6	DesiredEffect	4	correspond to	Measure (Effect attributed)
11	Measure	4	correspond to	Measure
12	MeasureType	4	correspond to	Measure (MeasureType attributed)
14	Performer	4	correspond to	Operational Node
24	Architectural Description	4	correspond to	Architecture
5	Data	3	correspond to	Data
8	Information	3	correspond to	Information
16	Project	3	correspond to	Project
17	Resource	3	correspond to	Resource
19	Service	3	correspond to	Activity (Service attributed)
23	Vision	3	correspond to	Measure (Vision attributed)
4	Condition	2	correspond to	Condition
7	Guidance	2	correspond to	Guidance
9	Location	2	correspond to	Location
10	Materiel	2	correspond to	Resource (Materiel attributed)
13	Organization	2	correspond to	Resource (Organization attributed)
15	PersonType	2	correspond to	Resource (Person attributed)
18	Rule	2	correspond to	Guidance
25	Constraint	2	correspond to	Condition
20	Skill	2	correspond to	Resource (Skill attributed)
21	Standard	2	correspond to	Guidance
2	Agreement	0	correspond to	Guidance
-	N/A	-	-	System Node
-	N/A	-	-	Function
-	N/A	-	-	JCA
-	N/A	-	-	UJTL

Table 7. Relation between DM2 CDM core concepts and Lee's CDM classes for CPM

And according to the metamodel development requirements, classes are related and named meaningfully and reflect operational requirement space and system solution space. Fig. 4 shows the resulted CDM for CPM methodology.

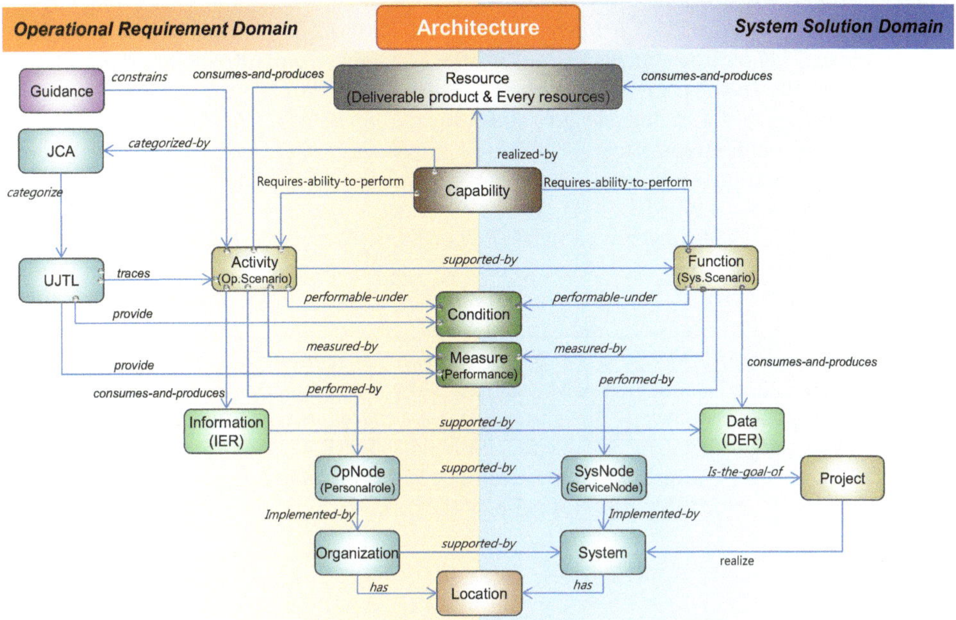

Fig. 4. Proposed CDM for CPM methodology

5. Conclusion

The purpose of this paper is to provide an abstracted metamodel for use in CPM effectively based on DoDAF V2.0. The proposed CPM methodology provides a process, tasks of the process, products, and model related technique which supports the generation of products in accordance with the methodology definition of ISO/IEC 24744. To promote the usability, the proposed methodology suggest a detailed CPM process. Additionally, in order to be an effective and efficient methodology, the CPM metamodel is developed in accordance with the MECE principles, systems engineering principles which was proposed earlier by Lee & Park's metamodel requirements. And to obtain the interoperability with DoDAF V2.0, the proposed CPM methodology is developed in accordance with DM2 CDM.

However, the current proposed abstracted metamodel remains on a theoretical and logical level and requires validation experimentally or in field applications. In the near future, the proposed metamodel must be validated for application use. However, the proposed CPM methodology is expected to be helpful in practice in the field.

6. References

DoD (Dec., 2006), *Capabilities-Based Assessment (CBA) User's Guide*, Ver.2, Sep. 25, 2011, Available from: www.dtic.mil/futurejointwarfare/strategic/cba_guidev2.pdf

DoD (Sep., 2008), *DoDD 7045.20 capability portfolio management*, Sep. 25, 2011, Available from: http://www.dtic.mil

DoD (Dec., 2008), *CJCSI 6212.01E Interoperability and supportability of information technology and national security systems*, , Sep. 25, 2011, Available from: https://acc.dau.mil

DoD (Aug., 2010), *The DoDAF Architecture Framework*, Ver. 2.02, Sep. 25, 2011, Available from: http://cio-nii.defense.gov/sites/dodaf20/products/DoDAF_v2-02_web.pdf

Federal Government (November 2007), *FEA Practice Guidance*, Available from: www.whitehouse.gov/sites/default/files/omb/assets/fea_docs/FEA_Practice_Guidance_Nov_2007.pdf

ISO (2007), *ISO/IEC 24744 Software Engineering - Metamodel for Development Methodologies*, ISO, Retrieved from: www.iso.org/iso/iso_catalogue.htm

Lee, J. & Park, Y. (2009), A Study on the Abstracted Metamodel of DoDAF 2.0 for CBA Methodology Execution, *International Conference on Software Engineering, 2009 10th ACIS Artificial Intelligences, Networking and Parallel/Distributed Computing*, ISBN: 978-0-7695-3642-2, Daegu, Korea, May 27- 29, 2009

Martin, J. (1997), Process concept, In : *Systems engineering guidebook: a process for developing systems and products*, pp. 51-56, CRC, ISBN 0-8493-7837-0, Boca Raton

Suh, N. (1990), Design axioms and corollaries, In: *The Principles of Design*, pp.47-48, Oxford University Press, ISBN 0-19-504345-6, New York

Part 2

New Systems Engineering Theories

11

Augmented Human Engineering: A Theoretical and Experimental Approach to Human Systems Integration

Didier Fass
ICN Business School and MOSEL – LORIA, Lorraine University, Nancy, France

1. Introduction

This chapter focuses on one of the main issues for augmented human engineering: integrating the *biological user's needs* in its methodology for designing human-artefact systems integration requirements and specifications. To take into account biological, anatomical and physiological requirements we need a validated theoretical framework. We explain how to ground augmented human engineering on the Chauvet mathematical theory of integrative physiology as a fundamental framework for human system integration and augmented human design. We propose to validate and assess augmented human domain engineering models and prototypes by experimental neurophysiology.

We present a synthesis of our fundamental and applied research on augmented human engineering, human system integration and human *in-the-loop* system design and engineering for enhancing human performance - especially for technical gestures, in safety critical systems operations such as surgery, astronauts' extra-vehicular activities and aeronautics. For fifteen years, our goal was to research and to understand fundamental theoretical and experimental scientific principles grounding human system integration, and to develop and validate rules and methodologies for augmented human engineering and reliability.

2. Concepts

2.1 Human being

A human being, by its biological nature - bearing in mind its socio-cultural dimensions, cannot be reduced to properties of mathematical or physical automaton. Thus, connecting up humans and artefacts is not only a question of technical interaction and interface; it is also a question of integration.

2.2 Human systems integration

As a technical and managerial concept (Haskins 2010), human systems integration (HIS) is an umbrella term for several areas of "human factors" research and systems engineering that include human performance, technology design, and human-interactive systems interaction

(Nasa 2011). Defining a system more broadly than hardware and software refer to human centred design (Ehrhart & Sage 2003). That issue requires thinking human as an element of the system and translating it qualitatively throughout design, development and testing process (Booher, 2003).

These are concerned with the integration of human capabilities and performances, from individual to social level into the design of complex human-machine systems supporting safe, efficient operations; there is also the question of reliability.

Human systems integration involves augmented human design with the objectives of increasing human capabilities and improving human performance[1] (Engelbart 1962) using behavioural technologies at the level of human-machine system and human machine symbiosis (Licklider 1960). By using wearable interactive systems, made up of virtual reality and augmented reality technologies or wearable robotics, many applications offer technical gesture assistance e.g. in aeronautics, human space activities or surgery.

2.3 Technical gesture assistance

Gesture is highly integrated neurocognitive behaviour, based on the dynamical organization of multiple physiological functions (Kelso, 2008)(de Sperati, 1997). Assisting gestures and enhancing human skill and performances requires coupling sensorimotor functions and organs with technical systems through artificially generated multimodal interactions. Thus, augmented human design has to integrate human factors - anatomy, neurophysiology, behaviour - and assistive cognitive and interactive technologies in a safe and coherent way for extending and enhancing the ecological domain of life and behaviour.

The goal of this type of human *in-the-loop* system design is to create entities that can achieve goals and actions (predetermined) beyond natural human behavioural, physical and intellectual abilities and skills – force, perception, action, awareness, decision…

2.4 Integrative design

Augmenting cognition and sensorimotor loops with automation and interactive artefacts enhances human capabilities and performance. It is extending both the anatomy of the body and the physiology of human behaviour. Designing augmented human beings by using virtual environment technologies requires integrating both artificial and structural elements and their structural interactions with the anatomy, and artificial multimodal functional interactions with the physiological functions (Fass, 2006). That needs a fitting organizational design (Nissen & Burton 2010).

Therefore, the scientific and pragmatic questions are: how to best couple and integrate in a coherent way, a biological system with physical and artifactual systems? How to integrate in a coherent way human and interactive artefact –more or less immersive and invasive, in a behaviourally coherent way by design? How augmented human engineering can anticipate and validate a technical and organizational design and its dynamics? How modelling and assessing such a design efficiency? How grounding HIS and augmenting human design on a validated theory? How assessing experimentally and measuring both performance and efficiency?

---

[1] Sensorimotor and cognitive

3. Augmented human domain engineering

Human-artefact systems are a special kind of *systems of systems*. They are made up of two main categories of systems. These two kinds of systems differ in their nature: their fundamental organization, complexity and behaviour. The first category, the traditional one, includes *technical* or *artifactual* systems that could be engineered. The second category includes *biological* systems: the human that could not be engineered. Thus, integrating human and complex technical systems in design is to couple and integrate in a behaviourally coherent way, a biological system (the human) with a technical and artifactual system. Augmented human engineering needs to model the human body and its behaviour to test and validate augmented human reliability and human systems integration (HSI).

3.1 Domain engineering

According to system engineering, taking into account user needs in the world of activities and tasks, designing system requirements is to find the system design, its three dimensional organizational dimensions of requirements - structural, geometrical and dynamical - and its three view plans of system design specifications -structure or architecture, behaviour - performance and efficiency, and evolution -adaptation, resilience capability…(Fig.1).

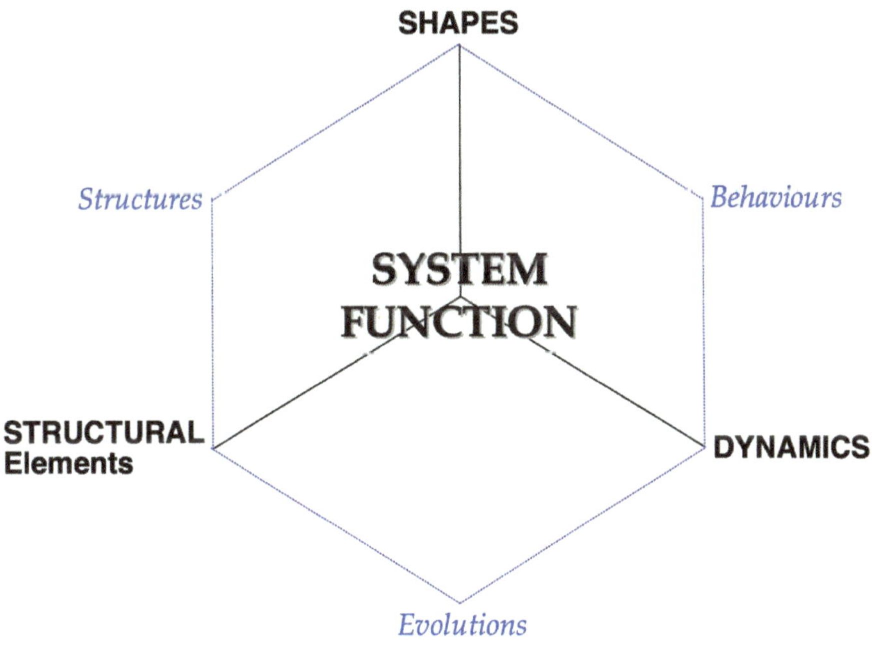

Fig. 1. Our overall system design general conceptual framework: System function results from the integrative organization of different structural elements, shapes and dynamics according there space and time scales relativity and specific qualitative and quantitative measurement units.

Thus, system engineering requires both expert skills and validated formal modelling methodologies. To some extent, the main difficulty is to build a system model from a collection of informal and sometimes imprecise, redundant and unstructured descriptions to the domain of expertise. A formal model could be relevant to highlight a hidden structure according to an intended function and its dynamics, or to apply operations or transformation on the system itself.

From domain engineering to requirements, our approach is situated inside Dines Bjoemer's framework (Bjoemer's 2006a, 2006b and 2009) based on the triptych: D, S -> R, where D is the domain of the problem and where requirements R are satisfied by the relation ->, which intends to mean *entailment*; so, S is a kind of model of our system built or expressed from D. If that triptych is able to express, in a synthetic manner, a situation related to the problem domain, a system model and the requirements, it remains at a global level and can thus be applied in different problem spaces and instances.

The domain provides a way to express properties and facts of the environment of the system under construction. The system model S is intended to summarize actions and properties of the system and it is a link between the requirements and the final resulting system. The relation -> is conceptualized as a deduction-based relation which can be defined in a formal logical system, and which helps to derive requirements from domain and model. This relation is sometimes called entailment and is used to ground the global framework. When one considers an application, one should define the application domain from the analysis and this may integrate elements of the world. The triptych helps for defining a global framework and offers the possibility to use tools that are useful for assessing the consistent relation between D, S and R; because we aim to use proof techniques for ensuring the soundness of the relation.

3.2 Human system integration

The major benefits of using augmented human modelling in design include reducing the need for physical development; reducing design costs by enabling the design team to more rapidly prototype and test a design; avoiding costly design 'fixes' later in the program by considering *human factors* requirements early in the design process; and improving customer communications at every step of product development by using compelling models and simulations. Thus, designing an artefact consists of organizing a coherent relation between structural elements and functions in a culture and context of usage. Modelling human beings consists of taking into account anatomical and physiological elements in the same model. It is to design functions by organizing a hierarchy of structural elements and their functions. Such models should be used to create models of individuals rather than using aggregated summaries of isolated functional or anthropometric variables that are more difficult for designers to use. Therefore augmented human modelling in design requires an integrative approach according to the three necessities we defined for human systems integration (Fass 2007).

3.3 Human system integration domain

Since technical systems are mathematically grounded and based on physical principles, HITLS needs to be considered in mathematical terms. There are several necessities to make HIS and augmented human reliable (Fass & e: Lieber 2009).

- Necessity 1 - Designing a HITLS is to couple two systems from different domains organized and grounded on different principles theory and framework: biological, physical, numerical.
- Necessity 2 - HITLS design is a global and integrative model based method ground on Chauvet's Mathematical Theory of Integrative Physiology and domain system engineering.
- Necessity 3 - Modelling augmented human and HSI is to organize the required hierarchically structural elements, shapes and their interactional dynamics according an architectural principles, behavioural needs of performance and efficiency and evolutionary needs.

Consequently, designing augmented human following human system integration is to organize hierarchically and dynamically human and artefact coupling. This requires a new domain engineering approach for requirements and specification based on biological user's needs and functions.

3.4 Augmented human engineering

Dealing with augmented human engineering is being able to situate and limit its domain for specifying the whole system - biological and artifactual integrated system- in accordance with the high-level and global requirements:

- *D:* The ecology of the augmented human: scientific validated principles of augmented human needs and functions;
- *R:* Augmented human teleonomy, augmented human economy and ethics;
- *S:* Biological, technical and organizational specifications of the human-artefact system - performance, efficiency, reliability, security, safety, stability.

4. Augmented human's needs

Who would even think about separating a living goldfish from its water and its fishbowl?

4.1 Epistemological needs

Converging technologies for improving human performances (Rocco & Brainbridge 2002), *augmented human,* need a *new epistemological and theoretical* approach to the nature of knowledge and cognition considered as an integrated biological, anatomical, and physiological process, based on a hierarchical structural and functional organization (Fass 2007). Current models for human-machine interaction or human-machine integration are based on symbolic or computational cognitive sciences and related disciplines. Even though they use experimental and clinical data, they are yet based on logical, linguistic and computational interpretative conceptual frameworks of human nature, where postulate or axiomatic replace predictive theory. It is essential for the robust modelling and the design of future rules of engineering for HIS, to enhance human capabilities and performance. *Augmented human* design needs an integrative theory that takes into account the specificity of the biological organization of living systems, according to the principles of physics, and a coherent way to organize and integrate structural and functional artificial elements (structural elements and functional interactions). Consequently, virtual environments design for *augmented human* involves a shift from a metaphorical, and scenario based design, grounded on *metaphysical* models and rules of interaction and cognition, to a predictive science and engineering of interaction and integration. We propose to ground HSI and *augmented human* design on an integrative theory of the human being and its principles.

4.2 Chauvet's mathematical theory of mathematical physiolgy (MTIP) needs

The mathematical theory of integrative physiology, developed by Gilbert Chauvet (Chauvet 1993a; Chauvet 1993b; Chauvet 1993c) examines the hierarchical organization of structures (i.e., anatomy) and functions (i.e., physiology) of a living system as well as its behaviour. MTIP introduces the principles of a functional hierarchy based on structural organization within spaces limits, functional organization within time limits and structural units that are the anatomical elements in the physical space. This abstract description of a biological system is represented in (fig. 2). MTIP copes with the problem of structural discontinuity by introducing functional interaction, for physiological function coupling, and structural interaction Ψ from structure-source s into structure-sink S, as a coupling between the physiological functions supported by these structures.

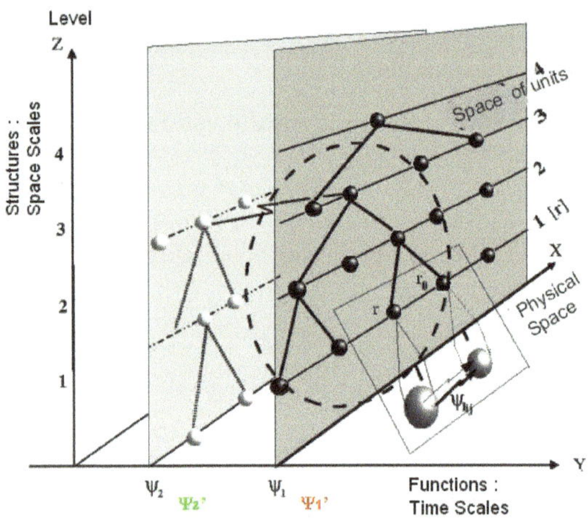

Fig. 2. Ω - 3D representation of a biological system based on the Chauvet's MTIP.

Chauvet had chosen a possible representation related to hierarchical structural constraints, and which involves specific biological concepts. MTIP consists in a representation: set of non-local interactions, an organizing principle: stabilizing auto-association principle (PAAS), and a hypothesis: any biological system may be described as a set of functional interactions that gives rise to two faces of the biological system, the potential of organization (O-FBS) and the dynamics in the structural organization, making an n-level field theory (D-FBS). Both are based on geometrical/topological parameters, and coupled via geometry/topology that may vary with time and space (state variables of the system) during development and adult phases. The structures are defined by the space scale Z, hence the structural hierarchy, the functions are defined by the time scale Y, hence the functional hierarchy.

MTIP shows three relevant concepts for grounding human system integration:

- Functional interaction: The first important hypothesis of the MTIP is that a biological system may be mathematically represented as a set of functional interactions of the type $s \overset{\psi}{\to} S$. Unlike interactions in physics, who are local and symmetric at each level of organization, biological or functional interactions are non-symmetrical, leading to directed graph, non local, leading to non local fields, and increase the functional stability of a living system by coupling two hierarchical structural elements. However, the main issue now is to determine whether there exists a cause to the existence of functional interactions, i.e. to the set of triplets' $s \overset{\psi}{\to} S$? What is the origin of the existence (the identification) of s, S and ψ that together make a component $s \overset{\psi}{\to} S$ of the system?
- PAAS: is a mathematical principle that makes of a framework, the MTIP, a veritable theory. The PAAS may be stated as follows: For any triple (s ψ S), denoted as $s \overset{\psi}{\to} S$, where s is the system-source, S the system-sink, and ψ the functional interaction, the area of stability of the system $s \overset{\psi}{\to} S$ is larger than the areas of stability of s and S considered separately. In other words, increasing in complexity the system $s \overset{\psi}{\to} S$, corresponds to increase in stability.
- Potential of functional organization: describes the ability of the system to combine functional interaction in a coherent way, in such a dynamic state of a maximum of stability and reorganization.

Therefore augmented human engineering needs designing artificial functional interactions – short sensorimotor artificial functions, which generate a maximum of stability for human-artefact systems in operational conditions. Thereby MTIP provide for us an abstract framework for designing human-artefact system and designing organizations for dynamic fit (Nissen & Burton 2011). These are the reasons why MTIP is a relevant candidate theory for grounding augmented human design.

5. Rational for a model of *augmented human*

As claims by Fass (Fass2006), since artifactual systems are mathematically founded and based on physical principles, HSI needs to be thought of in mathematical terms. In addition, there are several main requirements categories to make HIS and augmented human design safe and efficient. They address the technology - virtual environment-, sensorimotor integration and coherency.

Requirement 1: Virtual environment is an artifactual knowledge based environment

As an environment, which is partially or totally based on computer-generated sensory inputs, a virtual environment is an artificial multimodal knowledge-based environment. Virtual reality and augmented reality, which are the most well known technologies of virtual environments, are obviously the tools for the augmented human design and the development of human in-the-loop systems. Knowledge is gathered from interactions and dynamics of the individual-environment complex. It is an evolutionary, adaptive and integrative physiological process, which is fundamentally linked to the physiological functions with respect to emotions, memory, perception and action. Thus, designing an artifactual or a virtual environment, a sensorimotor knowledge based environment, consists of making biological individual and artifactual physical system consistent. This requires a neurophysiological approach, both for knowledge modelling and human in-the-loop design.

Requirement 2: Sensorimotor integration and motor control ground behaviour and skills

Humans use multimodal sensorimotor stimuli and synergies for interacting with their environment, either natural or artificial (vision, vestibular stimulus, proprioception, hearing, touch, taste…) (Sporn & Edelman 1998). When an individual is in a situation of immersive interaction, wearing head-mounted display and looking at a three-dimensional computer-generated environment, his or her sensorial system is submitted to an unusual pattern of stimuli. This dynamical pattern may largely influence the balance, the posture control (Malnoy & al. 1998), the spatial cognition and the spatial motor control of the individual. Moreover, the coherence between artificial stimulation and natural perceptual input is essential for the perception of the space and the action within. Only when artificial interaction affords physiological processes is coherence achieved.

Requirement 3: Coherence and HIS insure the human-artefact system performance, efficiency and domain of stability

If this coherence is absent, perceptual and motor disturbances appear, as well as illusions, vection or vagal reflex. These illusions are solutions built by the brain in response to the inconsistency between outer sensorial stimuli and physiological processes. Therefore, the cognitive and sensorimotor abilities of the person may be disturbed if the design of the artificial environment does not take into account the constraints imposed by human sensory and motor integrative physiology. The complexity of physiological phenomena arises from the fact that, unlike ordinary physiological systems, the functioning of a biological system depends on the coordinated action of each of the constitutive elements (Chauvet 2002). This is why the designing of a artificial environment as an augmented biotic system, calls for an integrative approach.

Integrative design strictly assumes that each function is a part of a continuum of integrated hierarchical levels of structural organization and functional organization as described above within MTIP. Thus, the geometrical organization of the virtual environment structure, the physical structure of interfaces and the generated patterns of artificial stimulations, condition the dynamics of hierarchical and functional integration. Functional interactions, which are products or signals emanating from a structural unit acting at a distance on another structural unit, are the fundamental elements of this dynamic.

As a consequence, the proposed model inside Chauvet's MTIP assumes the existence of functional interactions between the artificial and the physiological sensorimotor systems. This hypothesis has been tested through experiments described in the following section. This model in the framework of MTIP is formally described in figure 3, that is the 3D representation of the integrated augmented human design. The human (Ω) (fig.2.) is represented as the combination of the hierarchical structural (z) and functional (Y) organizations. X-Axis corresponds to the ordinary physical or Cartesian space. Each physiological function ψ is represented in the $x\psi y$ plane by a set of structural units hierarchically organized according space scales. Two organizational levels are shown: ψ_1 and ψ_2. The different time scales are on the y-axis, while space scales, which characterize the structure of the system, are on the z-axis. The role of space and time clearly appears. Ψ_{1ij} is the non-local and non-symmetric functional interaction.

Units at the upper levels of the physiological system represent the whole or a part of sensorial and motor organs. Augmented human (Ω') (fig.3.) design consists of creating an artificially extended sensorimotor loop by coupling two artifactual structural units *I'* and *J'*. Their integration into the physiological system is achieved by the functional interactions (i.e. sensorimotor) they generate. From sensors' outputs to effectors' inputs, the synchronized designed artificial system or process *S'* controls and adapts the integration of the functional interactions artificially created into the dynamics of the global and coherent system.

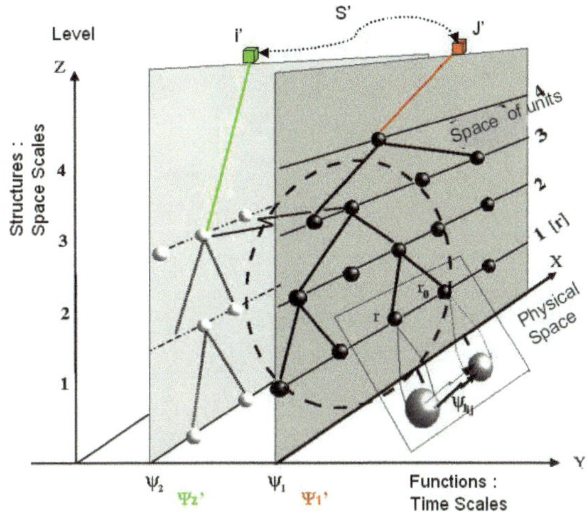

Fig. 3. Ω' – a representation of *augmented human*: artifactual loop coupling the biological system with an artifactual system to an artificial sensorimotor loop (Fass 2007).

This is our theoretical paradigm for augmented human modelling.

According MTIP we highlight three grounding principles for augmented human engineering and human-artefact system design[2]:

- Principle 1: functional interaction is an affordance, a sensorimotor and emotional coupling function depending on geometrical structure of the artifactual design, its architecture;
- Principal 2: the hierarchical structural and functional organization of the human-artefact system must allow behavioural performance and effectiveness inside the boundaries of an operation domain of stability.
- Principle 3: the degree of organization of a human-artefact design, its degree of functional complexity, must be compliant with the evolution of the human-artefact system situated in its operational environment, context, and domain of stability (safety, security and reliability).

---

[2] These theoretical principles of human system integration are consistent with the ten organizational HSI principles define by Harold Booher (Booher 2003) or the three HSI design principles defined by Hobbs et al. (Hobbs et al. 2008).

6. Experiments

The goals of this research are to search for the technical and sensorimotor primitives of augmented human design for gesture assistance by a wearable virtual environment, using virtual reality and augmented reality technologies, for human space activities, aeronautical maintenance and surgery. We have chosen as behavioural assessment adapts to a virtual environment, a neurophysiological method used in motor control researches to study the role of the body in human spatial orientation (Gurfinkel et al. 1993), and the representation of the peri-personnal space in humans (Ghafouri & Lestienne 2006).

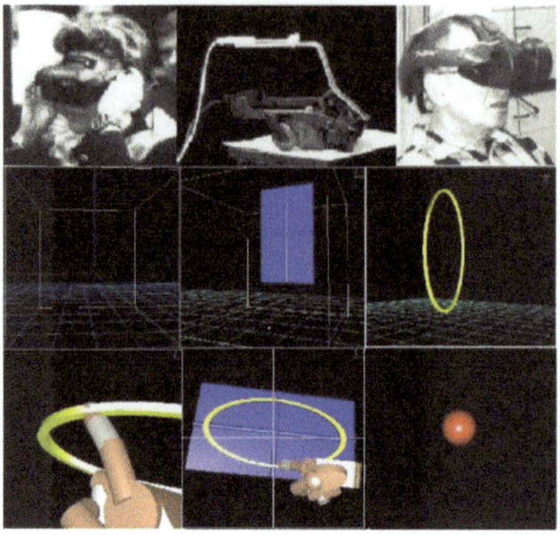

Fig. 4. Examples of different structural and functional primitives for virtual environment design.

6.1 Paradigm

The following method was developed for expert system engineering (knowledge based system) and to explore the knowledge nature as a behavioural property of coupling generated in the dynamics of the individual-environment interaction, either natural or artificial. We use gestures as a sensorimotor maieutic.

The gesture based method for virtual environment design and human system integration assessment is a behavioural tool inspired by Chauvet's theoretical framework, i.e.:

i. an integrated marker for the dynamical approach of augmented human design, and the search for interaction primitives and validation of organization principles; and
ii. an integrated marker for a dynamical organization of virtual environment integrative design.

By designing a artificial environment, a human *in-the-loop* system consists of organizing the linkage of multimodal biological structures, sensorimotor elements at the hierarchical level of the living body, with the artificial interactive elements of the system, devices and patterns of stimulation. There exists a "transport" of functional interaction in the augmented space of both physiological and artifactual units, and thus a *function* may be viewed as the final result of a set of functional interactions that are hierarchically and functionally organized between the artificial and biological systems.

6.2 Material and method

To find the main classes of virtual environments and highlight the dynamical principles of hierarchical organization of human systems integration and virtual environment design for assisting gesture, we set up a protocol according to a complex and incremental design (fig.4.). The experiments were performed in laboratory and a prototype was tested during a French National Space Centre (CNES) parabolic flight campaign.

Devices: Head mounted display I-Glasses® immersive or see-trough, Frastrack Pohlemus® electromagnetic motion tracking system, workstation with a specific software design for managing and generating the visual virtual environment in real-time.

Protocol: Our protocol is based on graphical gesture analysis, more specifically of the drawing of ellipses within 3D-spaces. It's inspired by neurophysiology of movement [20]. By selecting this experimental paradigm, the movement was considered as the expression of a cognitive process *per se*: the integrated expression of the sensorimotor three-dimensional space.

In laboratory, ellipses drawn without virtual environment are the control experiment. It consists of two main situations: open and closed eyes, touch or guided by a real wooden ellipse, and memorized without a model. To highlight the dynamical principles of organization for assisting gestures, we set up a protocol according to a complex and incremental VE design, allowing intuitive learning of both task and use of virtual environment. Ten volunteers (7 men and 3 women, 25 to 35 years old) were asked to performed graphical gestures (drawing of ellipses: eccentricity 0.87 - major axis 40cm and minor axis 20cm) in the three anatomical planes of reference for each step of incremental design (Fig. 5).

The first step of the protocol consisted of drawing ellipses wearing a turned off HMD to study the influence of HMD design and intrusiveness on sensorimotor integration and motor control. The last step of the virtual reality artefact combined allocentric and egocentric prototypic structural elements of artificial visual space, model of ellipses and their planes of movement, and a visual feedback of movement.

Parabolic Flights – hypergravity and weightlessness: to test our prototype (Fig. 6, 7 and 8), three right-handed trained volunteers were asked to draw ellipses (major axis 30 cm and minor axis 15cm) in two orientations of the three anatomical reference planes: vertical sagittal (VS) and transversal horizontal (TH). These drawing of ellipses were performed continuously and recorded during both the 1.8g ascents and the 0g parabola itself, feet in foot-strap (F) or in free-floating (FF), in two main situations: free gesture and assisted gesture wearing a visual virtual environment. Visual virtual environment was generated in immersion (RV) or in augmented reality (RA).

Data analysis: sixteen gesture-related variables are calculated from data produced during the parabola and recorded from the sensor worn on the tip of the index finger of the working hand: kinematics (Number of ellipses), Average velocity, Covariation Vt/Rt, Amplitude), position (Global position, Position / x axis, Position / y axis, Position / z axis), orientation (Global orientation, Orientation / sagittal plane, Orientation / frontal plane, Orientation / horizontal plane) and shape (Mean area, Eccentricity, Major axis variation, Minor axis variation) – indexes in Annex 1.

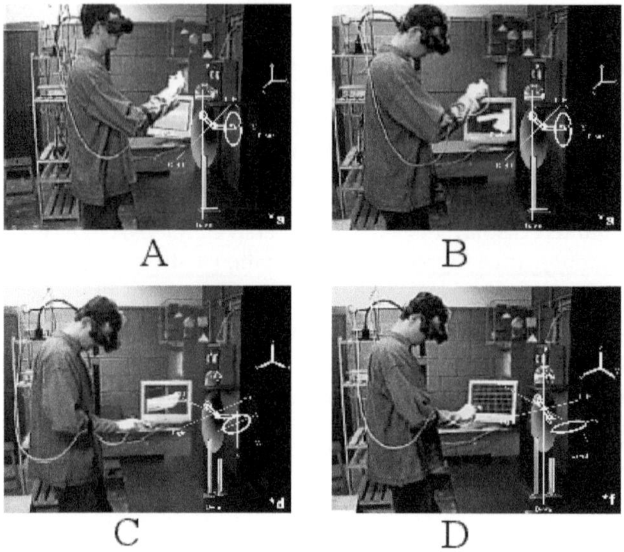

Fig. 5. Graphical gesture of ellipse drawing in the 3D space is performed and analysed in different configurations, more or less complex, of immersive virtual environment assisted drawing ellipses: A- SV ellipses and neutral and coloured background, B- SV ellipses and anthropomorphic visual feedback of movement (artificial hand), C- TF and model of ellipse insert in its plan of movement without visual feedback of movement, D- TH ellipses and abstract representation visual feedback of movement (ball).

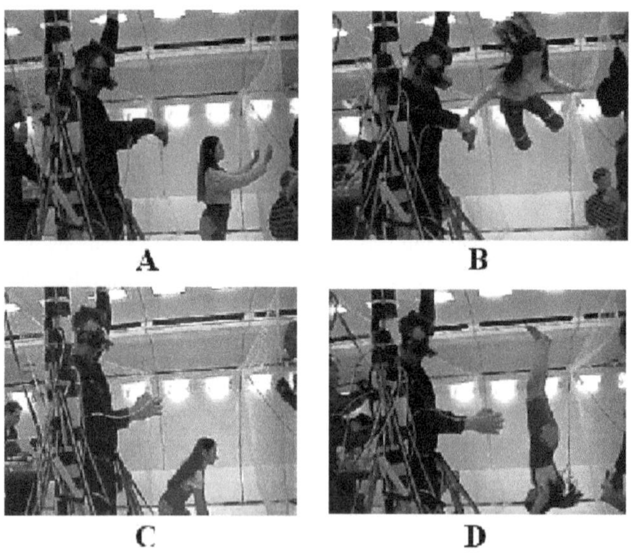

Fig. 6. Drawing of SV (A,B) and HT (C, B) ellipses with gesture assistance in hypergravity (1,8g - A, C) and microgravity (0g - B,D)

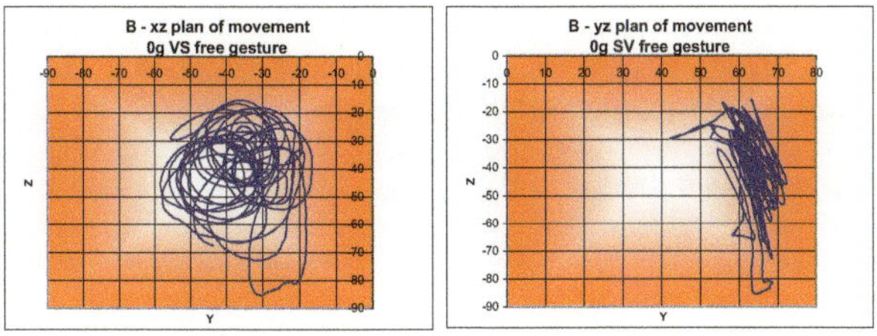

Fig. 7. Weightlessness (0g), example of ellipse drawing in vertical sagital orientation without assistance. We observe a total lost of shape and orientation accuracy.

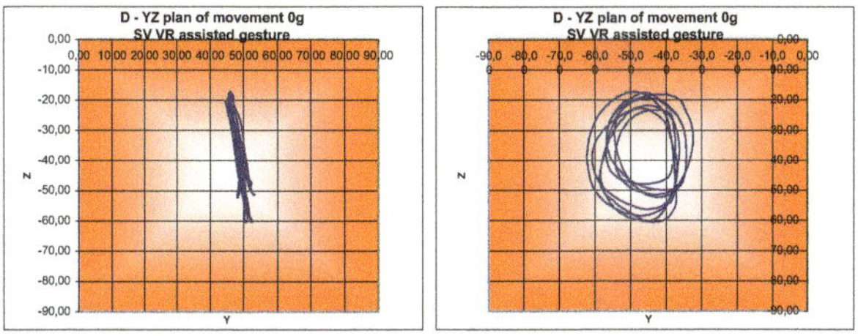

Fig. 8. Weightlessness (0g), example of ellipse drawing in vertical sagital orientation with assistance in vertical sagittal orientation. Even if the shape is not precise, orientation of movement is very accurate and stable (tacking into account the magnetic field distorsion) despite that loss of the gravitational referential and vestibular perturbations. Artificial visuomotor functional interaction coupling by virtual environment enhance stability according the Chauvet's MTIP theory and its principles of auto-associative stabilization.

Statistical analysis: We use a method of multidimensional statistical analysis. Principal component analysis and hierarchical classification are calculated with SPAD 4.0® to show the differential effects of hypergravity and microgravity on graphical gestures for each subject wearing or not the system. A second goal of this exploratory statistics is to assess the design of our prototype and the dynamics of the human virtual environment integration in weightlessness and on earth.

Results: The variable correlation circle (Fig. 9.) shows the first principal (F1) component is correlated in a negative manner with the position, kinematics and shape variables; especially with the global position F, the average velocity B and the mean area E. The second principal component (F2) is correlated in a negative manner with the variables of orientation M, J and K. Whereas K orientation variation in relation to the sagittal plane is fairly correlated with F1. Thereof, the more the average person is placed downward and on the left on the F1-F2 plane, the more their global orientation and orientation in relation to both the frontal and horizontal planes will be important (Annex 1).

Principal component analysis F1-F2 factorial plans (Fig. 10.): Axis 1 (42.70%) shows two sets of experimental status. The first set contains control status head free, touched ellipse, opened or closed eyes, visual guidance, and the virtual reality assisted gesture with visual feedback, ball or hand, and referential frames of action: plane of movement or ellipse model. The second set contains individuals without ellipse model; head free, opened or closed eyes and memorized, HMD off, no gesture feedback and no allocentric or egocentric referential frames. These positions of individuals on the axis 1 reveal the importance of visuo-haptic interactions for gesture in real or virtual environment. Inside that set, they are differences between real touched ellipses situations and " virtually touched". The visuo-haptic class contains two sub-classes (visuo-tactile and visuo-proprioceptive).

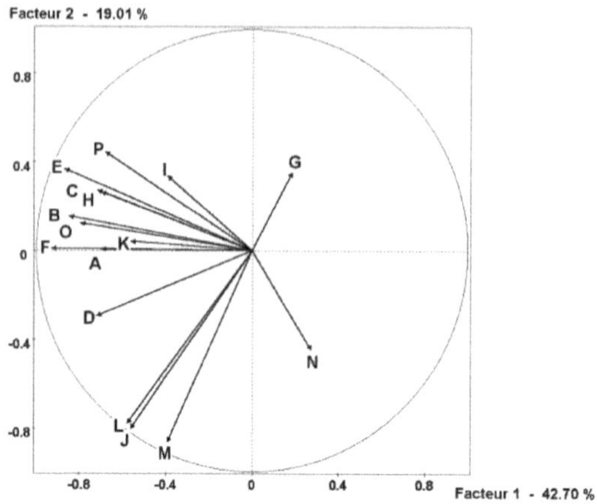

Fig. 9. Variables correlation circle.

Axis 2 (19.01%) shows difference functions of the orientation plane of movement. The distortion of the gesture spatial orientation is greater without visuo-haptic inputs, even with spatial frames of reference and models of action (ellipse model and plan of movement). These positions of individuals on the axis 2 reveal the importance of the gesture spatial orientation. Without visuo-haptic elements, situation of sagittal plane drawing ellipses are nearest to the gravity center of the factorial plane. Frontal and horizontal orientations influence motor behavior with contrary effect. The gesture distortion is greater in the horizontal plane. It also shows significant influence of HMD configurations and of gesture feedback representation. There are functional semiotic differences between ball and virtual hand with enhanced functional differences in absence of visuo-haptic elements. There are four noticeable statuses: 88A, 172a and 175a, without gesture feedback, induce similar behavior to situations with visuo-haptic interactions; 39f, drawing ellipses in the horizontal plane wearing HMD off immersive I-Glasses, induce the greatest distortion in motor control.

The multidimensional statistical analysis (Fig. 9 and 10) confirms the existence of structural and dynamical primitives of human system integration and virtual environment design, for assisting gestures the *a priori* main classes of virtual environment organizational elements. Their organizational and functional properties - the way to couple real and artificial sensori-motor functions - have a significant influence on the human *in-the-loop* system behavior. By enhancing and interacting with the sensorimotor loops, they are able to modify (disturbing or improving) the motor control, the gesture and, as a consequence, the global quality of human behavior. According to these experimental results, the interactions generated by the artefacts may be identified as functional interactions.

Thus we are able to show differential effects for each element of the incremental design of VE, and to assess the global design and dynamics of the human system integration. These experimental results will ground VE design modelling according to the hierarchical organization of theoretical integrative physiology.

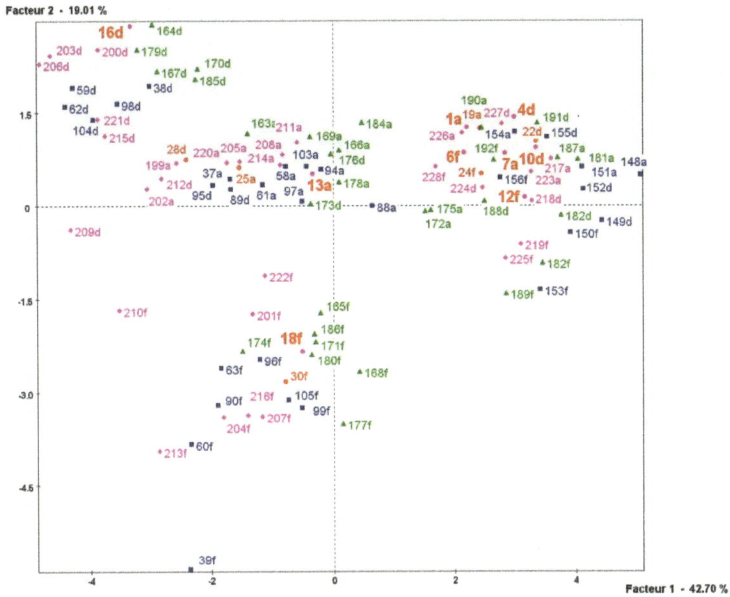

Fig. 10. Principal component analysis, F1-F2 factorial plans: outcome analysis of the virtual environment elements organization is done by observing statistical individuals (indexes Annex 2 and 3) position on the F1-F2 plan (representing 67.71% of the total inertia).

7. Conclusion and perspective

Designing a human-artefact system consists of organizing the linkage of multimodal biological structures, sensorimotor elements at the hierarchical level of the living body, with the artificial interactive elements of the system, devices and patterns of stimulation. There exists a "transport" of functional interaction in the augmented space of both physiological and artifactual units, and thus a *function* may be viewed as the final result of a set of functional interactions that are hierarchically and functionally organized between the artificial and biological system elements.

Structures or Architecture: spatial organization of the structural elements, natural and artificial, coupled by non-local and non-symmetric functional interactions according to PAAS. It is specifying the function(s) of the integrated system. Different organizations specify different architecture and their specific functions:

Behaviour: temporal organisation of the patterns of artificial functional interactions condition and specify the dynamics fit of augmented sensorimotor loops. It is determining augmented human behaviour.

Evolution: the spatiotemporal organization of the structural elements and the functional interactions they produce and processes specify functional stability of human-artefact system according to the *potential of functional organization* principle during the *life of augmented human.*

Contingent on ecology and economy, architecture, behaviour and evolution as specified, define and limit the *life domain of augmented human.*

MTIP is thus applicable to different space and time level of integration in the physical space of the body and the natural or artificial behavioural environment; from molecular level to socio-technical level; from drug design to wearable robotics, and to life and safety critical systems design.

Future work should address questions related to the development of formal models (Cansell & Méry 2008; Méry & Singh 2010) related to augmented human engineering. New questions arise when dealing with deontic or ethical questions that might be handled by an augmented human together with classical formal modelling languages based on deontic or modal languages.

Industrial scientific and pragmatic challenges rely on designing intelligent and interactive artifactual systems relating machines and human beings. This relationship must be aware of its human nature and its body: it is anatomy and physiology. The man-machine interface becomes an integrated continuation of the body between perception-action and sensory and motion organs. By integrating human body and behaviours, the automaton is embodied but this embodiment grounds on the user's body; it enhances capabilities and performances. Efficiency and reliability depend on respecting these fundamental necessities.

8. Acknowledgment

Our special thanks to Professor Dominique MÉRY head of MOSEL LORIA, University of Lorraine.

9. Annexes

9.1 Annex 1: Calculated variables

Index	Variables
A	Number of ellipse
B	Average velocity (cm/s)
C	Covariation Vt/Rt
D	Amplitude (cm)
E	Mean area (cm^2)
F	Global position
G	Position / x axis (cm)
H	Position / y axis (cm)
I	Position / z axis (cm)
J	Global orientation
K	Orientation / sagittal plane(d°)
L	Orientation / frontal plane(d°)
M	Orientation / horizontal plane(d°)
N	Eccentricity
O	Major axis variation
P	Minor axis variation

Table 1.

9.2 Annex 2: Training and control experimental status indexation

Control	INDEX		
	Situation		Gesture Orientation
Opened Eyes	touched ellipse	**1a**	VS
		4d	TF
		6f	TH
	visual guidance	**7a**	VS
		10d	TF
		12f	TH
	memorised	**13a**	VS
		16d	TF
		18f	TH
Closed Eyes	touched ellipse	**19a**	VS
		22d	TF
		24f	TH
	memorised	**25a**	SV
		28d	FT
		30f	HT

Table 2.

9.3 Annex 3: Assisted graphical gesture experimental status

Virtual Environment		INDEX			
	Visual environment	**I/O Immers**	**I/O N Immers.**	**Proview 60**	**Gesture orientation**
HMD off	no	**37a**	**163a**	**199a**	VS
	no	**38d**	**164d**	**200d**	TF
	no	**39f**	**165f**	**201f**	TH
No gesture feedback					
	Allocentric frames	**58a**	**166a**	**202a**	VS
	"	**59d**	**167d**	**203d**	TF
	"	**60f**	**168f**	**204f**	TH
	Egocentric frame	**61a**	**169a**	**205a**	VS
	"	**62d**	**170d**	**206d**	TF
	"	**63f**	**171f**	**207f**	TH
	Ellipse + Allo frames	**88a**	**172a**	**208a**	VS
	"	**89d**	**173d**	**209d**	TF
	"	**90f**	**174f**	**210f**	TH

	Ellipse + Allo+ Ego	**94a**	**175a**	**211a**	VS
	"	**95d**	**176d**	**212d**	TF
	"	**96f**	**177f**	**213f**	TH
Gesture Feedback					
Ball	simple	**97a**	**178a**	**214a**	VS
	"	**98d**	**179d**	**215d**	TF
	"	**99f**	**180f**	**216f**	TH
	Ellipse + all references	**148a**	**181a**	**217a**	VS
	"	**149d**	**182d**	**218d**	TF
	"	**150f**	**183f**	**219f**	TH
Hand	simple	**103a**	**184a**	**220a**	VS
	"	**104d**	**185d**	**221d**	TF
	"	**105f**	**186f**	**222f**	TH
	Ellipse + all references	**151a**	**187a**	**223a**	VS
	"	**152d**	**188d**	**224d**	TF
	"	**153f**	**189f**	**225f**	TH
	"				
Vision and touch	Ellipse and hand	**154a**	**190a**	**226a**	VS
	"	**155d**	**191d**	**227d**	TF
	"	156f	192f	228f	TH

Table 3.

10. References

Bjorner, D. (2006a), Software Engineering 1 Abstraction and Modelling. *Theoretical Computer Science*, An EATCS Series. Springer. ISBN: 978-3-540-21149-5.

Bjorner, D. (2006b), Software Engineering 2 Specification of Systems and Languages. *Theoretical Computer Science*, An EATCS Series. Springer, 2006. ISBN: 978-3-540-21150-1.

Bjorner, D. (2009), Domain Engineering Technology Management, Research and Engineering. COE Research Monograph Series, Vol. 4, JAIST.

Booher, H.R. (2003), Introduction: Human Systems Integration, In: *Handbook of human systems integration*, Booher H.R., pp. 1-30, Wiley, ISBN: O-471-02053-2.

Cansell, D. and Méry, D. 2008. The Event-B Modelling Method: Concepts and Case Studies, in "Logics of specifi- cation languages", D. BJØRNER, M. C. HENSON (editors), Monographs in Theoretical Computer Science, Springer, p. 47-152.

Chauvet, G. A. (1993) Hierarchical functional organization of formal biological systems: a dynamical approach. I. An increase of complexity by self-association increases the domain of stability of a biological system. *Phil Trans Roy Soc London B*, Vol. 339 (March 1993), pp. 425-444, ISSN: 1471-2970.

Chauvet, G. A. (1993), Hierarchical functional organization of formal biological systems: a dynamical approach. II. The concept of non-symmetry leads to a criterion of evolution deduced from an optimum principle of the (O-FBS) sub-system. *Phil Trans Roy Soc London B*, Vol. 339 (March 1993), pp. 445-461, ISSN: 1471-2970.

Chauvet, G. A. (1993),"Hierarchical functional organization of formal biological systems: a dynamical approach. III. The concept of non-locality leads to a field theory describing the dynamics at each level of organization of the (D-FBS) sub-system." *Phil Trans Roy Soc London B*, Vol. 339 (March 1993), pp. 463-481, ISSN: 1471-2970.

Chauvet, G.A. (2002), On the mathematical integration of the nervous tissue based on the S-Propagator formalism: I Theory, *J.Integr.Neurosci.*, Vol. 1, No. 1, 31-68, ISSN: 0219-6352.

Delattre P. (1985), *Système, structure, fonction, évolution : essai d'analyse épistémologique (2ème édition)*, Maloine, ISBN: 2.224-01055-9, Paris, France.

de'Sperati, C. and Viviani, P. (1997) The relationship between curvature and velocity in two-dimensional smooth pursuit eye movements. *Journal of Neurosciences*, Vol. 17, No. 10 (May 1997), pp. 3932-3945, ISSN: 0270-6474.

Ehrhart L.S. and Sage A.P. (2003), User-centrerd systems engineering framework, In: *Handbook of human systems integration*, Booher H.R., pp. 295-373, Wiley, ISBN: O-471-02053-2.

Engelbart, D.C. (1960), Augmenting human intellect: a conceptual framework, AFOSR-3233 Summary Report, Stanford Research Institute, Menlo Park, California 94025, USA, October 1962, http://www.dougengelbart.org/.

Fass, D. (2006). Rationale for a model of human systems integration: The need of a theoretical framework. Journal of Integrative Neuroscience, Vol. 5, No. 3 (September 2006), pp. 333-354, ISSN: 0219-6352.

Fass, D. (2007), Integrative Physiological Design: A Theoretical and Experimental Approach of Human Systems Integration, In Harris D.: *Engin. Psychol. and Cog. Ergonomics, HCII 2007, LNCS, LNAI 4562* (July 2007), pp. 52–61. Springer-Verlag Berlin Heidelberg, ISSN: 0302-9743.

Fass, D. 2005, Virtual Environment a Behavioral and Countermeasure Tool for Assisted Gesture in Weightlessness: Experiments during Parabolic Flight, *Journal of Gravitational Physiology*, Vol. 12, No. 1; July 2005, pp.19-20, ISSN: 1077-9248

Fass, D. and Lieber, R. (2009), Rationale for human modelling in human in the loop systems design, Proceedings of *3rd Annual IEEE International Systems Conference, SysCon*, ISBN: 978-1-4244-3462-6, Vancouver, Canada, Mars 2009.

Ghafouri M, Lestienne FG. (2006), Contribution of reference frames for movement planning in peripersonal space representation. *Experimental Brain Research*, Vol. 169, No. 1 (February 2006), pp. 24-36, ISSN: 0014-4819.

Gurfinkel, VS., Lestienne, F., Levik, YU.S., Popov K.E. et Lefort, L. (1993), Egocentric references and human spatial orientation in microgravity : II body-centered coordinates in the task of drawing ellipses with prescribed orientation. *Experimental Brain Research*, Vol. 95 (August 1993), pp. 343-348, ISSN: 0014-4819.

Haskins C. (Ed.), January 2010, *Systems Engineering Handbook: a guide for processes and activities*, International Council on Systems Engineering (INCOSE).

Hobbs A. N., Adelstein B. D., O'Hara J. , & Null C.H. (2008), Three principles of human-system integration, *Proceedings of the 8th Australian Aviation Psychology Symposium*, Sydney, Australia, April 8-11, 2008

Licklider, J. C. R., Man-Computer Symbiosis, *IRE Transactions on Human Factors in Electronics*, Vol. HFE-1 (March 1960), pp. 4-11, ISSN: 0096-249X.

Kelso, JA. (2008). An Essay on Understanding the Mind: The A.S. Iberall Lecture, *Ecological Psychology*, Vol. 20, No. 2 (April 2008), pp. 180-208, ISSN: 1040-7413.

Malnoy F., Fass. D. and Lestienne F. (1998), Vection and postural activity: application to virtual environment design, *La lettre de l'IA*, Vol. 134-136, 121-124.

Méry, D. and Singh N.K. (2010), Trustable Formal Specification for Software Certification, T. Margaria and B. Steffen (Eds.): ISoLA 2010, Part II, LNCS 6416, pp. 312–326, ISSN: 0302-9743.

Nasa - Human System Integration Division, What is human system integration?, September 2011, http://humansystems.arc.nasa.gov

Nissen M.E. and Burton R.M. (2011), Designing organizations for dynamic fit: system stability, manoeuvrability, and opportunity loss, *IEEETransactions on Systems, Man and Cybernetics-Part A: Systems and Humans*, Vol. 41, No. 3, pp. 418-433, ISSN: 1083-4427.

Roco, M.C. and Brainbridge, W.S. (2002), Converging technologies for improving human performance. National Science Foundation, June 2002.
www.wtec.org/ConvergingTechnologies/Report/NBIC_report.pdf.

Sporns, O. and Edelman G., (1998), Bernstein's dynamic view of the brain: the current problems of modern neurophysiology (1945), *Motor Control*, Vol. 2 (October 1998), pp. 283-305, ISSN: 1087-1640.

Viviani P, Burkhard PR, Chiuvé SC, Corradi-Dell'Acqua C, Vindras P. 2009. Velocity control in Parkinson's disease: a quantitative analysis of isochrony in scribbling movements, Exp Brain Res. , Vol. 194, No. 2 (April 2009), pp. 259-83. Epub 2009 Jan 20. Erratum in: Exp Brain Res. 2009 Apr; 194(2):285.

12

Usage of Process Capability Indices During Development Cycle of Mobile Radio Product

Marko E. Leinonen
Nokia Oyj
Finland

1. Introduction

Mobile communication devices have become a basic need for people today. Mobile devices are used by all people regardless of the race, age or nationality of the person. For this reason, the total number of mobile communication devices sold was almost 1.6 billion units worldwide during 2010 (Gartner Inc., 2011). Manufacturability and the level of quality of devices need to be taken into account at the early stages of design in order to enable a high volume of production with a high yield.

It is a common and cross-functional task for each area of technology to build the required level of quality into the end product. For effective communication between parties, a common quality language is needed and process capability indices are widely used for this purpose.

The basis for the quality is designed into the device during the system specification phase and it is mainly implemented during the product implementation phase. The quality level is designed in by specifying design parameters, target levels for the parameters and the appropriate specification limits for each parameter. The quality level of the end product during the manufacturing phase needs to be estimated with a limited number of measurement results from prototype devices during the product development phase. Statistical methods are used for this estimation purpose. A prototype production may be considered to be a short-term production compared to the life cycle of the end product and a long-term performance process is estimated based on the short-term production data. Even though statistical process control (SPC) methods are widely used in high volume production, the production process may vary within statistical control limits without being out of the control leading product to product variation between product parameters.

Easy to use statistical process models are needed to model long-term process performance during the research and development (R&D) phase of the device. Higher quality levels for the end product may be expected, if the long-term variation of the manufacturing process is taken into account more easily during the specification phase of the product's parameters.

2. Product development process

An overview of a product development process is shown in Figure 1 (based on Leinonen, 2002). The required characteristics of a device may be defined based on a market and competitor analyses. A product definition phase is a cross-functional task where marketing and the quality department and technology areas together define and specify the main functions and target quality levels for features of the device. A product design phase includes system engineering and the actual product development of the device. The main parameters for each area of technology as well as the specification limits for them are defined during the system engineering phase. The specification limits may be 'hard' limits which cannot be changed from design to design, for example governmental rulings (e.g., Federal Communications Commission, FCC) or standardisation requirements (e.g., 3GPP specifications) or 'soft' limits, which may be defined by the system engineering team.

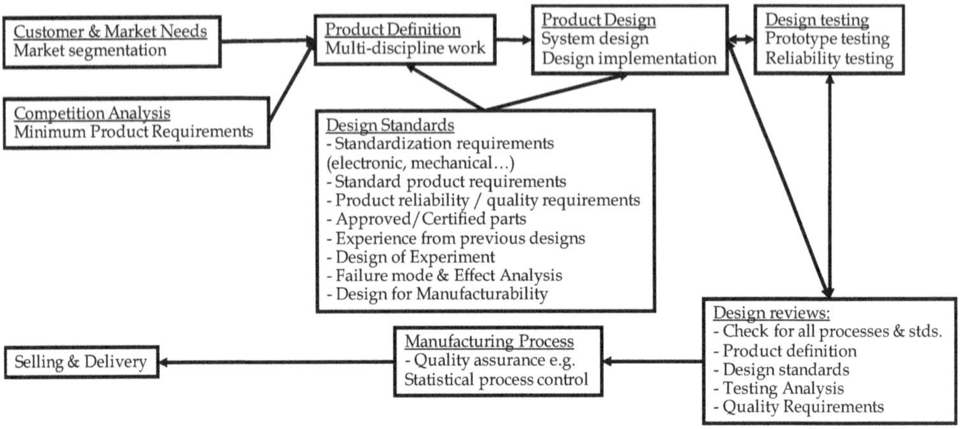

Fig. 1. An overview of a product development process

The main decisions for the quality level of the end product are done during the system engineering and product design phases. Product testing is a supporting function which ensures that the selections and implementations have been done correctly during the implementation phase of the development process. The quality level of the end product needs to be estimated based on the test results prior to the design review phase, where the maturity and the quality of the product is reviewed prior the mass production phase. New design and prototype rounds are needed until the estimated quality level reaches the required level. Statistical measures are tracked and stored during the manufacturing phase of the product and those measures are used as a feedback and as an input for the next product development.

2.1 Process capability indices during the product development

An origin of process capability indices is in the manufacturing industry where the performance of manufacturing has been observed with time series plots and statistical process control charts since 1930s. The control charts are useful for controlling and monitoring production, but for the management level a raw control data is too detailed and thus a simpler metric is needed. Process capability indices were developed for this purpose and the first metric was introduced in early 1970s. Since then, numerous process capability indices are presented for univariate (more than twenty) and multivariate (about ten) purposes (Kotz & Johnson, 2002). The most commonly used process capability indices are still C_p and C_{pk} which are widely used within the automotive - an electrical component - the telecommunication and mobile device industries. An overview of the use of process capability indices for quality improvement during the manufacturing process (based on Albing, 2008; Breyfogle 1999) is presented in Figure 2.

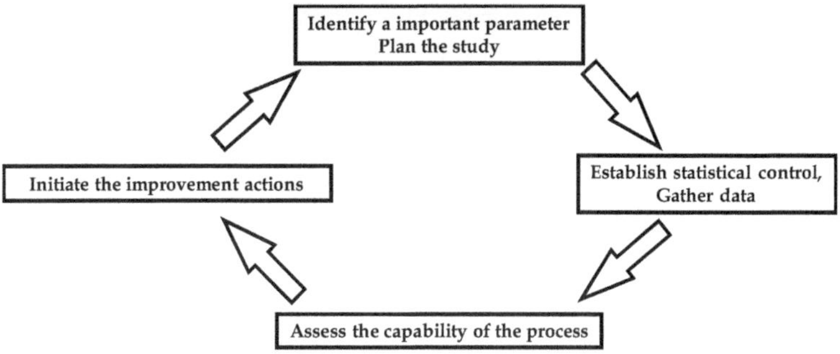

Fig. 2. An improvement process for production related parameters

The usage of process capability indices has been extended from the manufacturing industry to the product development phase, where the improvement of the quality level during product development needs to be monitored, and process capability indices are used for this purpose. The main product development phases, where product capability indices are used, are shown in Figure 3.

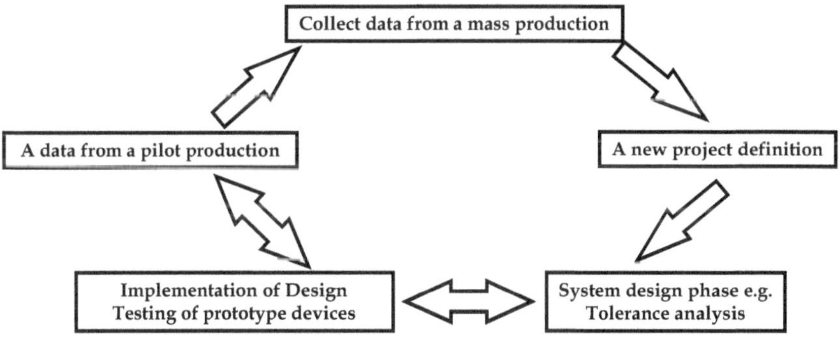

Fig. 3. Product development steps where process capability indices are actively used

An advantage of process capability indices is that they are unitless, which provides the possibility of comparing the quality levels of different technology areas to each other during the development phase of the mobile device, for the example mechanical properties of the device may be compared to radio performance parameters. Additionally, process capability indices are used as a metric for quality level improvement during the development process of the device. The following are examples of how process capability indices may be used during the product development phase:

- A common quality level tool between R&D teams during the product definition phase and business-to-business discussions
- An estimate for the expected quality level of the end product during the R&D phase

- A robustness indicator of design during the R&D phase and product testing
- A decision-making tool of the quality level during design reviews
- A process capability indicator during the mass production phase
- A tool to follow the production quality for quality assurance purposes

Process capability indices can be calculated with some statistical properties of data regardless of the shape of a data distribution. The shape of the data needs to be taken into account if the product capability index is mapped to an expected quality level of the end product. Typically, normal distributed data is assumed for simplicity, but in real life applications a normality assumption is rarely available, at least in radio engineering applications. One possibility to overcome the non-normality of the data is to transform the data closer to the normal distribution and to calculate the process capability indices for the normalised data (Breyfogle, 1999); however, this normalisation is not effective for all datasets. An alternative method is to calculate the process capability indices based on the probability outside of the specification limits and to calculate the process capability index backwards.

2.2 RF system engineering during the product development

RF (Radio Frequency) engineering develops circuitries which are used for wireless communication purposes. RF system engineering is responsible for selecting the appropriate RF architectures and defining the functional blocks for RF implementations. System engineering is responsible for deriving the block level requirements of each RF block based on specific wireless system requirements, e.g., GSM or WCDMA standards and regulatory requirements such as FCC requirements for unwanted radio frequency transmissions.

RF system level studies include RF performance analyses with typical component values as well as statistical analyses with minimum and maximum values of components. The statistical analyses may be done with statistical software packages or with RF simulators in order to optimise performance and select the optimal typical values of components for a maximal quality level. RF block level analyses with process capability indices are studied in Leinonen (1996) and a design optimisation with process capability contour plots and process capability indices in Wizmuller (1998). Most of the studied RF parameters are one-dimensional parameters which are studied and optimised simultaneously, such as the sensitivity of a receiver, the linearity of a receiver and the noise figure of a receiver.

Some product parameters are multidimensional or cross-functional and need a multidimensional approach. A multiradio operation is an example of a multidimensional radio parameter, which requires multidimensional optimisation and cross-technology communication. The requirements for the multiradio operation and interoperability need to be agreed as a cross-functional work covering stake holders from product marketing, system engineering, radio engineering, testing engineering and the quality department. The requirement for multiradio interoperability - from the radio engineering point of view - is a probability when the transmission of the first radio interferes with the reception of a second radio. The probability may be considered as a quality level, which may be communicated with a process capability index value and which may be monitored during the development process of the device. A multiradio interoperability (IOP) may be presented with a two-dimensional figure, which is shown in Figure 4 (based on Leinonen, 2010a). Interference is present if the signal condition is within an IOP problem area. The probability of when this situation may occur can be calculated with a two dimensional integral, which includes the probabilities of radio signals and a threshold value. The actual threshold value for the transmission signal level is dependent on - for example - an interference generation mechanism, an interference tolerance of the victim radio and the activity periods of radios.

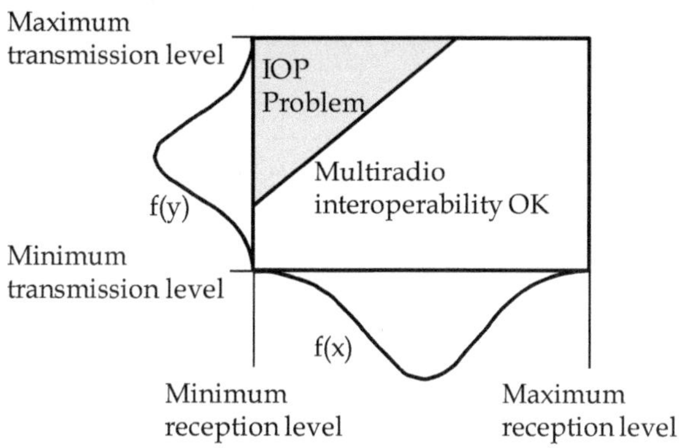

Fig. 4. Illustration of multiradio interoperability from the RF system engineering point of view

3. Overview of process capability indices

Process capability indices are widely used across different fields of industry as a metric of the quality level of products (Breyfogle, 1999). In general, process capability indices describe a location of a mean value of a parameter within specification limits. The specification limits can be 'hard' limits, which cannot be changed from product to product, or 'soft' limits, which are defined during the system engineering phase based on the mass production data of previous or available components, or else the limits are defined based on numerical calculations or simulations.

The most commonly used process capability indices within industry are so-called 'first generation' process capability indices C_p and C_{pk}. The C_p index is (Kotz S. & Johnson, 1993)

$$C_p = \frac{USL - LSL}{6\sigma}, \tag{1}$$

where USL is an upper specification limit and LSL is a lower specification limit, and σ is a standard deviation unit of a studied parameter. C_{pk} also takes the location of the parameter into account and it is defined (Kotz S. & Johnson, 1993)

$$C_{pk} = \min\left(\frac{USL - \mu}{3\sigma}, \frac{\mu - LSL}{3\sigma}\right), \tag{2}$$

where μ is a mean value of the parameter. The process capability index C_{pk} value may be converted to an expected yield with a one-sided specification limit (Kotz S. & Johnson, 1993)

$$\text{Yield} = \Phi\left(3C_{pk}\right), \tag{3}$$

where Φ is a cumulative probability function of a standardised normal distribution. A probability outside of the specification limit is one minus the yield, which is considered to be a quality level. A classification of process capability indices and expected quality levels are summarised in Table 1 (Pearn and Chen, 1999; Leinonen, 2002). The target level for C_{pk} in high volume production is higher than 1.5, which corresponds to a quality level of 3.4 dpm (defects per million)

Acceptable level	C_{pk} value	Low limit	High limit
Poor	$0.00 \leq C_{pk} < 0.50$	500000 dpm	66800 dpm
Inadequate	$0.50 \leq C_{pk} < 1.00$	66800 dpm	1350 dpm
Capable	$1.00 \leq C_{pk} < 1.33$	1350 dpm	32 dpm
Satisfactory	$1.33 \leq C_{pk} < 1.50$	32 dpm	3.4 dpm
Excellent	$1.50 \leq C_{pk} < 2.00$	3.4 dpm	$9.9*10^{-4}$ dpm
Super	$C_{pk} \geq 2.00$	$9.9*10^{-4}$ dpm	

Table 1. A classification of the process capability index values and expected quality level

The C_{pk} definition in equation 2 is based on the mean value and the variation of the data, but alternatively the C_{pk} may be defined as an expected quality level (Kotz S. & Johnson, 1993)

$$C_{pk} = -\frac{1}{3}\Phi^{-1}(\gamma), \tag{4}$$

where γ is the expected proportion of non-conformance units.

Data following a normal distribution is rarely available in real life applications. In many cases, the data distribution is skewed due to a physical phenomenon of the analysed parameter. The process capability analysis and the expected quality level will match each other if the shape of the probability density function of the parameter is known and a statistical analysis is done based on the distribution. The process capability index C_{pk} has been defined for non-normally distributed data with a percentile approach, which has now been standardised by the ISO (International Standardisation Organisation) as their definition of the C_{pk} index. The definition of C_{pk} with percentiles is (Clements, 1989)

$$C_{pk} = \min\left\{\frac{\text{USL-M}}{U_p - \text{M}}, \frac{\text{M-LSL}}{\text{M} - L_p}\right\}, \tag{5}$$

where M is a median value, U_p is a 99.865 percentile and L_p is a 0.135 percentile.

A decision tree for selecting an approach to the process capability analysis is proposed in Figure 5. The decision tree is based on the experience of the application of process capability indices to various real life implementations. The first selection is whether the analysed data is a one-dimensional or a multidimensional. Most of the studied engineering applications have been one-dimensional, but the data is rarely normally distributed. A transform function, such as a Cox-Box or a Johnson transformation, may be applied to the data to convert the data so as to resemble a normal distribution. If the data is normally distributed, then the results based on equations 2 and 3 will match each other. If a probability density function of the parameter is known, then the process capability analysis should be done with the known distribution. Applications for this approach are discussed in Chapter 4. The process capability analysis based on equation 5 is preferred for most real-life applications.

In general, the analysis of multidimensional data is more difficult than one-dimensional data. A correlation of the data will have an effect to the analysis in the multidimensional case. The correlation of the data will change the shape and the direction of the data distribution so that the expected quality level and calculated process capability index do not match one another. A definition of a specification region for multidimensional data is typically a multidimensional cube, but it may alternatively also be a multidimensional sphere, which is analysed in Leinonen (2010b). The process capability analysis may be done with analytical calculus or numerical integration of multidimensional data, if the multidimensional data is normally distributed (which is rarely the case). Transformation functions are not used for non-normally distributed multidimensional data. A numerical integration approach for process capability analysis may be possible for non-distributed multidimensional data but it may be difficult with real life data. A Monte Carlo simulation-based approach has been preferred for non-normally distributed multidimensional data. The process capability analysis has been done based on equation 3, where simulated probability out of the specification region is converted to a corresponding C_{pk} value. The Monte Carlo simulations are done with computers, either with mathematical or spread sheet software based on the properties of the statistical distribution of the data.

Process performance indices P_p and P_{pk} are defined in a manner similar to the process capability indices C_p and C_{pk}, but the definition of the variation is different. P_p and P_{pk} are defined with a long-term variation while C_p and C_{pk} are defined with a short-term variation (Harry & Schroeder, 2000). Both the short-term and the long-term variations can be distinguished from each other by using statistical control charts with a rational sub-grouping of the data in a time domain. The short-term variation is a variation within a sub-group and the long-term variation sums up short-term variations of sub-groups and a variation between sub-group mean values, which may happen over time. Many organisations do distinguish between C_{pk} and P_{pk} due to similar definitions of the indices (Breyfogle, 1999).

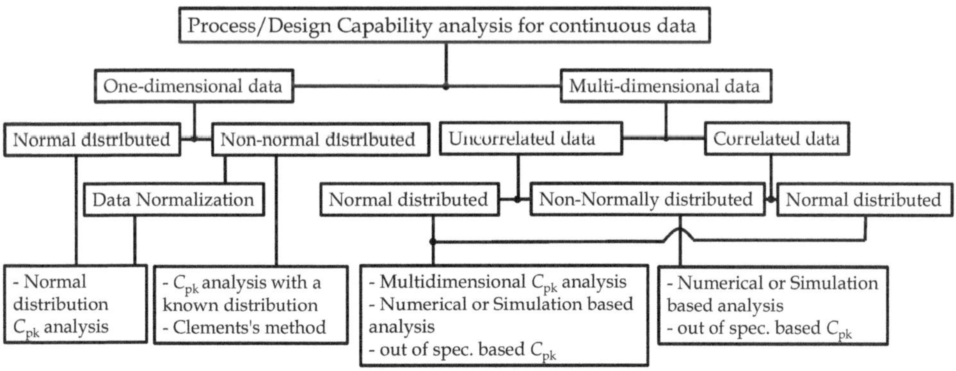

Fig. 5. Process capability analysis selection tree

3.1 Statistical process models for manufacturability analysis

An overview of the usage of process capability analyses during the product development process is shown in Figure 6. Data from a pilot production is analysed in R&D for development purposes. These process capability indices provide information about the maturity level of the design and the potential quality level of the design. The pilot production data may be considered as a short-term variation of the device as compared with a mass production (Uusitalo, 2000). Statistical process models for sub-group changes during a mass production process are needed in order to estimate long-term process performance based on the pilot production data. A basic assumption is that the manufacturing process is under statistical process control, which is mandatory for high volume device production. A mean value and a variation of the parameters are studied during mass production. The mean values of parameters change over time, since even if the process is under statistical process control, statistical process control charts allow the process to fluctuate between the control limits of the charts.

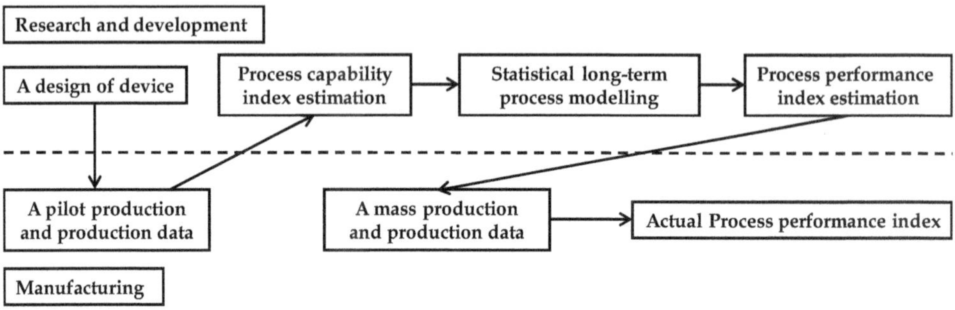

Fig. 6. Long-term process performance estimation during product development

An ideal process is presented in Figure 7, where the mean value and the variation of the process are static without a fluctuation over time. There are some fluctuations in real life processes and those are controlled by means of statistical process control. SPC methods are based on a periodic sampling of the process, and the samples are called sub-groups. The frequency of sampling and the number of samples within the sub-group are process-dependent parameters. The size of the sub-group is considered to be five in this study, which has been used in industrial applications and in a Six Sigma process definition. The size of sub-group defines control limits for the mean value and the standard deviation of the process. The mean value of sub-groups may change within +/- 1.5 standard deviation units around the target value without the process being out of control with a sub-group size of five. The variation of the process may change up to an upper process control limit (B4) which is 2.089 with a sub-group size of five.

The second process model presented in Figure 8 is called a Six Sigma community process model. If the mean value of the process shifts from a target value, the mean will shift 1.5 standard deviation units towards the closer specification limit and the mean value will stay there. The variation of the process is a constant over time in the Six Sigma process model, but it is varied with a normal and a uniform distribution in Chapter 3.2.

The mean value of the process varies over time within control limits, but the variation is a constant in the third process model presented in Figure 9. The variation of the mean value within the control limits is modelled with a normal and a uniform distribution.

The mean value and the variation of the process are varied in the fourth process model presented in Figure 10. The changes of the mean value and the variation of sub-groups may be modelled with both a normal and a uniform distribution. The normal distribution is the most common distribution for modelling a random variation. For example, tool wear in the mechanical industry produces a uniform mean value shift of the process over time.

A short-term process deviation is calculated from the ranges of sub-group and a long-term variation is calculated with a pooled standard deviation method over all sub-groups (Montgomery, 1991). If the number of samples of the sub-group is small - i.e., less than 10 - the range method in deviation estimation is preferred due to the robustness of outlier observations (Bissell, 1990). For control chart creation, 20 to 25 sub-groups are recommended (Lu & Rudy, 2002). It is easier and safer to use a pooled standard deviation method for all the data in an R&D environment for the standard deviation estimation to overcome time and order aspects of the data.

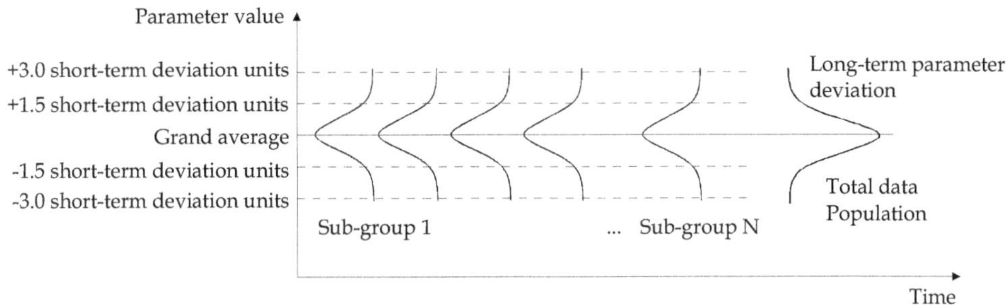

Fig. 7. An ideal process model without mean or deviation changes

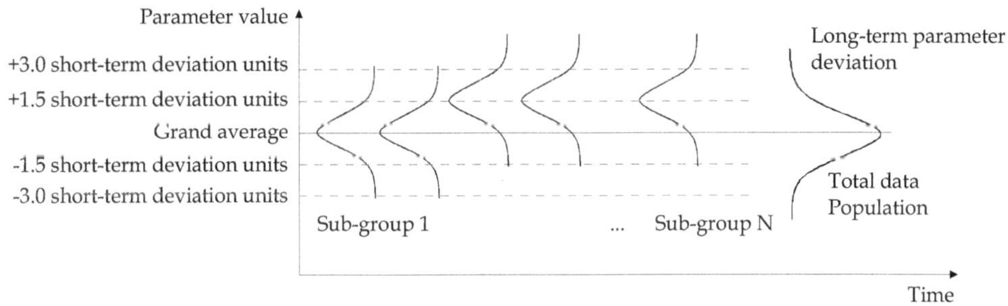

Fig. 8. A Six Sigma process model with a constant mean value shift of sub-groups

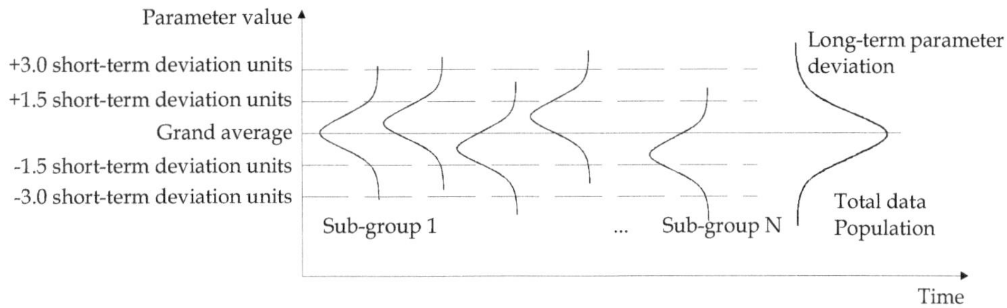

Fig. 9. A process model with a variable mean value and a constant variation of sub-groups

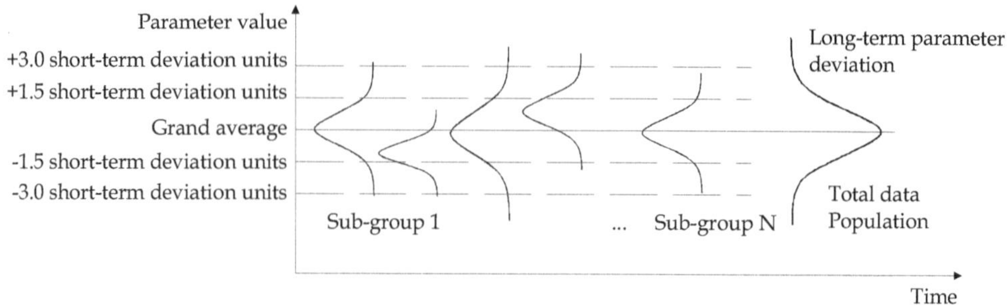

Fig. 10. A process model with a variable mean value shift and variations of sub-groups

3.2 Process model effect to one-dimensional process performance index

Long-term process performance may be estimated based on short-term process capability with a statistical process model. The easiest model is a constant shift model, which is presented in Figure 8. The mean value of sub-groups is shifted with 1.5 deviation units with a constant variation. The process performance index is (Breyfogle, 1999)

$$P_{pk} = \min\left(\frac{USL-\mu}{3\sigma} - 0.5, \frac{\mu - LSL}{3\sigma} - 0.5\right), \tag{6}$$

where σ is a short-term standard deviation unit.

A constant variation within sub-groups with a varied mean value of sub-groups is presented in Figure 9. It is assumed that the variation of the mean value of sub-groups is a random process. If the variation is modelled with a uniform distribution within statistical control limits (+/- 1.5 standard deviation units), then long-term process standard deviation is

$$\sigma_{\text{Long term}} = \sqrt{\sigma^2 + \frac{(1.5-(-1.5))^2}{12}\sigma^2} = \frac{\sqrt{7}}{2}\sigma \approx 1.330\sigma \tag{7}$$

and a corresponding long-term process performance index P_{pk} is

$$P_{pk} = \min\left(\frac{USL-\mu}{3\cdot 1.330\sigma}, \frac{\mu-LSL}{3\cdot 1.330\sigma}\right) \approx \min\left(\frac{USL-\mu}{3.99\sigma}, \frac{\mu-LSL}{3.99\sigma}\right). \tag{8}$$

The second process model for the variation of the mean values of the sub-groups of the process presented in Figure 9 is a normal distribution. The process is modelled so that the process control limits are assumed to be natural process limits or the process is within the control limits with a 99.73% probability. Thus, the standard deviation of the mean drift is 0.5 standard deviation units and the total long-term deviation with normal distributed sub-group mean variation is

$$\sigma_{\text{long term}} = \sqrt{\sigma^2 + (0.5\sigma)^2} = \sigma\sqrt{1+0.25} = 1.118\sigma \tag{9}$$

A corresponding long-term process performance index P_{pk} is

$$P_{pk} = \min\left(\frac{USL-\mu}{3\cdot 1.118\sigma}, \frac{\mu-LSL}{3\cdot 1.118\sigma}\right) \approx \min\left(\frac{USL-\mu}{3.35\sigma}, \frac{\mu-LSL}{3.35\sigma}\right). \tag{10}$$

The effects of the process models to the process performance indices are summarised in Figure 11. The Six Sigma process is defined in that the process capability 2.0 corresponds with the process performance 1.5. The same relationship for process capability 2.0 can be seen if the sub-group means are varied with a uniform distribution. If the process capability is less than 2.0, then the process performance index based on a normal distribution model is clearly higher than with other process models. This may be taken into account when specification limits are defined for the components during the R&D phase. A tolerance reserved for manufacturability may be reduced if a normal distribution may be assumed for the process model instead of the uniform distribution or the constant mean shift, based on previous experience. The process capability Cp value 2.0 is mapped to a process performance index Ppk 1.66 with a normal distribution, and only to 1.50 with the constant mean shift and the uniform distribution models. The estimated quality levels for the process with a process capability C_p value 2.0 are 3.4 dpm with the constant mean shift, 2.9 dpm with the uniform distribution and 0.048dpm with the normal distribution.

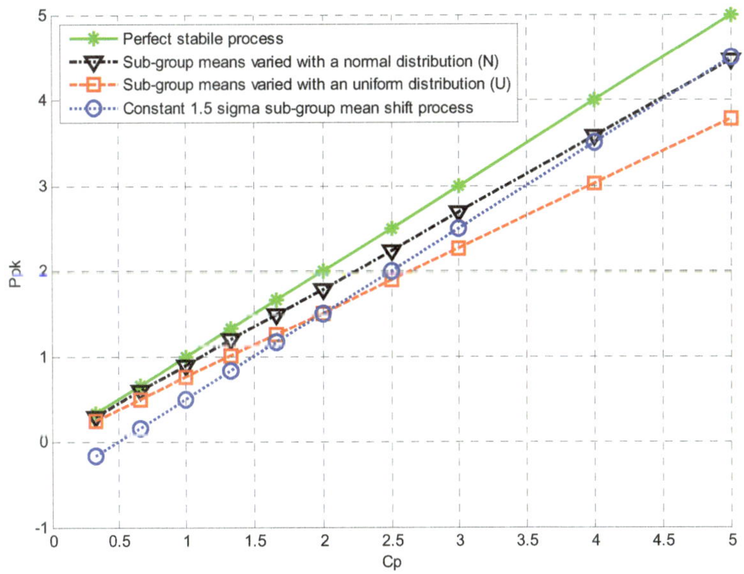

Fig. 11. The effect of the statistical model of sub-group mean value to the process performance index

A realistic statistical process model is presented in Figure 10, where both the mean value and the variation of the sub-groups are varied within the control limits of the control charts for both mean values (Xbar-chart) and variations (s-chart). The effects of the variation within the sub-groups are modelled with both a normal and a uniform distribution. the effect of the variation distribution for the variation within the sub-groups is calculated for a process with a constant mean value, and the combined effect of the variation of the sub-group means and sub-group variations are simulated.

Firstly, the mean value of the process is assumed to be a constant, and a long-term standard deviation is calculated by combining within sub-groups and between sub-groups' variations. The variation within sub-groups is modelled to be one and a standard deviation between sub-groups is defined so that a probability exceeding an UCL (Upper Control Limit) of the s-chart is 0.27 per cent, or the UCL limit is three standard deviation units away from the average value. The UCL (or B4 value) value for the s-chart is 2.089 when a sub-group size 5 is used and a Lower Control Limit (LCL) is zero. The long-term variation can be calculated by

$$\sigma_{\text{Long term}} = \sqrt{\sigma^2 + \left(\frac{(2.089-1)}{3}\right)^2 \sigma^2} \approx 1.064\sigma \tag{11}$$

A corresponding long-term process performance index P_{pk} is

$$P_{pk} = \min\left(\frac{USL-\mu}{3\cdot 1.064\sigma}, \frac{\mu-LSL}{3\cdot 1.064\sigma}\right) \approx \min\left(\frac{USL-\mu}{3.19}, \frac{\mu-LSL}{3.19}\right). \tag{12}$$

The second process model is a uniform distribution for the variation between sub-groups. The uniform distribution is defined so that the variation may drift between the control limits of the s-chart, where the UCL is 2.089 and the LCL is zero. The variation within the sub-group is assumed to be normally distributed with a standard deviation of one. The long-term variation is

$$\sigma_{\text{Long term}} = \sqrt{\sigma^2 + \frac{(2.089-0)^2}{12}\sigma^2} \approx 1.168\sigma \tag{13}$$

A corresponding long-term process performance index P_{pk} is

$$P_{pk} = \min\left(\frac{USL-\mu}{3\cdot 1.168\sigma}, \frac{\mu-LSL}{3\cdot 1.168\sigma}\right) \approx \min\left(\frac{USL-\mu}{3.50}, \frac{\mu-LSL}{3.50}\right). \tag{14}$$

The combined effects of the variations of the sub-group mean value and the variation are simulated with Matlab with ten million observations ordered into sub-groups with five observations within each sub-group. The results of the combined effects of variations of the mean and variation of the sub-groups are presented in Figure 12. The results based on a normal distribution process model for the mean value are closest to the perfect process. The results based on a uniform distribution process model for variation give the most pessimistic quality level estimations.

New equations for process performance indices with various statistical process models are presented in Table 2. It is assumed that the upper specification limit is closer to the mean value in order to simplify the presentation of equations without losing generality. The top left corner equations are used in the literature for process performance indices and others are based on the results from Figures 11 and 12. Short term data models the long term process performance based on these equations. These equations may be used with measured data from the pilot production or during the system engineering phase when component specifications are determined. The short-term data during the system engineering phase may be generated based on Monte Carlo–simulations. A system engineer may test the effects of different statistical process models to the specification limit proposals with these simple equations and estimate a quality level.

	Constant variation	Normal distributed variation between sub-groups	Uniformly distributed variation between sub-groups
Perfect stabile process	$\frac{USL-\mu}{3.00\sigma}$	$\frac{USL-\mu}{3.19\sigma}$	$\frac{USL-\mu}{3.50\sigma}$
Constant 1.5s deviation units mean shift	$\frac{USL-\mu}{3.00\sigma}-0.50$	$\frac{USL-\mu}{3.19\sigma}-0.47$	$\frac{USL-\mu}{3.50\sigma}-0.41$
Normally distributed sub-group mean shift	$\frac{USL-\mu}{3.35\sigma}$	$\frac{USL-\mu}{3.51\sigma}$	$\frac{USL-\mu}{3.92\sigma}$
Uniformly distributed sub-group mean shift	$\frac{USL-\mu}{3.99\sigma}$	$\frac{USL-\mu}{4.10\sigma}$	$\frac{USL-\mu}{4.45\sigma}$

Table 2. Equations to include statistical process model effects for one-dimensional P_{pk}

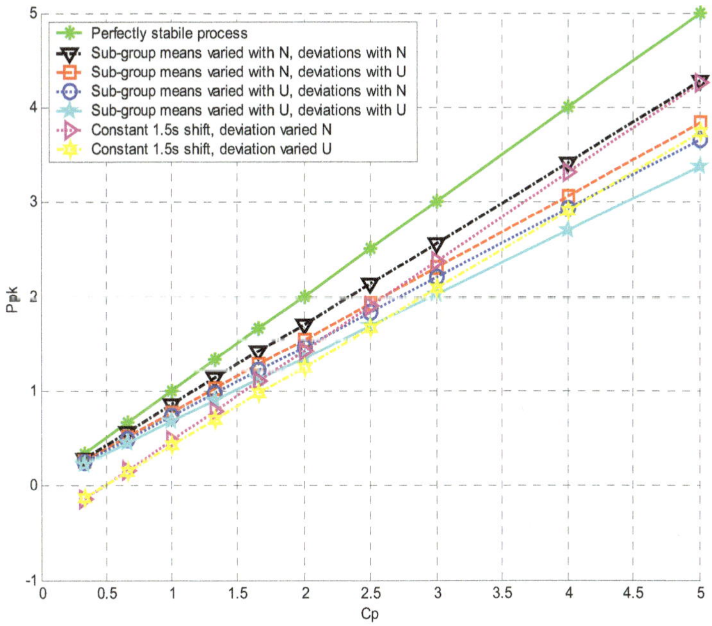

Fig. 12. The combined effects of statistical processes models to the process performance index

3.3 Multidimensional process capability indices

The research into multivariable process capability indices is limited in comparison with one-dimensional ones due to a lack of consistency regarding the methodology for the evaluation of the process's capability (Wu, 2009). In the multidimensional case, the index gives an indication about the problem, but the root cause of the indicated problem needs to be studied parameter by parameter. In general, multidimensional process indices are analogous to univariate indices when a width of variation is replaced with a volume. A multivariable counterpart of C_p is C_p (Kotz & Johnson, 1993)

$$C_p = \frac{\text{Volume of specification region}}{\text{Volume of region containing 99.73\% of values of X}}, \tag{15}$$

where volume of specification is

$$\prod_{i=1}^{v} (\text{USL}_i - \text{LSL}_i).$$

where USL_i and LSL_i are upper and lower specification limits for i:th variable. For multidimensional C_{pk} there is no analogous definition as with single dimensional C_{pk}. For multidimensional cases, a probability outside of the specification can be defined and it can be converted backwards to a corresponding C_{pk} value which can be regarded as a generalisation of C_{pk}. (Kotz & Johnson, 1993). A definition for a multidimensional C_{pk} is (Kotz & Johnson, 1993)

$$C_{pk} = -\frac{1}{3}\Phi^{-1}(\text{expected propotion of non-conformance items}) \tag{16}$$

3.4 Process model's effect on two-dimensional process capability indices

Statistical process models of long-term process variation for the two-dimensional case are similar to those presented in Figures 6 through to 9. An additional step for two-dimensional process capability analysis is to include a correlation of two-dimensional data into the analysis. The correlation of the data needs to be taken into account in both the process capability index calculation and statistical process modelling.

A two-dimensional process capability analysis for a circular tolerance area has been studied in reference to Leinonen (2010b). The circular tolerance area may be analysed as two separate one-dimensional processes or one two-dimensional process. One-dimensional process indices overestimate the quality level for circular tolerance since one-dimensional tolerances form a square-type tolerance range. Additionally, correlation of the data cannot be taken into account in analysis with two separate one-dimensional process indices.

In order to overcome the problems of one-dimensional process indices with a circular tolerance, a new process capability index has been proposed (Leinonen, 2010b), as shown in Figure 13. The one-dimensional C_{pk} process capability indices for X and Y dimensions are marked with and, respectively. The one-dimensional specification limits for the X and Y axis are shown in Figure 13 and the circular tolerance area has the same radius as one-dimensional specifications. A two-dimensional process capability index estimates the process capability based on a probability outside of the circular specification limit. One-dimensional process capability indices overestimate the process capability of the circular tolerance area and they may be regarded as upper bounds for the two-dimensional process capability.

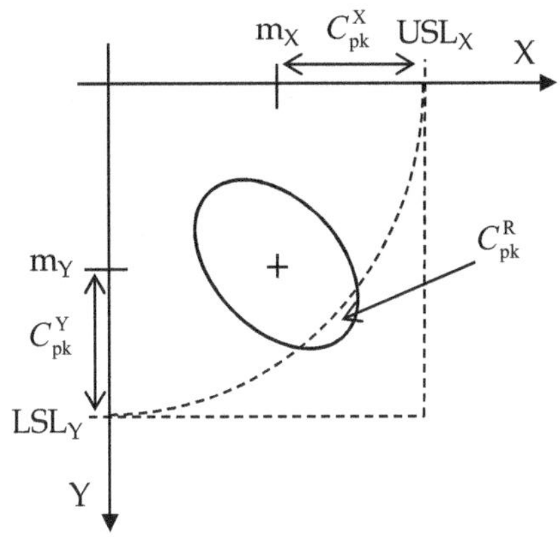

Fig. 13. C_{pk} definitions with circular specification limits

The analysed two-dimensional data distribution is a non-central elliptical normal distribution, and the probability inside of a circular acceptance limit can be calculated (Leinonen, 2010b)

$$p_4 = \int_{-R}^{R} \int_{-\sqrt{R^2-x^2}}^{\sqrt{R^2-x^2}} \frac{1}{2\pi\sigma_X\sigma_Y\sqrt{1-\rho^2}} e^{-\frac{1}{2(1-\rho^2)}\left(\left(\frac{x-m_X}{\sigma_X}\right)^2 -2\rho\left(\frac{x-m_X}{\sigma_X}\right)\left(\frac{y-m_Y}{\sigma_Y}\right)+\left(\frac{y-m_Y}{\sigma_Y}\right)^2\right)} dxdy\,. \tag{17}$$

The process capability index is specified with a probability outside of the circular and it may be calculated based on (16)

$$C_{pk}^{R} = \frac{1}{3}\Phi^{-1}(1-p_4)\,. \tag{18}$$

A long-term process performance for a two-dimensional process with a circular tolerance may be modelled with a similar statistical model (a normal and a uniform distribution) which were used in Chapter 3.2 for the one-dimensional case. However, the correlation of the mean shift of sub-groups is added. Two assumptions are analysed: the first is that there is no correlation between variations of sub-group mean values and the second is that the sub-group mean values are similarly correlated than the individual observations.

The analysed numerical cases are based on Leinonen (2010b) and these are summarised in Table 3. A graphical summary of the numerically analysed two-dimensional Cases 1, 2 and 3 is shown in Figure 14.

The location of the data set is in the first quadrant of the plane in Cases 1, 2 and 3, while the location is on the X-axis in Case 4. The variation of the data is the same in both directions in Cases 1 and 4, while the variation is non-symmetrical in Cases 2 and 3. The location of the data set is defined with mean values m_X and m_Y, and the variation with s_X and s_Y. One-dimensional process capability indices and are calculated for each case and the smaller one is regarded as the one-dimensional C_{pk} value.

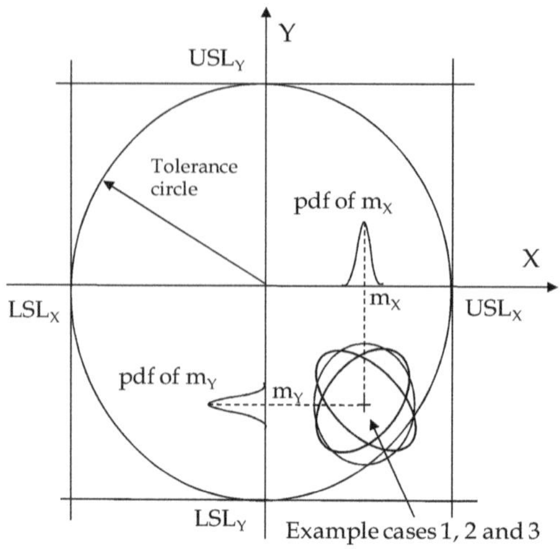

Fig. 14. A graphical representation of a two-dimensional process capability case-study

	Case 1	Case 2	Case 3	Case 4
USL	0.45	0.45	0.45	0.45
LSL	-0.45	-0.45	-0.45	-0.45
m_X	0.225	0.225	0.225	0.225
m_Y	-0.2	-0.2	-0.2	0.0
s_X	0.05	0.025	0.05	0.05
s_Y	0.05	0.05	0.025	0.05
Distribution shape, main direction	Circle	Ellipse, y-axis	Ellipse, x-axis	Circle
	1.50	3.00	1.50	1.50
	1.67	1.67	3.33	3.00
C_{pk} = min (,)	1.50	1.67	1.50	1.50

Table 3. Input data for a two-dimensional process capability case study

The effects of statistical process models of the variation of the mean values of sub-groups in the two-dimensional process performance index are simulated with Matlab with ten million observations ordered into sub-groups with five observations within each sub-group. The same process performance index name is used for both indices, whether based on the short- or the long-term variation.

A significant effect of data correlation to the process's capability may be seen in Figure 15, which summaries the analysis of the example in Case 1. The X-axis is the correlation factor ρ of the data set and the Y-axis is the value. The process capability index is calculated with a numerical integration and simulated with a Monte Carlo-method without any variation of the sub-groups for reference purposes (Leinonen, 2010b). The one-dimensional C_{pk} value is 1.5, and it may be seen that the two-dimensional process performance is maximised and approaching 1.5 when the correlation of data rotates the orientation of the data set in the same direction to that of the arch of the tolerance area.

The statistical process models have a noticeable effect on the expected quality level. If the mean values of the sub-groups are varied independently, with a normal distribution in both the X and Y directions, the effect varies between 0.05 and 0.25. If the mean values vary independently with a uniform distribution in both directions, then the process model has a significant effect up to 0.45 to the with a correlation factor value of 0.6. If the maximum differences in values are converted to the expected quality levels, then the difference ranges from 13 dpm to 2600 dpm. The uniform distribution model suppresses the correlation of data more than normal distribution, and for this reason the long-term process performs worse. If the sub-group's mean values are varied with normal distribution and correlated with the same correlation as the observations, then the long-term performance is a shifted version of the original process's performance and the effect of the correlated process model on average is 0.1 units.

The results for Case 2 are shown in Figure 16. The variation in the X-axis direction is a half of the variation of the Y-axis direction and the one-dimensional C_{pk} value is 1.66. The two-dimensional index approaches the one-dimensional value when the correlation of data increases. If the correlation is zero, then the circle tolerance limits the process performance to 1.2 as compared with the one-dimensional specification at 1.66. If the mean values of the sub-groups are varied independently, either with a normal or a uniform distribution, the process performs better than with the correlated process model. In this case, the correlated mean shift model changes the distribution so that it points out more from the tolerance area than the uncorrelated models. It may be noted that when the correlation changes to positive, then the normal distribution model performs closer to the original process than the uniform distribution model.

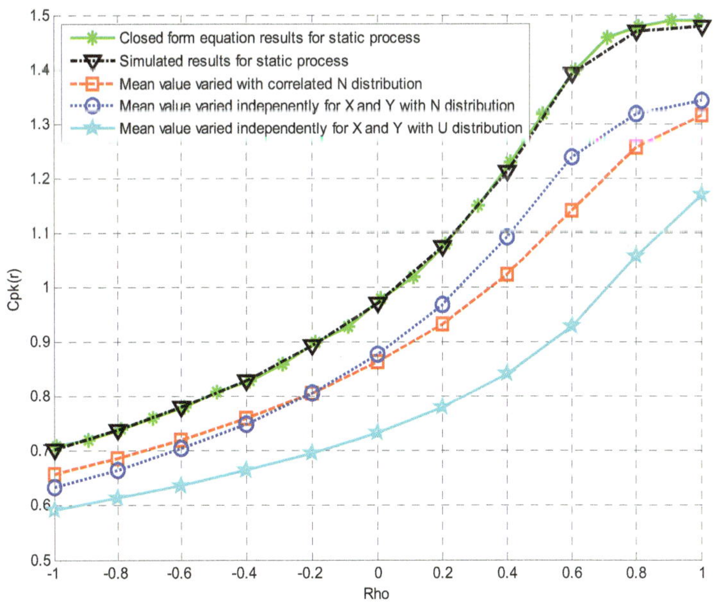

Fig. 15. The effect of the variation of the mean value of sub-groups on a two-dimensional , Case 1

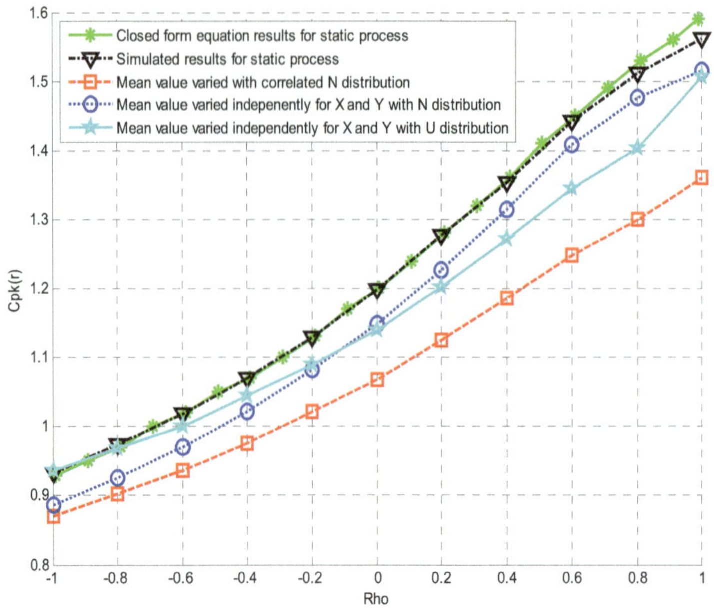

Fig. 16. The effect of the variation of the mean value of sub-groups on a two-dimensional , Case 2

The example of Case 3 shows half of the variation in the Y-axis direction as compared with the X-axis direction, and the one-dimensional C_{pk} value is 1.50. The results for Case 3 are presented in Figure 17. If the sub-group mean values are varied with correlated normal distributions, then the process capability with negative correlations is the best since the correlated process model maintains the original correlation of the data. The uncorrelated normal distribution has an overall data correlation between -0.8 to 0.8, and the uncorrelated uniform distribution has a correlation between -0.57 and 0.57. The uncorrelated uniform distribution model has an effect from 0.25 up to 0.42 of the value.

The results for the Case 4 are presented in Figure 18. The example provided by Case 4 has a symmetrical variation and the distribution is located along the X-axis. For these reasons, the correlation has a symmetrical effect on the two-dimensional process performance indices. The one-dimensional C_{pk} value is 1.50 and the close form equation result without the correlation has a value of 1.45. Both normally distributed process models have a value of 1.31 with the correlation factor at zero. The correlated process model differs from the uncorrelated one with high correlation factor values. The uniform distribution model clearly has the biggest impact on the estimated quality level up to 0.3. The process performance indices maintain the order of the quality level estimations over the correlations due to the symmetrical distribution and location.

As a conclusion, it is not possible to derive similar easy-to-use process capability indices, including the effects of the statistical process models of two-dimensional process performance indices as compared with one-dimensional ones based on the results presented in Chapter 3.2.

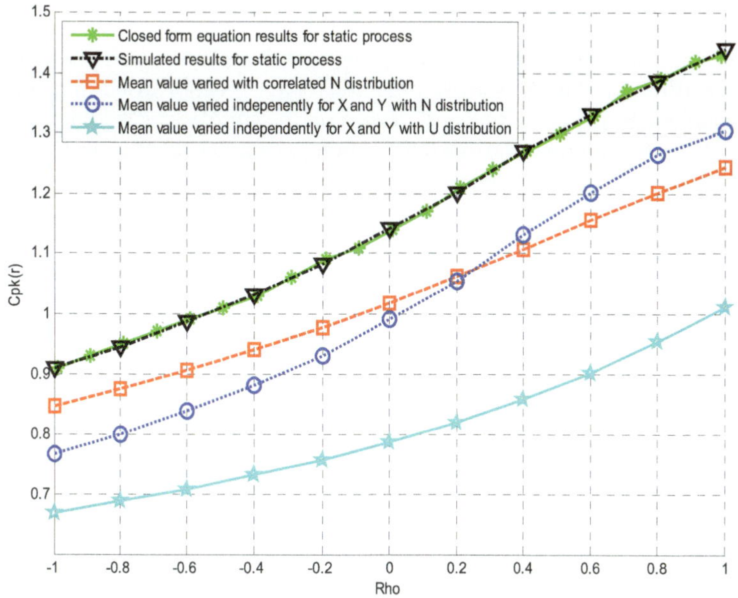

Fig. 17. The effect of the variation of the mean value of sub-groups on a two-dimensional , Case 3

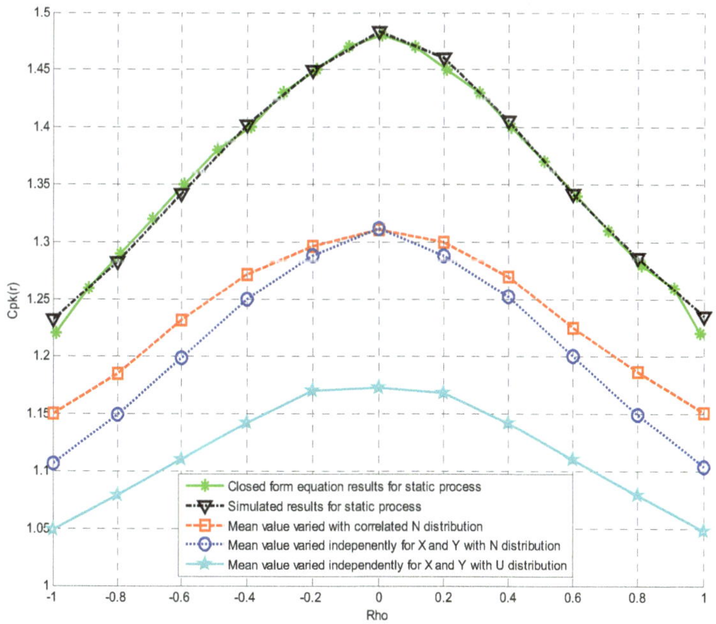

Fig. 18. The effect of the variation of the mean value of sub-groups on a two-dimensional , Case 4

4. Usage of process capability indices in radio engineering

Most of the parameters which are studied during the RF system design phase do not follow a normal distribution. Monte Carlo-simulations have been carried out for the most important RF block level parameters and - based on the simulations results - none of the RF block level parameters follow a normal distribution (Vizmuller, 1998). This is due to fact that the dynamic range of signal levels in radio engineering is huge and typically a logarithm scale is used for signal levels. Unfortunately, in most cases the signal levels do not follow a normal distribution on such a scale. In order to perform a process capability analysis properly for radio engineering parameters, the analysis should be done according to specific distributions, as shown in Figure 5. If a production quality level estimation of an RF parameter is done based on a process capability index with a normal distribution assumption, then the quality level may be significantly under- or overestimated. The problem is that the underlying distributions for all important RF parameters are not available or known and the analyses are based on measured results. The problem with a measurement-based approach is that the properties of the data distributions may change during the development cycle of the device.

Another problem with a measurement-based approach for process capability analysis is that a measurement error of an RF parameter may change the properties of the data distribution. The measurement error based on the RF test equipment on the process capability indices has been studied (Moilanen, 1998). Based on the study of the effect of RF, test equipment needs to be calibrated out and the analysis should be done with actual variation which is based on product-to-product variation. An actual number of RF measurements cannot be reduced based on mathematical modelling, since most RF parameters do not follow the normal distribution and the accuracy of the modelling is not good enough for the purposes of design verification or process capability analysis (Pyylampi, 2003).

Some work has been done in order to find the underlying functions for some critical RF parameters. The statistical properties of the bit error rate have been studied and a statistical distribution of it would follow an extreme value function on a linear scale or else it would follow a log-normal distribution on a logarithm scale with a DQPSK modulation (Leinonen, 2002). In order to validate this result in real life, an infinitive measurement result and measurement time would be needed. It has been shown that, based on measurement results, a peak phase error of a GSM transmission modulation would follow - statistically - a log-normal distribution (Leinonen, 2002). The statistical distribution of a bit error rate of a QPSK modulation has been studied and, with a limited measurement time and measurement results, the distribution of the bit error rate is a multimodal distribution (Leinonen, 2011). The multimodal distribution has a value of zero and a truncated extreme value function distribution part on a linear scale or else a truncated extreme value function distribution on a logarithm scale. Based on the previous results, the process capability analysis of the bit-error rate based on known statistical distribution functions has been studied (Leinonen, 2003, 2011).

Process capability indices give an indication of the maturity level of the design even though the process capability indices may over- or underestimate the expected quality level. The maturity levels of multiple designs may be compared to each other, if the calculation of the indices has been done in a similar manner.

Process capability indices are used as a communication tool between different parties during the development process of the device. Different notation for the process capability index may be used in order to create differences between a process capability index based on a normal distribution or those based on a known or non-normal distribution assumption. One proposal is to use the C^*_{pk} notation if the process capability index is based on non-normal distribution (Leinonen, 2003).

Typically, the studied parameters during the RF system engineering and R&D phases are one-dimensional parameters, and multiradio interoperability may be considered to be one of the rare two-dimensional RF design parameters. Multiradio interoperability in this context is considered to be purely defined as a radio interference study, as shown in Figure 4. Multiradio interoperability may be monitored and designed in the manner of a process capability index (Leinonen, 2010a). A new capability index notation MRC_{pk} has been selected as a multiradio interoperability index, which be defined in a manner similar to the process capability index in equation 16, at least for communication purposes. In order to make a full multiradio interoperability system analysis, all potential interference mechanisms should be studied. A wide band noise interference mechanism has been studied with an assumption that the noise level is constant over frequencies (Leinonen, 2010a). Typically, there is a frequency dependency of the signal level of the interference signals and new studies including frequency dependencies should be done.

The effects of statistical process models on normally distributed one- and two-dimensional data has been studied in 3.2. and 3.4. Unfortunately, most of RF parameters are, by nature, non-normally distributed and thus previous results may not apply directly. More studies will be needed in order to understand how simple statistical process models will affect non-normally distributed parameters. If the manufacturing process could be taken into account more easily during the system simulation phase, either block level or component level specifications could - potentially - be relaxed. If the manufacturing process cannot be modelled easily, then the block level and component level specifications should be done in a safe manner which will yield the over-specification of the system. If the system or solution is over-specified, the solution is typically more expensive than the optimised solution.

5. Conclusion

In high volume manufacturing, the high quality level of the product is essential to maximise the output of the production. The quality level of the product needs to be designed during the product development phase. The design of the quality begins in the system definition phase of product development by agreeing upon the most important parameters to follow during the development phase of the device. Block level and component level specifications are defined during the system engineering phase. The actual specifications and how they are specified are main contributors towards the quality level of the design.

The maturity and potential quality level of the design may be monitored with process capability indices during the product development phase. The process capability indices had originally been developed for the quality tools for manufacturing purposes. Multiple parameters may be compared to each other, since process capability indices are dimensionless, which is an advantage when they are used as a communication tools between technology and quality teams.

Components may be defined using previous information regarding the expected variation of the parameter or based on the calculation and simulation of the parameters. If the component specifications are only defined when based on calculations and simulations, then the variability of the manufacturing of the component and a variability of a device's production need to be taken into account. The manufacturability margin for parameters needs to be included, and one method for determining a needed margin is to use statistical process variation models. Statistical process control methods are used in high volume production and they allow the actual production process to vary between the control limits of statistical control charts. The control limits of the control charts are dependent on a number of samples in a sample control group, and the control limits define the allowable process variation during mass production. A constant mean shift process model has been used in a Six Sigma community to model mass production variation. The effects of a constant process shift model and normal distribution- and uniform distribution-based process models are compared with each other and with the one-dimensional normally distributed data. Based on the simulation results, the constant shift and the uniform distribution models expect a similar quality level with a process capability index value of 2, while at a lower process capability level a constant shift process estimates the lowest quality level. The normal distribution model of the manufacturing process expects a higher quality level than other process models with a one-dimensional parameter. New equations for one-dimensional process capability indices with statistical process models based on calculations and simulations have been presented in the Chapter 3.2.

Process capability indices have been defined according to multidimensional parameters which are analogous to one-dimensional process capability indices. One of the main differences between one- and two-dimensional process capability index analyses is that a correlation of the data with two-dimensional data should be included into the analysis. Another difference is the definition of the specification limit, which may be rectangular or circular or else a sub-set of those. A rectangular tolerance area may be considered if the two-dimensional data is uncorrelated, and the specifications may be considered to be independent of each other. Otherwise, the tolerance area is considered to be circular. The effects of statistical process models for two-dimensional process capability indices with a correlated normal distribution with a circular tolerance area have been studied. The correlation of the data has a significant effect on the expected quality level based on the simulation results. The location and the shape of the data distribution have an additional effect when statistical process models are applied to the data. Easy to use equations which take the statistical process models into account with two-dimensional data cannot be derived due to multiple dependences in terms of location, shape and the correlation of the data distribution.

Most radio performance parameters are one-dimensional and they are not distributed with a normal distribution, and so the process capability analysis should be carried within known statistical distributions. A process capability analysis based on a normality assumption may significantly under- or overestimate the expected quality level of the production. The statistical distributions of some RF parameters are known - e.g., the bit error rate - but more work will be needed to define the others. Also, a multiradio interoperability may be considered to be a two-dimensional parameter which may be analysed with process capability indices.

6. References

Albing, M. (2008). Contributions to Process Capability Indices and Plots, Doctoral thesis, Luleå University of Technology, ISSN 1402-1544

Bissell, A. F. (1990). How reliable is Your Capability Index? Applied Statistics, Vol. 39, No. 3, pp.331-340

Breyfogle, F. (1999). *Implementing Six Sigma: Smarter Solutions Using Statistical Methods*, John Wiley & Sons, ISBN 0-471-29659-7, New York, USA

Clements, J. A. (1989). Process Capability Calculations for Non-normal Distribution, *Quality Progress*, pp. (95-100), ISSN 0033-524X

Gartner Inc., (February 2011) Competitive Landscape: Mobile Devices, Worldwide, 4Q10 and 2010, 11.8.2011, abstract available from: http://www.gartner.com/it/page.jsp?id=1543014

Kotz, S. & Johnson, N. L. (2002). Process Capability Indices - A review, *Journal of Quality Technology*, Vol. 34, No 1, pp.2-19

Kotz, S. & Johnson, N. L. (1993). *Process capability indices*, Chapman & Hall, ISBN 0-412-54380-X, London, UK

Leinonen, M. E. (1996). The Yield Analysis of RF blocks with C_{pk} method, Diploma Thesis, University of Oulu, p.60, in Finnish

Leinonen, M. E. (2002). The Statistical Analysis of RF Parameters in GSM Mobile Phones, Licentiate Thesis, University of Oulu, p.121.

Leinonen, M. E. (2003). Process Capability Index Usage as a Quality metric of Digital Modulation Receiver, *Proceedings of URSI/IEEE XXVIII Convention of radio science & IV Finnish Wireless Communication Workshop*, pp.50–53, Oulu, Finland, October 16-17, 2003

Leinonen, M. E. (2010,a). Multiradio Interoperability index, *Proceedings of 3rd European Wireless Technology Conference*, pp.145-148, ISBN: 972-2-87487-018-7, Paris, France, September 27-28, 2010

Leinonen, M. E. (2010,b). The Effect of Data Correlation to Process Capability Indices in Position Tolerancing, *Proceedings of ISSAT2010 Conference*, pp.21 25, ISBN: 978 0 9763486-6-5, Washington, USA, August 5-7, 2010

Leinonen, M. E. (2011). Process Capability Index Analysis of Bit Error Rate of QPSK with Limited Observations, *to be presented at 41st European Microwave Conference*, Manchester, United Kingdom, October 9-14, 2011

Lu, M. & Rudy R. J. (2002). Discussion. *Journal of Quality technology*, Vol. 34, No 1, pp.38–99

Harry, M. & Schroeder, R. (2000) *Six Sigma the Breakthrough Management Strategy Revolutionizing the World's Top Corporations*, Random House, ISBN 0-385-49438-6, New York, USA

Moilanen, T. (1998). The Error Determination which Comes from Measurement Equipment during Mobile Phone RF Measurements, Engineering Thesis, Polytechnic of Ylivieska, 63 p., in Finnish

Montgomery, D. C. (1991). *Introduction to Statistical Quality Control, 2nd ed.*, John Wiley & Sons, Inc., USA

Pearn, W.L. and Chen, K. S. (1999) Making decisions in assessing process capability index C_{pk}. *Quality and Reliability engineering international*, Vol. 15, pp.321-326

Pyylampi, K. (2003). The Mathematical Modelling of RF parameters in GSM phone, Diploma Thesis, University of Oulu, p.63, in Finnish

Uusitalo, A. (2000). The Characterization a Long Term Process Deviation for RF Parameters in high volume Mobile Phone Production, Engineering Thesis, Polytechnic of Oulu, p.45, in Finnish

Vizmuller, P. (1998). *Design Centering Using Mu-Sigma Graphs and System Simulation*, Artech House, ISBN 0-89006-950-6, Norwood, MA

Wu C.-H., Pearn W. L. & Kotz S. (2009) An Overview of theory and practice on process capability indices for quality assurance. *Int. Journal Production Economics*, 117, pp.338 - 359

13

Systems Engineering and Subcontract Management Issues

Alper Pahsa
Havelsan Inc.
Turkey

1. Introduction

One of the major problems that the Systems Engineering processes come across is how to deal with the subcontractors. Assuring the activities of subcontractors are convenient and compliant with the systems engineering standard procedures and criteria is an important for the systems engineering program management. One of the most challenging jobs for a systems engineering team is to understand the needs and requirements of the customer, the constraints and variables that are established and the limits of business conduct that are acceptable for the particular job under contract. This understanding should directly reroute to the people who work under the subject contract of the customer. All of the requests, criteria and generic standards of customer needs associated with the subcontractor are directly written in the subcontracts statement of work or tasking contract too.

The process of the dealing with the subcontractors is the responsibility of the systems integrators in order to ensure the whole systems engineering process is followed. The systems integrator has the responsibility of helping the subcontractor take functional point of view of the organization and of all procurement process. It is the responsibility of the systems integrator to aid the subcontractor in erecting a parallel technical auditing process.

So what does all of this mean to project management team responsible for issuing contracts and subcontracts that enable systems engineering team solutions to meet the requirements of the systems integration project? It should be clear that the unclear instructions as part of the subcontracts issued to subcontractors with metrics spelled out by which are able to gauge both technical and on-time performance should be clear.

Systems engineering teams incorporate this into the terms and conditions of the subcontracts. It must be careful to avoid the pass-through of non deterministic risk factors, as if the systems engineering team lose control of these once they are in the hands of others. Pass-through of known risk elements is natural, and a revision activity must be in place such that it is able to keep track of the progress in the resolving the items with the risk.

Systems Engineering Teams discussed how to implement and maintain an audit trail throughout the systems integration process and how to perform and record the details of the quality assurance process. Each of these activities carries special important on how it is implemented in systems integration approach with subcontractors that is engaged for assistance with the project or for procurement of hardware and software.

Just as the customer provides the facilities with a set of requirements that it believes to be representative of the actual needs of the user. The corporation must prepare a detailed set of valid requirements for subcontractors. Absence of strategic plan on the part of a subcontractor should result in imposition of the systems integration organization strategic plan, especially those parts that related to audit trail maintenance; risk identification, formulation, and resolution; and such management process and procedures as we feel are essential for satisfactory performance the contract or subcontract. In the following sections initially Systems Engineering process is explained. Secondly Program Management process is explained then the process of the subcontract management and the activities related to Issues with contractor and subcontractor management will be given. In this section a systems engineering and the program management teams' perspectives for subcontract management issues are explained. Then the concerns related to subcontractor management process are given and finally conclusion section is drawn for the subcontract management in sense of systems engineering process is given.

2. Systems engineering

Systems engineering defined as "An interdisciplinary approach to evolve and verify an integrated and life cycle balanced set of systems product and process solutions that satisfy customer needs. Systems engineering: (a) encompasses the scientific and engineering efforts related to the development, manufacturing, verification, deployment, operations, support, and disposal of systems products and processes; (b) develops needed user training equipments, procedures, and data (c) establishes and maintains configuration management of the systems; (d) develops work breakdown structures and statements of work; and (e) provides information for management decision making." Figure 1 displays the Systems Engineering process outline (David E. S. et al, 2006).

The basic Systems Engineering process needs successful products and/or process. It is largely an iterative process that provides overarching technical management of systems from the stated need or capability to effective and useful fielded systems. During the process, design solutions are distributed evenly to the stated needs through the constraints imposed by technology, budgets, and schedules (INCOSE, 2011).

Systems engineering should support acquisition program management in defining what must be done and gathering the information, personnel, and analysis tools to define the mission or program objectives. This includes gathering customer inputs on "needs" and "wants", systems constraints (costs, technology limitations, and applicable specifications/legal requirements), and systems "drivers" (such as capabilities of the competition, military threats, and critical environments). The set of recommended activities that follow are written for a complex project that meets a stated mission or goal, but the word "product" can be substituted to apply these steps to commercial products, for example (Associate CIO of Architecture, 2002),.

Based on the acquisition strategy, the technical team needs to plan acquisitions and document the plan in developing Systems Engineering Management Plan (SEMP). The SEMP covers the technical teams before contract award, during contract performance, and upon contract completion. Included in acquisition planning are solicitation preparation, source selection activities, contract phase-in, monitoring contractor performance, acceptance of deliverables, completing the contract, and transition beyond the contract. The SEMP focuses on interface activities with the contractor, including technical team involvement with and monitoring of contracted work. Often overlooked in project staffing estimates is the amount of time that technical team members are involved in contracting-related activities. Depending on the type of procurement, a technical team member involved in source selection could be consumed nearly full time for 6 to 12 months. After contract award, technical monitoring consumes 30 to 50 percent, peaking at full time when critical milestones or key deliverables arrive (Shamieh, 2011).

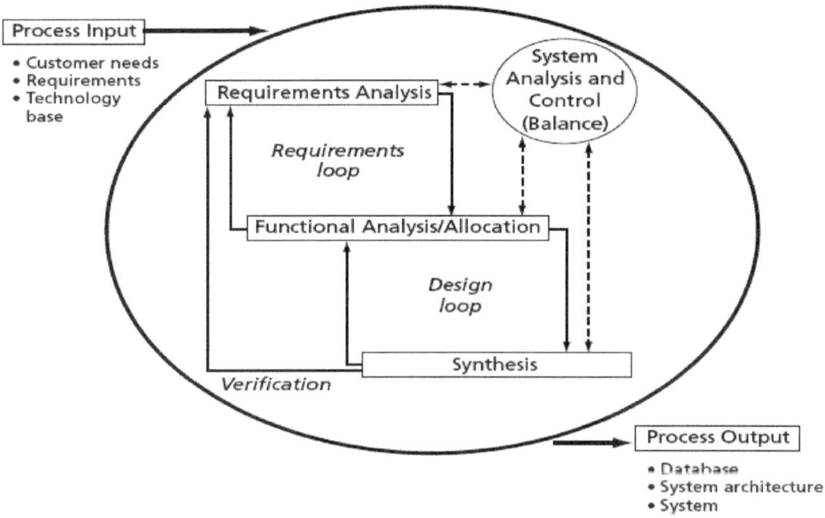

Fig. 1. Systems Engineering Process

The technical team is intimately involved in developing technical documentation for the acquisition package. The acquisition package consists of the solicitation (e.g., Request for Proposals (RFPs) and supporting documents. The solicitation contains all the documentation that is advertised to prospective contractors (or offers). The kkey technical sections of the solicitation are the SOW (or performance work statement), technical specifications, and contract data requirements list. Other sections of the solicitation include proposal instructions and evaluation criteria. Documents that support the solicitation include a procurement schedule, source evaluation plan, Government cost estimate, and purchase request. Input from the technical team will be needed for some of the supporting documents. It is the responsibility of the contract specialist, with the input from the technical team, to ensure that the appropriate clauses are included in the solicitation. All of the features related to solicitation are important for a subcontractor for fully understanding the content of the work that is aimed to realize. Figure 2 shows the process of the contract requirement development process (NASA, 2007).

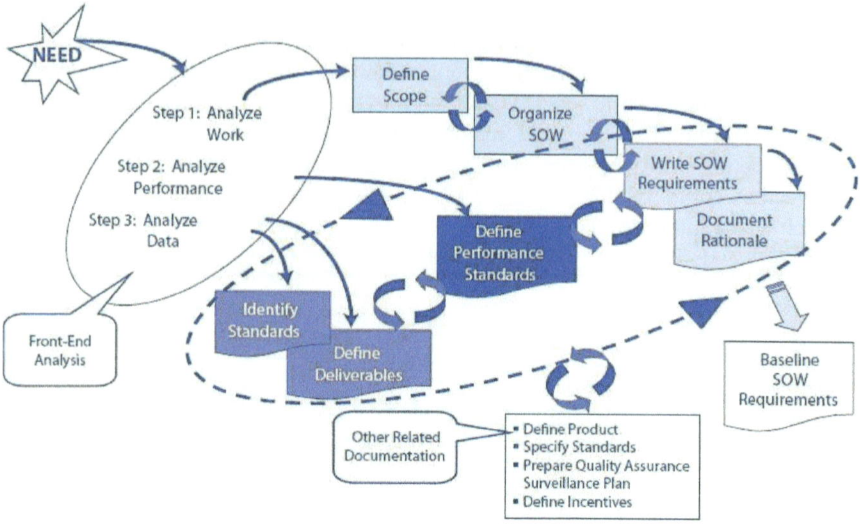

Fig. 2. Contract Requirements Development Process

3. Program management

Program management has been defined as "the management of a series of related projects designed to accomplish broad goals, to which the individual projects contribute, and typically executed over an extended period of time". Program management is very different from corporate administrative management that involves an ongoing oversight role. Program management usually has the more specific task of completing a project or set of projects for which there is a common goal and a finite termination point. The program manager has the responsibility of planning the project, controlling the project's activities, organizing the resources, and leading the work within the constraints of the available time and resources (Associate CIO of Architecture, 2002).

Project planning involves mapping the project's initial course and then updating the plan to meet needs and constraints as they change throughout the program. In the planning process, an overall plan, called an "acquisition strategy," is formulated by analyzing the requirements; investigating material solutions (designs); and making technical, cost, and performance trade-offs to arrive at the best solution. A formal acquisition plan details the specific technical, schedule, and financial aspects of a specific contract or group of contracts within a specific phase of a program. Functional plans detail how the acquisition strategy will be carried out with respect to the various functions within the program (i.e., systems engineering, test and evaluation, logistics, software development). Schedules that are continually updated are used to ensure that various milestones along a timeline are being met. Budgeting, another aspect of project planning, involves developing an initial cost estimate for the work to be performed, presenting and defending the estimate to parties responsible for budget approvals, and expending the funding.

Control of the project's activities is primarily concerned with monitoring and assessing actual activities and making sure they align with program goals. Monitoring involves conducting program reviews, measuring actual costs with planned costs, and testing incremental aspects of the program. It also includes managing the internal aspects of a program (e.g., the current contract) and monitoring external organizations (Government etc.) that may have a stake in the program's outcome. From time to time, a program assessment is needed to determine if the overall requirement is still being addressed, adequate funds are available, the risks are being managed, and the initial acquisition strategy is sound. Leading the work, given time and resource constraints, involves not only the previously mentioned tasks, but also directing that tasks be carried out and maintaining consensus within and outside the program. The program manager must give direction to his or her organization and take direction from organizations outside of his or her direct control. Maintaining a consensus requires making sure that the competing goals of internal and external organizations remain in balance and are working toward the desired goal (David E. S. et al, 2006).

There exists an agreement between the systems engineer and the contract management team. Systems engineer supports the development and maintenance of the agreement between the project office and the contractor that will perform or manage the detail work to achieve the program objectives. This agreement has to satisfy several stakeholders and requires coordination between responsible technical, managerial, financial, contractual, and legal personnel. It requires a document that conforms to the acquisition regulations, program product breakdown structure documentation and the systems architecture. The figure given below shows the contractual process (David E. S. et al, 2006):

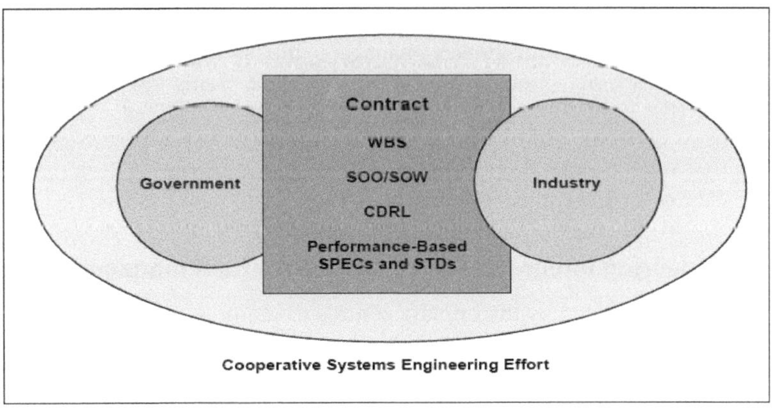

Fig. 3. Contractual Process

The role of technical managers or systems engineers is crucial to satisfying these diverse concerns. Their primary responsibilities include:

- Supporting or initiating the planning effort. The technical risk drives the schedule and cost risks which in turn should drive the type of contractual approach chosen,
- Prepares or supports the preparation of the source selection plan and solicitation clauses concerning proposal requirements and selection criteria,

- Prepares task statements
- Prepares the Contract Data Requirements List (CDRL),
- Supports negotiation and participates in source selection evaluations,
- Forms Integrated Teams and coordinates the government side of combined government and industry integrated teams,
- Monitors the contractor's progress, and Coordinates government action in support of the contracting officer (Global Intergy Corporation, 2002).

4. Subcontracting in System Engineering program

When a Systems Engineering program/project includes a contracting service/product, a challenge is occurred in the Systems Engineering people minds: "Do it in our company or purchase it?" As "everything is program/project...", it is always seem to Systems Engineer team that the option of doing the service/product in the company would be more manageable and cause less trouble. In fact this option is an illusion and reinforced by closed project/program contracting experiences. These experiences are large and were not successful and over which the system engineering process had little control.

However, it is already known that small or large project/programs there are many benefits of subcontracting the program/project. These benefits caused by purchasing the service/product from the product that is already available. Because of the lack of information or interest in a certain technology, carrying out a program/project without subcontracting external services/product, many problems would frequently occur for the program activities. Moreover, it must be well known that the training to hire outsources is important.

The reasons for failing in subcontracting activities are started with lack of a well-defined process to guide the systems engineering team. Purchasing services/product, despite being a rather routine task in program/project manager's life, is a high-risk endeavour and, usually, an empirical activity. Nonetheless, there are many items are bought during the program life cycle (De Mello Filho, 2005). When the program management acquisition team purchase hardware or some material, they are performing a search procedure for certain characteristics that will be evaluated during the acquisitions. This procedure or acquisition activity is defined in classical engineering terms as the procurement process.

5. System Engineering integration roles for subcontract management

When a project being managed by the primary contractor requires a wide range of skills and experience, it may require subcontracting with other companies. It is the prime contractor project manager's responsibility to ensure that the teaming partners and subcontractors are held to the same quality standards as the prime contractor as specified in the Project Plan.

- Statements of work for the subcontractors must clearly reflect the project requirements and state what activities and reviews are expected in their performance. Primary activities that the prime contractor will address with subcontractors include:
- Acceptance criteria.
- Subcontractor Project Plan, Quality Plan, Quality Assurance Plan.
- Quality assessments of subcontractor performance.
- Subcontractor assessments, audits, preventive and corrective action plans (Associate CIO of Architecture, 2002).

The systems integrators company teams such as the systems engineering and contract management have the responsibility of helping the subcontractor take a functional point of view of the organization and of all procurement efforts.

It is the responsibility of the systems integrator to aid the subcontractor in grows a parallel technical auditing process. In addition to interaction matters already discussed, there are several other points to be made (U.S. DoT, 2009). These include the following items:

No favoured treatment for specific vendors. It is only human perhaps for clients and systems integrators to have favourite vendors. These are companies or individuals within certain companies that have provided excellent service in the past. Perhaps a previous acquisition more than met specifications or was unusually trouble-free. Sometimes a particular marketing organization has gone the extra mile to be of assistance in an emergency. It is only natural under such circumstances that this favourable past impression might bias the client or systems integrator. Indeed, the concept of the favoured client is a common one in the private sector. But this attitude is illegal and improper in government procurements. We want to emphasize here that we are not talking about collusion or conspiracy to defraud the government. It is entirely possible that, on occasion, biased behaviour could benefit the government. That is of no matter. It is illegal and not to be condoned.

Timely, Accurate Client Reports. Technical personnel, engineers, computer scientists and the like, tend not to support active, timely reporting on progress to clients. They follow the mushroom growers to client interactions-"keep them in the dark and cover them with manure." That approach may work when things are moving well, but it runs the risk of forfeiting client confidence in troubled times. It seems better to report progress accurately and in a timely fashion, so that if slippages occur they are minor when first mentioned. Naturally the systems integrator should make every effort to stay on schedule, and if the schedule slips or a problem surfaces, the systems integrator should present the recommended solution at the same time the problem is first mentioned.

Prudential Judgement. Suppose the systems integrator has reason to believe that the client is unable or unwilling to handle setbacks in an objective manner. The parable of the king who "killed messengers who brought him bad news" would not remain current in our folklore if it did not have a basis in reality. Thus, reports of delays and difficulties should be brought to the attention of top management rather than directly to the client. This is the sort of prudential judgement call that should be handled by the top management within your organization rather than someone at the operating level. It is suggested that the matter be brought to the attention of top management within the organization as soon as possible and in a calm, factual manner.

Management of subcontractors is of special importance for systems integration involving large, complex engineered systems. It is highly likely that multiple subcontractors will be employee by the prime contractor. Prudent management of these subcontracts is critical to the success of the systems integration program (Grady 1994, 2010).

There are a number of key activities that must be completed by the systems integrator to assure integration of the products provided b the subcontractors prior to test and delivery of the final configuration. Some of the more important activities that must be accomplished include the following (Grady 1994, 2010):

- Organize overall team support for the subsystems integration and test activity, including personnel from various subcontractors.
- Validate incremental deliveries as these are made by subcontractor.
- Prepare the various subsystems for the test and evaluation prior to integration to assure performance meets the stated specifications.
- Integrate hardware/software (HW/SW) subsystems from subcontractors with systems developed by the corporation and legacy systems.
- Monitor test activity and assure that all tests conform to the systems testing regimens agreed to by the client.
- Provide for both Alpha and Beta site tests.
- Conduct necessary post-test activities to review outcomes with all concerned parties.
- Conduct formal reviews and review documentation.
- Provide for failure recovery and error correction in the event subcontractors are unable to meet design specifications.

The corporation must be able to demonstrate that it has gone about its business in a legal, objective, unbiased fashion. In large procurements it is often the case that outside contractors will be let for validation and verification and to develop and administer an audit trail relative to the prime contractor. The necessity for an external enterprise to create and follow a technical audit trail arises not so much from the need to respond to potential procurement difficulties as it does from a need to be able to demonstrate that an objective and unbiased procurement process was utilised. In the figure given below systems integration acquisition strategy is given (INCOSE, 2004):

Systems Integration Acquisition Strategy	
Specification Component	Auditing Component
I. *Functional Architecture Concept* Establish the general technical capabilities (i.e., the client functional needs).	*Validation Test Document Conceptual Discussion of:* 1. Traceability 2. Conflict resolution 3. Risk analysis and management 4. Consistency 5. Ambiguity evaluation 6. Testability 7. Constraints 8. Feasibility
II. *Technical Architecture Plan* Define the configuration categories.	*Validation Test and Audit Plan* For each configuration category, name and describe the relevant characteristics that delimit the requirement.
III. *Technical Component Specifications* Define and select the configuration components.	*Validation Test and Audit Implementation* For each configuration component, set down explicit functional and quantitative tests.
IV. *Contract(s)*	*Establish the Operational Requirements for Validation and Audit*

Fig. 4. Generic Technical Acquisition Strategy for a Systems Integration Viewpoint

The Validation Test Document will contain a conceptual discussion of items such as the following (NASA, 2007):

- Traceability
- Potential conflicts and resolution procedures
- Risk analysis and management
- Consistency of requirements
- Potential ambiguities in evaluation procedures
- Testability

1. *Traceability.* The fundamental requirement for the auditing component is traceability. This is classic requirement in all of engineering and in scientific efforts. All work must be written up on a regular basis in a laboratory notebook, dated signed, and witnessed. In engineering construction only registered professional engineers inspect and approve drawings. This seems to be a reasonable precaution when lives may be at stake when using the finished product. While the traceability and validation aspect of computer software is not as formal and rigid as in conventional engineering, the trend is undoubtedly in that direction.
2. *Potential Conflicts and Resolution Procedures.* At the Validation Test Document level, we do not identify specific technical conflicts and their solutions. At this highest level we expect only to see outlined the recommended procedure for resolving technical conflicts. This procedure should be formal, with a special form to be filled out if the conflict is not resolved at the first level discovered. Informal resolution of potential conflicts is the purpose of frequent peer reviews of the systems while it is under construction. Yourdon (1988) recommends this in his data flow method of design. But the idea of frequent peer reviews is a general tool and should be adopted in some form of team design and analysis. Peer review meetings should probably occur at least weekly, with any conflicts not resolved at that time being written up and forwarded to the first level of management. This should not be viewed as an additional burdensome administrative load; rather, it is simply what a group leader would do automatically in a management-by-exception environment.
3. *Risk Analysis and Management.* Risk analysis and management is also derived to the subcontractor from the systems integrator. Risk analysis and management process should be thought to the Subcontractors and with frequent peer reviews and coordinated meetings risks should be identified and managed to resolve.
4. *Consistency of Requirements.* Consistency of requirements would seem to be essentially similar to the previous issue of conflict resolution procedures and it may be taken as so if convenient. We separate the two simply to indicate that consistency of requirements can be checked at the general level, whereas conflicts sometimes occur in an unfortunate application of requirements that are not necessity inconsistent in them.

(a) Potential Ambiguities in Evaluation Procedures. In effect, a conflict is an error of omission. It is almost impossible to write a set of specifications for complex systems that is totally without conflict and ambiguity. Be that as it may, it is the job of systems integrators to produce a set of specifications that reduce ambiguity to a minimum, while at the same time remaining within the bounds of reasonableness as far as complexity goes.

(b) Testability. Testability is an absolutely necessary attribute or feature of a specification. If a specification is not testable, it is not really realistic. It is the job of the installation team or the validation component of the systems integrator effort to require a feasible test scheme for

each proposed specification. Some specifications can be validated or tested by simple observation. One can count the entry ports or disk drives or what have you. But other specifications are intrinsically impossible to complete until after final installation and break-in of the systems. The second level of the audit component is the Validation and Audit Plan. At this level the generic Validation Test Document produced in the first phase is refined and sharpened. For each configuration category, name and describe the relevant characteristics that delimit the requirement.

Then in the third audit component, Validation and Test audit Implementation, for each configuration component set down explicit functional and quantitative tests. At the fourth and final audit level, within the contract request for proposal, establish the operational requirements for validation and audit.

(c) Audit Reports and Sign-off. It is known how the auditing process proceeds. The procedures just discussed above establish the requirements for a complete audit trail, but only if the requirements are actually followed. Often, in practice, reality is far from the theoretical ideal. For example, program evaluation and review technique (PERT) and critical path method (CPM) charts are merely useless impedimenta if not maintained on a timely basis. We also know that documentation sometimes lags production by several cycles. Similarly, audit reports and sign-off will not be kept up to date and functional unless management insists. This is especially so in dealing with subcontractors and one can see why this is so. A subcontractor is paid to produce one or more deliverables. Paper records of any kind seem to some subcontractors to b a non functional and unnecessary.

For each of the activities, components "at risk" are identified, the risk aspects are analysed, the steps to avoid the risk and the ensuing consequences are taken, and management of the risk initiated, an internal processes and procedures are developed to address components at risk. In addition, the risk detection and identification plan is modified to incorporate similar occurrences of such risk, if these are not already to address components at risk. In addition, the risk detection and identification plan is modified to incorporate similar occurrences of such risk, if these are not already included in the plan. The risk management plan, as part of the overall strategic plan for the systems integration program, begins wit an analysis of the requirements at the onset of the program to ascertain if there are requirements statements that could jeopardize successful completion of the program (NASA, 2007).

The risk management plan continues with risk assessment for each of the phases of the systems integration life cycle. One of the most vexing problems in risk management is the early identification of potential causes of risk. This is especially true in the development of large, complex life-support systems and for large systems integration programs that are heavily dependent on the integration of legacy systems and newly developed requirements. What has made this problem particularly difficult has been necessity of using qualitative processes in an attempt to identify risk areas and risk situations. Risk detection and identification should commence with the issuance of requirements and the development of specifications. It is often assessing the risk assessment process is delayed until development of systems designs or even until procurements of major subsystems. This is fundamentally an untenable situation, since by this point in a program, investments of resources and personnel have been made, designs have been developed, and it is much too late to achieve an economical and efficient recovery without significant rework. This impact and ripple effect due to program elements at risk becomes known only after discovery of the nature and character of risk, thus jeopardizing the entire development program (Grady, 1994, 2010).

Consider the instance of systems and hardware and software requirements that may be at risk. If these requirements are found to be ambiguous, in conflict, incomplete, or changing too much (Requirement volatility), they may be considered to be a cause of risk to successful completion of the program. Any of these sources may in and of itself, be sufficient to jeopardize the entire program if not resolved.

6. Issues related with subcontractor arrangements

In the ideal world, a systems integrator group that has systems engineering management and program management group manages its subcontractors, each subcontract contains all the right requirements, and resources are adequate. In the real world, the technical team deals with contractors and subcontractors that are motivated by profit, subcontracts with missing or faulty requirements, and resources that are consumed more quickly than expected (Grady 1994, 2010). These and other factors cause or influence two key issues in subcontracting:

- Limited or no oversight of subcontractors and
- Limited access to or inability to obtain subcontractor data.

These issues are exacerbated when they apply to second-(or lower) tier subcontractors. Scenarios other than those above are possible. Resolutions might include reducing contract scope or deliverables in lieu of cost increases or sharing information technology in order to obtain data. Even with the adequate flow down requirements in (sub) contracts, legal wrangling may be necessary to entice contractors to satisfy the conditions of their (sub) contracts. Activities during contract performance will generate an updated surveillance plan, minutes documenting meetings, change requests, and contract change orders. Processes will be assessed, deliverables and work products evaluated, and results reviewed (De Mello Filho, 2005).

Systems engineering companies, who use an internal pool of technical resources to develop the entire program/project in their organization, need independent control and audit to their process. System's owners who select to use their internal resources and capabilities of their organization to perform the development process should obey the Systems Engineering Management process defined a Systems Engineering Process guidebook such as "INCOSE System Engineering Handbooks". Internal agreements in the organization should be written and signed between the customer and the systems development team as though they were procured from the outside. Moreover there should be independent review (by another division such as quality control assurance teams, agency, or independent consultant) of products and activities. In fact the development is done internally, an independent review team is recommended to provide a sanity check on the development process. This will create a healthy and clear perspective in the project and help to identify and manage project risks (De Mello Filho, 2005).

If the company uses an independent subcontractor in their development program/project then the control on the subcontracted service is performed by the contractor system integrator. Independent control of the system integrator carries the same responsibility as the independent control/audit of the consultants or agencies that use to select internal system development resources in the program/project development. However subcontracting a service/product then brought different problems with itself. Distributing the Systems Engineering process of the system integrator to the subcontractor, sharing the program schedule and program risks related to the subcontracted activity to the subcontractor are the important headlines in the subcontract activity.

7. Conclusion

One of the major problems that the Systems Engineering processes come across is how to deal with the subcontractors in order to assure activities of subcontractors are convenient and compliant with the systems engineering standard procedures and criteria. One of the most challenging job for a systems engineering team is to understand the needs and requirements of the customer and the constraints and variables that are established and the limits of business conduct that are acceptable for the particular job under contract. This understanding should directly reroute to the people who work under the subject contract of the customer. All of the requests, criteria and generic standards of customer needs associated with the subcontractor are directly written in the subcontracts statement of work or tasking contract too.

Systems Engineering Teams discussed how to implement and maintain an audit trail throughout the systems integration process and how to perform and record the details of the quality assurance process. Each of these activities carries special important on how it is implemented in systems integration approach with subcontractors that is engaged for assistance with the project or for procurement of hardware and software. Just as the customer provides the facilities with a set of requirements that it believes to be representative of the actual needs of the user. The corporation must prepare a detailed set of valid requirements for subcontractors. Absence of strategic plan on the part of a subcontractor should result in imposition of the systems integration organization strategic plan, especially those parts that related to audit trail maintenance; risk identification, formulation, and resolution; and such management process and procedures as we feel are essential for satisfactory performance the contract or subcontract.

8. References

Associate CIO of Architecture (2002), "Departmental Information Systems Engineering (DISE) Volume 1 Information Systems Engineering Lifecycle", DISE-V1-F1-013102, (2002)

David E. S., Michael B., Obaid Y. (2006), "Systems Engineering and Program Management-Trends and Costs for Aircraft and Guided Weapons Programs", United States Air Force, ISBN 0-8330-3872-9

De Mello Filho M. C., (2005),"A Whitepaper from IBM's Rational Software Division-Managing Subcontractors with Rational Unified Process", Rev 2.0, 2005, IBM

Global Intergy Corporation (2002), "Phase 1 Systems Engineering Management Plan, Volume No 1 Appendix A-A Process Review of the Accepted Standards and Best Practices for Developing Systems Engineering Process Models"

Grady O. J. (2010), "System Synthesis: Product and Process Design", Taylor & Francis, 2010

Grady O. J., (1994), "System Integration", CRC Press, 1994

INCOSE (2004), "Systems Engineering Handbook-A "What to" Guide for All SE Practitioners", INCOSE-TP-2003-016-02, Version 2a, (2004)

NASA (2007), "NASA Systemss Engineering Handbook", NASA/SP-2007-6105 Rev 1, 2007

Shamieh C., IBM (2011), "Systems Engineering for Dummies, IBM Limited Edition", Wiley Publishing, Inc, ISBN: 978-1-118-10001-1, (2011)

U.S. Department of Transportation (DoT) (2009), "Systems Engineering Guidebook for Intelligent Transportation Systems", Version 3, November 2009

14

A System Engineering Approach to e-Infrastructure

Marcel J. Simonette and Edison Spina
University of São Paulo, Escola Politécnica,
Computer and Digital Systems Department,
Brazil

1. Introduction

Electronic infrastructures (e-Infrastructures) are the basic resources used by Information and Communication Technologies. These resources are heterogeneous networks, which together constitute a large computing and storage power, allowing resources, facilities and services to be provided to the creation of systems in which communication and business operations are almost immediate, with implications in business organization, task management and human relations, forming a kind of patchwork of technologies, people and social institutions.

e-Infrastructure are present in several areas of knowledge, and they are helping the competitiveness of economies and societies. However, in order to continue with this paradigm, e-Infrastructures must be used in a sustainable and continuous way, respecting the humans and the social institutions that ultimately use them, demand their development and fund their paradigm.

This work presents an approach to deal with the interactions between e-Infrastructure technologies, humans and social institutions, ensuring that the emergent properties of this system may be synthesized, engaging the right system parts in the right way to create a unified whole, greater than the sum of its parts. The social components of this system have needs. The answers to these needs must not be associated with the engineering old philosophy of "giving the customers what they want", as the technology alone does not have a purpose; it is only a technological artifact. Technology has a purpose only when one or more humans use it to perform a task. This human presence in a e-Infrastructure System make it a complex system, because humans are diverse - multi cultural, multi generational multi skilled. This diversity can lead to differences between what is expected (planned) and the actual System behavior, and this variation is called complexity in this study.

Soft System Methods emerged as a way of addressing complex and fuzzy problems, the objectives of which may be uncertain. Soft methods are aimed at systems in which human and social institutions are present, these methods have an underlying concept and theory of systems, with which the Systems Engineering approach can focus on solving the customer's problem and provides all the customer needs, not only on what has been required (Hitchins, 2007).

e-Infrastructure design should have a holistic approach, seeking steps that ensure functional and failsafe systems, respecting humans and social institutions dimensions. This chapter is about Systems Engineering in the design of e-Infrastructure Systems, using Soft System Methods to develop a Systemic Socio-technical approach, crucial in order to identify the correct quality factors and expectations of the social infrastructure in an e-Infrastructure. The following sections, Dealing with Complexity, and e-Infrastructure as a Socio-technical System, introduce background information related to System Engineering and Socio-technical Systems. Next, the Soft System Method Approach section is about design process of systems in which human and socio institutions are present; in this section, the Consensual Methods are highlighted, and a perspective to a method selection is presented. Next, this chapter presents a Case Study, the design of an e-Infrastructure to be used by ALCUE Units of the Vertebralcue Project, from the ALFA III Program of the European Commission. A Conclusion section is followed by Acknowledgment and References.

2. Dealing with complexity

Problems arise in many ways, several problems are complex, difficult to be understood and analyzed; problems the solution of which is often only a "good enough" response, based on previous experience, common sense, and subjective judgment. Sometimes, the response to this kind of problem is just a change in the problem domain, so that the problem disappears.

Addressing problems is part of human nature. Humans have already faced numerous problems in history, and, especially after the Scientific Revolution, the approach adopted to deal with problems is to divide them into smaller parts, prioritizing and addressing the parts thought to be the most important first. Unfortunately, sometimes this approach fails, especially when it is necessary to deal with multiple aspects of a problem at the same time. When an aspect is prioritized, either it is not possible to have an understanding of emergent properties that may exist, or the problem can change in nature, emerging with another format. Neither scenario allow the identification of the existing complexity in the original problematic situation. Systems Engineers need to deal with complexity, identifying the interrelationships that exist in problematic situations, especially those related with human demands.

3. e-infrastructure as socio-technical system

The operation of e-Infrastructures depends both on the technology involved (developed by several engineering disciplines), and humans and social institutions interfaces (social interfaces), i.e., the operation depends on technological and social infrastructures. People, social institutions and technology result in a Socio-technical System, which has a social infrastructure and a technological infrastructure (Hitchins, 2007; Sommerville, 2007).

Although the Traditional Engineering methods with their reductionist approach, successfully address technological components and Human Factors (Chapanis, 1996; Nemeth, 2004; Sadom, 2004), these methods have difficulties in the treatment of the social infrastructure of e-Infrastructures Systems, both for addressing people and social institutions, which are often seen only as part of a context, without directly belonging to the System, treating human and social dimensions as constants, or some-times, ignores them (Bryl et al. 2009; Fiadeiro, 2008; Hollnagel & Woods, 2005; Nissenbaum, 2001; Ottens et al., 2006).

The social infraestrucure actors of an e-Infrastructure are more than system components, a part of the context, they want to optimize their decisions, considering their own subsystems, proposes and interests (Houwing et al., 2006).

4. Soft system method approach

There are several Systems Engineering approaches to address a solution to a problem. Nevertheless, Hitchins (2007) argues that the approach that makes use of Soft System Methods is the one that investigates the problem to be treated, looking for practical experiences and interactions with the problematic situation, trying to develop an understanding about the nature of problem symptoms and to propose solutions.

The use of Soft System Methods - a Soft Systems Approach - both allows the System Engineer to understand the problem domain, and helps him with the identification of social and human dimensions present in the problem domain. The former is because the activity to understand the problem domain is essentially an activity in which the components are human activities, and the second because there is an intrinsic complexity for accurately identifying human and social dimensions all along the System life.

The approach to go beyond Human Factors, and deal with the humans dimensions, is the use of the Soft System Approach with an evolutionary approach strategy. This approach deals with the interaction between Reality and Thought, and the interaction between Problem and Solution, it is represented at Figure 1 and was proposed by Soares (1986) as a way to understand, design, and implement solutions to a problematic situation.

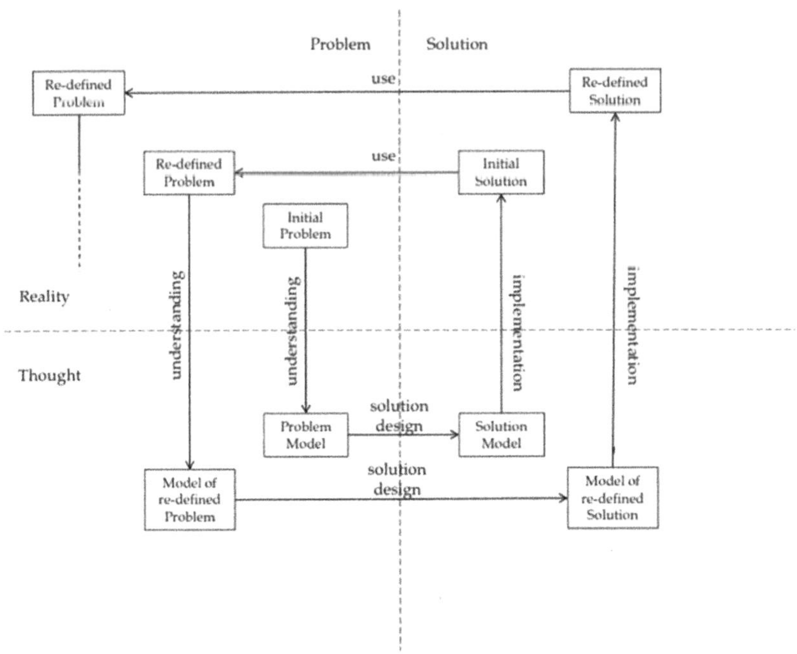

Fig. 1. Representation of the Evolutionary Spiral Approach.

From the two interactions - Reality x Thought and Problem x Solution, there are four actions that generate a cycle to treat a problem. These actions are: (i) *Understanding*: when the System Engineer develops an understanding, an abstract representation of the real problem, (ii) *Design*: when the System Engineer creates a response to the problem that satisfies the Problem in the Thought dimension, (iii) *Implementation*: the construction of the response to the problem in terms of Reality, (iv) *Use:* set up of a response to the Problem, in the environment of the Problem.

The set up of a response to a Problem may cause changes in Reality, emerging scenarios not previously determined, giving rise to new demands and a redefinition of the Problem. The treatment sequence of the problems leads to an Evolutionary Spiral as in Figure 1.

However, different from Soares, the authors of this chapter consider Solution not only as a response to a problem, but also as an overcoming restrictions, improvements in an existing Reality through actions to treat the problematic situation. Solution is an indicative of an improvement, a response that satisfies, but does not always solve, the problem, i.e., a response to the problem that is the best at that moment.

Although the identification of human and social dimension all along the System life is important to System success; the first action of the process - *Understanding* - is crucial.

4.1 Consensual methods

Understanding the Problem in the Reality dimension (Fig. 1) is the first step to determine the System construction possibilities. A proposal to develop this understanding and reduce users' dissatisfaction - respecting the human and social dimensions - is the use of Consensual Methods

Consensual Methods are not only about getting a consensus about a problem to be treated, it is also about getting the Systems Requirements from the people that have interests in the System. The consensual processes deal with the human activities involved in identifying the requirements and the human and social dimensions, reducing the discrepancy between the expected Systems features and the ones that will be perceived by the users.

Next, the Consensual Methods used by the authors in their work are listed. Hitchins (2007) stated that these methods are specifically meant to the front end of the Systems methodology, they are: Brainstorming, Nominal Group Technique, Idea Writing, Warfield's Interpretive Structural Modeling, Checkland's Soft System Methodology, Hitchins' Rigorous Soft Method.

4.1.1 Brainstorming

This method is an approach in which a selected group of people is encouraged by a moderator to come up with ideas in response to a topic or a triggering question.

4.1.2 Nominal Group Technique (NGT)

This method is similar to Brainstorming. A moderator introduces a problematic situation to a group of people and asks participants to write down their ideas about the problem on a sheet of paper. After a suitable time for people to generate their ideas, all participants read their ideas and the moderator, or an assistant, write them in a flip chart. With all the ideas written, the moderator conducts a discussion about these ideas, and then the participants are invited to rank all ideas. An idea-ordered list is generated and this constitutes the ideas that have been produced by the group as whole.

4.1.3 Idea writing

This method takes TGN a little farther. The moderator introduces the theme, and the participants are asked to write their ideas, suggestions, etc., on a piece of paper. After two or three minutes, the moderator asks each participant to pass his sheet on to another person, to pass the sheet to the second person on the left, for example. The one who receives the sheet can see the ideas already written, which may lead him (her) to a new set of ideas. After a short time, the moderator asks for the sheet recirculation, this time, to a different number of people. The process is repeated for about 30 minutes, or until the moderator notes that most people do not have any more ideas. There are two purposes in this strategy: encouraging ideas emergence within the working group and hiding the origin of a particular idea. The lists of ideas are worked later through Brainstorming or TGN to generate an action plan.

4.1.4 Interpretive Structural Modelling (ISM)

This method is similar to a computer-assisted learning process that enables individuals or groups to map complex relationships between many elements, providing a fundamental understanding and the development of action courses to treat a problem. An ISM session starts with a set of elements (entities) to which a relationship must be established. These entities are identified using any other method. The result of ISM is a kind of graph, where the entities are nodes and the relations are edges. The whole process can be time-consuming, especially when there are many divergences among the group members. Therefore, this time is important. It is essential for participants to understand and to recognize the each other' arguments, reaching a consensus.

4.1.5 Checkland's Soft Systems Methodology (SSM)

This method promotes the understanding of a problematic situation through the interaction between the people involved in the problematic situation. It promotes the agreement of the multiple problem views and multiple interests, and may be represented by a seven-stage model. Stages one and two explore the problematic situation (unstructured) and express it in a rich picture. Stage three is the root definition of the relevant systems describing six aspect of the problem, which are called CATWOE, they are: Customers, Actors, Transformation process, World view, Owner and Environment constrains. In stage four, the conceptual models of the relevant systems are developed, and, in stage 5, the conceptual model is compared with the perceptions of the real situation. In stage six, an action plan is developed for the changes, which are feasible and desirable; and in stage seven, the action plan is implemented. As a method developed from the Soft Systems Thinking, SSM does not produce a final answer to the problematic situation, it seeks to understand the problem situation and find the best possible response (Checkland, 2000).

4.1.6 Hitchins' Rigorous Soft Method (RSM)

As SSM, this method is based on the General-Purpose Problem-Solving Paradigm and is context free. The people who are experiencing a problem, and have knowledge about it, provide information about it in meetings with a coordinator. This investigation, which searches for dysfunction sources related to the problem, can create a lot of information and data. Differently from SSM, RSM employs tools and methods for treating, organizing and processing information; the action of "process" implies a gradual reduction of the problematic situation by ordering the data, transforming them into information for the treatment of the problem. RSM has seven steps: (1) *Nominate Issue & Issue domain,* in which the problem issues are indentified and a description of the situation is made; (2) *Identify Issue Symptoms & Factors,* that identifies the symptoms of the problem, and the factors that make them significant to be explored; (3) *Generate implicit systems,* each symptom implies the existence of at least one implicit system in the problem situation; (4) *Group into Containing System:* at this step, the implicit systems are aggregated to form clusters, one cluster for each symptom, named containing system, which can generate a hierarchy of systems, highlighting issues related to the problem; (5) *Understanding Containing Systems, interactions, imbalances*: at this step, the interactions between the containing systems are evaluated; (6) *Propose Containing Systems Imbalance resolution:* this step uses the differences between an ideal world, where the symptoms do not exist, and the real world, to propose Socio-technical solutions to the imbalances identified in the previous step; (7) *Verify proposal against original symptoms*: at this step, the system model are tested to see if they would, if implemented, eliminate the symptoms identified at step two and the imbalance found at step six. This model could also be tested for cultural acceptability by the people that are experiencing the problem (Hitchins, 2007).

4.2 Perspectives of consensual method selection

The diversity of people involved in an e-Infrastructure System development is a reality that Engineering must deal with. Zhang (2007) states that it is impractical to limit the diversity of people involved in a process to get a consensus about a problem to be treated. However, the methods to develop Systems requirements are under the Engineer's control.

Kossiakoff & Sweet (2003) stated that the function of System Engineering is to guide the Engineering of complex Systems, and that System Engineering is an inherent part of Project Management - the part that is concerned with guiding the Engineering effort itself. Kossiakoff and Sweet also propose a System Engineering life cycle model that corresponds to significant transitions in Systems Engineering activities, and it is the model adopted as the life cycle framework to this work. It has three broad stages: (i) *Concept Development Stage*: with the Needs Analysis, Concept Exploration and Concept Definition phases; (ii) *Engineering Development Stage*: with: Advanced Development, Engineering Design and Integration & Evaluation phases; (iii) *Post development* with the Production and Operation & Support phase.

The use of Consensual Methods to get a consensus about the problematic situation is a System requirements elicitation process. Consequently, a Consensual Method is a technique to implement the *Concept Development Stage;* thus, to be adherent to the System life cycle, the Consensual Methods must also provide information to other phases that are dependent on the requirement definition process. The information that is demanded by the following phases, and its purpose, is presented in Table 1.

The authors' experience in dealing with Consensual Methods has allowed the development of a comparison context, which considers if a Method complies with the demands of the Primary Purpose and the Inputs of each phase listed in Table 1.

	Main Activity	Primary Purpose	Inputs
Advanced Development	Risk Abatement	Identification and reduction of development risks.	System functional specification and defined system concept
Engineering Design	Component Engineering	Ensuring that individual components faithfully implement the functional and compatibility requirements.	System design specification and validated development model
Integration & Evaluation	System Integration	Ensures that all interfaces are fit and component interactions are compatible with functional requirements.	Test & Evaluation Plan and Engineered Prototype
Production	Production Process	Diagnosing the source of problems and finding effective solution.	Production specification and production systems
Operation & Support	Logistic Support System	Continuous training programs for operators and maintenance personnel.	Operation & Maintenance documents and installed operational system

Table 1. List of System Engineering life cycle phases after the Concept Development stage.

In Table 2, the adherence of each Consensual Method to System Engineering life cycle model phases is summarized. The first cell of the left column is a label that presents the level of adherence.

+++: Method recognizes the phase issues and provides means to deal with it; ++: Method supports the phase issues but not as strongly as before; +: Method addresses the phase need but weakly or indirectly; -: Method does not address the phase issues.	Brainstorming	Nominal Group Technique (NGT)	Idea Writing	Interpretive Structural Modeling (ISM)	Soft Systems Methodology (SSM)	Rigorous Soft Method (RSM)
Advanced Development	+++	+++	+++	+++	+++	+++
Engineering Design	++	++	++	+++	++	+++
Integration & Evaluation	-	-	-	+	++	+++
Production	++	++	++	-	++	+++
Operation & Support	-	-	-	+	+	+++

Table 2. Table of Method Selection.

Table 2 is illustrative, rather than comprehensive. It is based on empirical findings from the authors' experience. It provides a practical starting point for organizing an approach to identify the Consensual Method that complies with the demands of the System life cycle.

5. Case study: e-Infrastructure for an ALCUE unit

From the Perspective of Method Selection, RSM is the Consensual Method that provides more information for the phases of the System life cycle. As a Consensual Method, it promotes the consensus among people about the problem issues, so that people feel welcomed by the process. Of course, as Hitchins (2007) argues, people who feel dissatisfied with this approach are those who have no interest in consensus, who want to impose their worldview.

As a Case Study, the RSM is used to understand the problem of developing an e-Infrastructure to an ALCUE Unit, a kernel concept of Vertebralcue Project from the ALFA III Program of the European Commission. This Case Study also assessed whether the information obtained by RSM may actually contribute to other life system stages, according to the Perspective of Comparison of Consensus Methods.

5.1 The issue and its domain

KNOMA is designing an ALCUE Unit, and desires to develop and maintain an e-Infrastructure to support it.

As usually occurs in Engineering practice, the demand comes to the Engineer with words that are known by the people involved with the problematic situation, which the Engineer is still unaware of.

5.1.1 Issue

The concern about the e-Infrastructure to be developed and maintained is about what needs to be done. However, this depends on the features needed for an ALCUE Unit, which are not clear.

5.1.2 Domain

The Knowledge Engineering Laboratory (KNOMA) is a research laboratory of the Department of Computer Engineering and Digital Systems (PCS) of the School of Engineering (EPUSP) of the University of São Paulo (USP), and acts as a partner in projects sponsored by the European Commission (EC), including Vertebralcue from the ALFAIII Program of the EC.

Each project partner should develop and implement an ALCUE Unit (VERTEBRALCUE, 2011). These Units must operate independently from each other; however, they must be linked as "vertebras" of the framework, strengthening the academic cooperation networks that already exist between the project partners institutions, providing structural support for new partnerships and corporations networks. The Vertebralcue Project board stated that each ALCUE Units operate as an Information Center, broadcasting information about both the intuition and the region it belongs to. Likewise, the Unit must receive information from partner institutions for internal disclosure.

The ALCUE Unit operation deal with information and policy, as an academic collaborative process consists of multiple academic partners working together for information exchange and development of policy cooperation. In this operation process, there are interests of multiple actors: students, professors, researchers, and academic and social institutions. In the scenario of ALCUE Unit as an information center, there may be a distortion of information due to political interests, which can occur with pressures related to the disclosure of information or not. Uncertainty, diversity, quality and quantity of information are factors that can lead to a variation between the expected (planned) for a ALCUE Unit and the actual situation, perceived by the people who interact with the Unit, this variation is called complexity in this study.

5.2 Symptoms and Issue factors

The e-Infrastructure required for an ALCUE Unit depends on the purposes of the people who interact with the Unit. In order to indentify these purposes, meetings have been held with diverse groups of people who had interest in an ALCUE Unit. Furthermore, the Vertebralcue Project documentation and documents about the EPUSP academic cooperation was studied.

5.2.1 A Socio-technical System

e-Infrastructures are Socio-technical Systems. The technology in these Systems does not have a purpose by itself; this technology must meet the purpose of the people and institutions that interact with it. The difficulty in identifying the purpose of an ALCUE Unit can be seen by the description of the domain of the problematic situation.

The existence of a relationship between ALCUE Units and academic cooperation networks is evidence that there are different people's and institutions' interests in the System. This diversity of institutions and people, possibly with different cultures, makes it difficult to identify the specific System goals. Consequently, the identification of e-Infrastructure technological requirement is also made difficult.

5.2.2 Information center

The demand for an ALCUE Unit to be an Information Center is vague. As an Information Center, the Unit must both generate and disclose the information, and receive information and publish it. Nevertheless, before defining how the information will be received or generated, and how access will be provided to this information, it is necessary to identify what information is of interest to the people involved with the ALCUE Unit and what information is of interest to the academic cooperation networks. All this information has been identified by a Brainstorming session with the topic: "What subjects related to academic cooperation would you like to know?"

The Brainstorming session identified the following subjects: (i) Equivalence of titles between higher education institutions; (ii) Graduate and Undergraduate courses offered by institutions, including information about the disciplines and curriculum; (iii) Training programs and continuous education programs offered by institutions; (iv) Distance Learning; (v) Scholarships and funding of studies and research in institutions; (vi) Qualifications of faculty and researchers; and (vii) Mobility and exchange between institutions for faculty, students and researchers.

This list was not definitive; it was a first sample of what a group of people with interest in an ALCUE Unit had thought to be relevant at that stage of the problem treatment. Figure 2 presents the Brainstorming diagram that was created during the session. Diagrams were used in the Brainstorming session to improve communication and association of ideas.

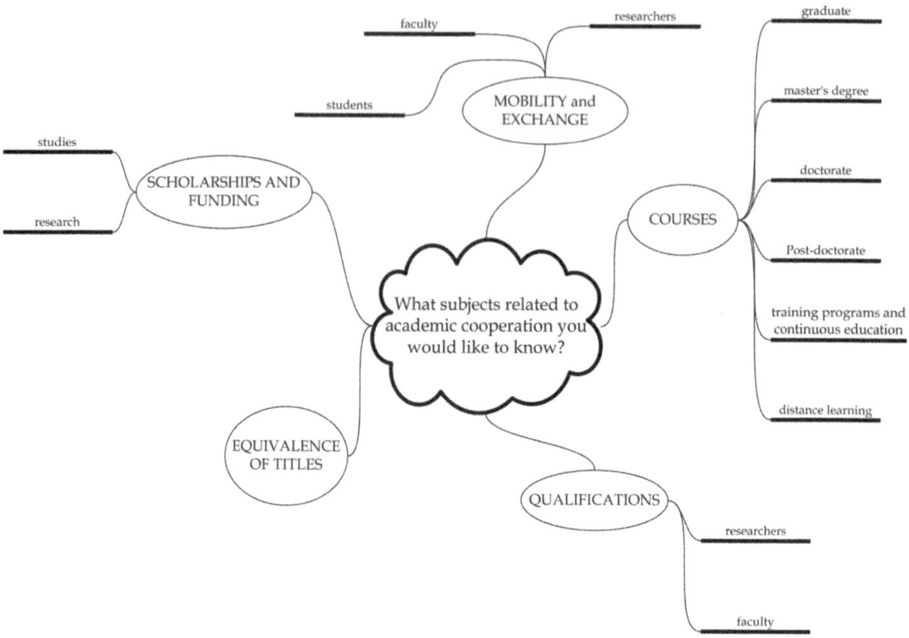

Fig. 2. Brainstorming diagram.

5.2.3 The relationships

The information, generated or received by the ALCUE Unit, occurs within a context with several institutions that have interests in academic cooperation. In order to identify some institutions, the Nominal Group Technique was used with the subjects that were identified in the Brainstorming session as a starting point. The Nominal Group session resulted in Table 2, in which the first column shows the identified institutions; the second column indicates if the institution is a funding institution, a support foundation, an academic institution, or an international cooperation institution. The third column was not identified in that session; it was identified only in the workshop that followed that session, and presents the characteristic of each type of institution.

The list of the institutions indentified in the Nominal Group session was used in a workshop, which aimed to build an institution chart and identify the relationship and information flow between them. In that workshop, the Interpretative Structural Modeling was used, and the work group decided to group institutions according to their characteristics - the results of which are present in the third column in Table 3. Figure 3 presents the institutions relationship and the information flow that was identified in the workshop.

INSTITUTION	TYPE	CHARACTERISTIC
Private Companies	Funding	Provides scholarships and grants, financial or not, for scientific and technological research.
European Commission	Funding	
Fundação de Amparo a Pesquisa do Estado de São Paulo - (FAPESP)	Funding	
Financiadora de Estudos e Projetos (FINEP)	Funding	
Fundação de Apoio à Universidade de São Paulo - (FUSP)	Support Foundation	Provides scholarships that are associated to research projects also provide institutional support to projects.
Fundação para o Desenvolvimento Tecnológico da Engenharia - (FDTE)	Support Foundation	
Universidade de São Paulo - (USP)	Academic	Belonging to the USP structure
Escola Politécnica da Universidade de São Paulo - (EPUSP)	Academic	
Departamento de Engenharia de Computação e Sistemas Digitais da EPUSP - PCS	Academic	
Laboratório de Engenharia do Conhecimento do PCS-EPUSP - (KNOMA)	Academic	
Comissão de Relações Internacionais da EPUSP - CRInt-POLI	International Cooperation	
Comissão de Cooperação Internacional (CCInt)	International Cooperation	
ALCUE Units	Academic Cooperation	Support academic networks at various levels: regional, national and international.

Table 3. Institutions with interests in academic cooperation.

5.2.4 Threats, opportunities, weaknesses and strengths

When the System Engineer deals with a problem such as the design of e-Infrastructure Systems to support the ALCUE Unit, he must not only be concerned about the needs to have the System operating according to the demands at the moment when he understands the problem domain. If the Engineer only considers these needs, the product of the design may be a System in which the changes and the evolutions required to meet new demands will be impossible. Therefore, to identify future scenarios for the ALCUE Unit, a situational analysis tool was used: the TOWS Matrix. This Matrix is a tool that allows the formulation of a strategy for the future by examining the present.

In a single workshop, the ALCUE Unit internal factors - Strengths and Weaknesses - and external factors - Threats and Opportunities - were identified and the relationship between them were established. Table 4 presents the result of this workshop: the TOWS Matrix.

5.3 Implicit systems

The Symptoms and Issue Factors imply the existence of Implicit Systems[1] in problematic situations. At this point in the RSM process, the needs of the ALCUE Unit that indicate the existence of Implicit Systems in the e-Infrastructure System are indentified.

Usually, skilled System Engineering can indentify Implicit Systems by the analysis and synthesis of the content in Figure 3, a rich picture - as in SSM - and the content in Table 4, the TOWS Matrix. The Implicit Systems identified by the authors are:

- System to store information: all the information obtained or generated should be stored for later access;
- System to support static disclosure: a system that allows access to information when people want it;
- System to support dynamic disclosure: a system that sends information to people who are interested in receiving them;
- System to support relationship networks: a system that allows the construction and operation of social and thematic networks;
- System for obtaining[2] information from FUSP: a system that accesses an interface at FUSP to retrieve information;
- System for obtaining information from FAPESP: a system that accesses an interface at FAPESP to retrieve information;
- System for obtaining information from Private Companies: a system that accesses an interface at a Private Company to retrieve information. There may be a different system for each Company that wishes to disclose information;
- System for obtaining information from FDTE: a system that accesses an interface at FDTE to retrieve information;
- System for obtaining and sending information to CRInt-POLI: a system that accesses an interface at CRInt-POLI to send and retrieve information;
- System for obtaining and sending information to CCInt: a system that accesses an interface at CCInt to send and retrieve information;
- System for obtaining and sending information to other ALCUE Units: a system that accesses an interface at another ALCUE Unit to send and retrieve information. There may be a different system for each ALCUE Unit.

---

[1] The authors consider that Implicit Systems are sub-systems of the e-Infrastructure System, but the term Implicit Systems is used to follow the RSM pattern.

[2] Another possibility would be to have Implicit Systems that receive information from these sources, which was discarded by the authors, because this involves a demand for work in the partner institution.

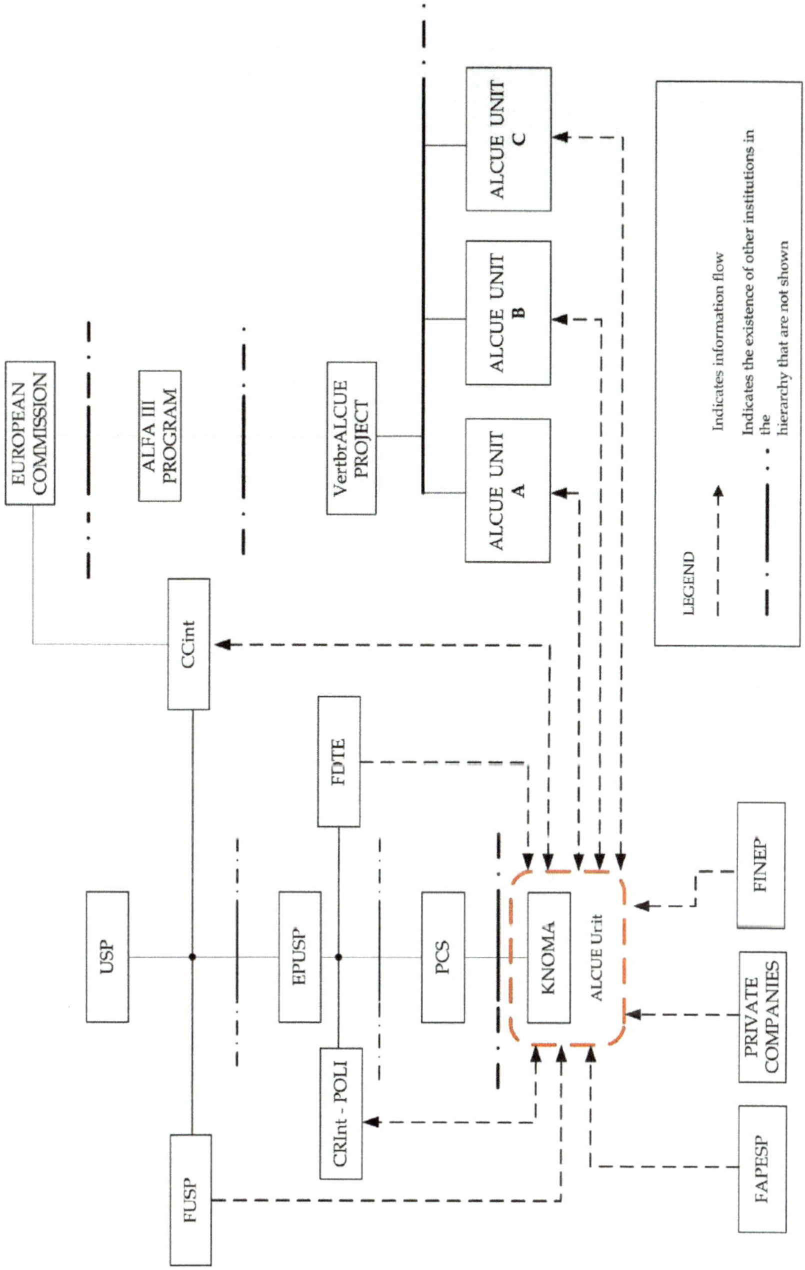

Fig. 3. Relationship between institutions.

		Internal Factors	
	TOWS Matrix	**STRENGTHS (S)** **S1**. The Unit is an access point that consolidates information from various institutions on academic networks of cooperation, interchanges, scholarships and research funding. **S2**. The Unit is a place of permanent dissemination of key policies and projects of academic cooperation. **S3**. The Unit can disseminate institutional contacts of several universities, not only of the institutions involved in the Vertebralcue Project.	**WEAKNESSES (W)** **W1**. Inefficient cooperation between institutions that promote academic cooperation. **W2**. The people with interest in academic cooperation do not know the real dimension and importance of the actual cooperation process and agreements. **W3**. There is a delay between the identification of the needs of the academic community and the identification of programs that already exist, or the creation of a new program. No tools are available to accelerate this process.
External Factors	**OPPORTUNITIES (O)** **O1.** Strengthening relationships between universities. **O2.** Strengthening relations between universities and society. **O3**. It is possible to use academic knowledge to develop and to implement an Information Center. **O4**. It is possible to access and to use advanced networking infrastructure, such as RNP (Rede Nacional de Ensino e Pesquisa) in Brazil, CLARA in Latin America, and Géant in Europe.	ALCUE Unit Strengths are based on information. Both for its potential to consolidate information from multiple sources such as the ability to be a permanent location to be used to access this information. It is possible to use the academic knowledge to develop techniques to organize and maintain the information. The social networks can be used for information dissemination, strengthening the relationship between universities and society. The communication structure can use the network infrastructure that interconnects several institutions in different countries for information dissemination. In short: **Strengths** S1 and S2 can be used to make good use of **Opportunities** O2, O3, and O4. **Strenghts** S2 and S3 can be used to make good use of **Opportunities** O1, O3, and O4.	The development of social networks and thematic networks that are specific for representatives of institutions can promote and make the relationship between institutions more efficient. On the other hand, the mere existence of social and thematic networks does not guarantee agility in identifying new needs, or agility in the treatment of these needs. No action has been identified to overcome **Weaknesses** W3, and take advantages of the **Opportunities**. In short: **Weaknesses** W1 and W2 can be overcome by the development of social and thematic networks specific for institutions representatives, to take advantage of **Opportunities** O1, O3, and O4. No actions can be taken to overcome **Weaknesses** W3, and take advantages of the **Opportunities**
	THREATS (T) **T1**. There is a concentration of few countries in academic interchanges. **T2**. There is an overlapping between the activities of the Unit and other institutions, which already have academic Information Systems. **T3**. There are other institutions that only take care of the dissemination of academic scholarships and funding. **T4**. Maintenance costs	All the Strengths can be used for information dissemination, contributing to more institutions and being aware of opportunities to participate in academics networks of cooperation, and scholarships and funding to students, faculty and researchers interchange. The Unit must create and maintain contact with other institutions, providing room for information dissemination. How the Strengths can be used in relation to any of the other Threats has not been identified. Notably, a way to ensure the ALCUE Unit sustainability after the end of the Vertebralcue Project was not identified. In short: **Strengths** S1, S2, and S3 can be used to protect against **Threat** T1. None of the **Strengths** can be used against **Threats** T2, T3, and T4	To overcome the **Weaknesses** that do not help in dealing with threats, the Unit can create a network of contacts for the promotion, development, and operation of cooperation and academic interchanges. This network can promote and coordinate the knowledge exchange about cooperation projects. The improvement of information exchange and dissemination of knowledge can reduce the concentration of interchanges between the same institutions. No action has been identified that could overcome **Weaknesses** W3, and take advantages of the **Opportunities.** In short: **Weaknesses** W1 and W2 can be used to help with dealing with **Threat** T1. No actions can be taken to overcome **Weaknesses** W3, to help in dealing with **Threats.**

Table 4. TOWS Matrix for ALCUE Unit

5.4 Containing systems

The authors have decided not to use any special technique of clustering to group the Implicit Systems in containing sets. Therefore, the Implicit Systems have been grouped together according to partners indentified in their own characteristics, in order to get sets of systems grouped by the symptoms of the ALCUE Unit e-Infrastructure. The resulting Containing Systems are:

- **Storage System:** System that contain as elements the following Implicit System:
  - System to store information.
- **Disclosure Support System:** System that contain as elements the following Implicit System:
  - System to support static disclosure;
  - System to support dynamic disclosure;
  - System to support relationship networks.
- **Information Gathering System:** System that contain as elements the following Implicit System:
  - System for obtaining information from FUSP;
  - System for obtaining information from FAPESP;
  - System for obtaining information from Private Companies;
  - System for obtaining information from FDTE.
- **Information Gathering/Dispatch System:** System that contain as elements the following Implicit System:
  - System for obtaining and sending information to CRInt-POLI;
  - System for obtaining and sending information to CCInt;
  - System for obtaining and sending information to other ALCUE Units.

The systems identified represent a perspective about the problematic situation in an ideal world. This means that they do not necessarily have to be designed and implemented in the real world. Furthermore, it does not mean that they are the only systems in the problematic situation. During the following phases of the System life cycle, new symptoms may appear that were not determined in this phase of the method execution, which can lead to a redefinition of the issue or the emergence of new issues. The sequence of treatments for these symptoms follows the concept of the previously mentioned Evolutionary Spiral.

5.5 Interactions and imbalances of containing systems

The interactions between Containing Systems always occur when there is an information related demand. These interactions are represented in Figure 4, in which the arrow indicates the direction in which information is being sent.

Following the concept of the Evolutionary Spiral (Fig. 1), a new workshop was held with the aim of assessing the interactions identified in reality dimension. At that meeting, it was identified:

- The **Disclosure Support System** contains the Implicit System that supports relationship networks, and this Implicit System also generates information to be stored.

- Two distinct Containing Systems - **Information Gathering System** and **Information Gathering/Dispatch System** - have Implicit Systems with the same characteristic: obtaining information in as institution. This scenario indicates a duplication of systems, even if the institutions are of different types, as identified in Table 2.

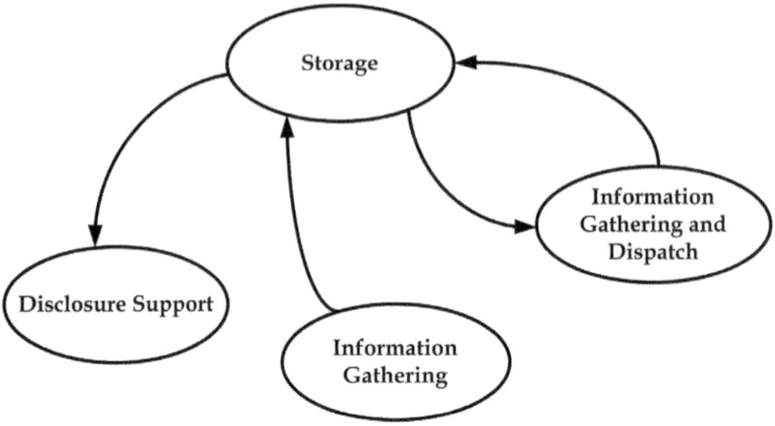

Fig. 4. Containing Systems Interaction.

5.6 Treatment for Imbalance and impact of the proposal

The new symptoms, identified in the workshop commented above, were considered in a new proposal for the Containing Systems, in which the **Information Gathering System** was merged with the **Information Gathering/Dispatch System.** The proposal also considered the symptom that the **Disclosure Support System** demands interactions with the **Storage System,** generating information that should also be accessed later by the system. This new scenario is depicted in Figure 5, where the arrows indicate the direction in which information is being sent.

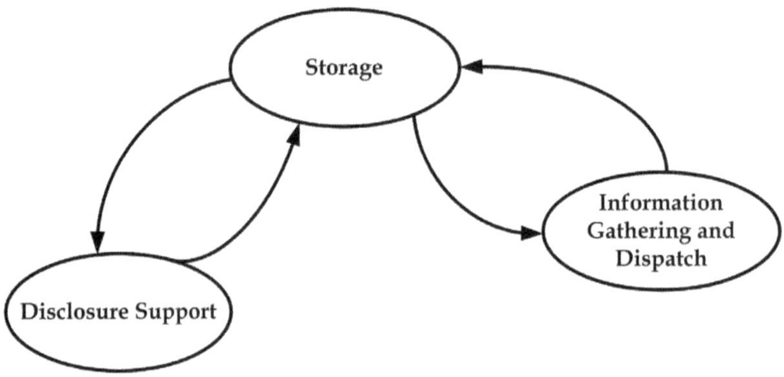

Fig. 5. Containing Systems Interaction, after the treatment of symptoms.

	Main Activity	Primary Purpose	Inputs	Contribution to phase
Advanced Development	Risk Abatement	Identification and abatement of development risks.	System functional specification and defined system concept	The rich picture (figure 3) and the contents of Matrix TOWS (Table 4) assist in understanding the concept of the system and its functionality.
Engineering Design	Component Engineering	Ensuring that individual's components faithfully implements the functional and compatibility requirements.	System design specification and validated development model	The Implicit Systems, the Containing Systems, and the rich Picture (Figure 3) assist the validation process.
Integration & Evaluation	System Integration	Ensure that all interfaces are fit and components interactions are compatible with functional requirements.	Test & Evaluation Plan and Engineered Prototype	The relationship between the Containing Systems (Figure 5), the rich picture (Figure 4), and the content of Matrix TOWS (Table- 4) assist the development of test plan and evaluation.
Production	Production Process	Diagnose the source of problems and find effective solution.	Production specification and production systems	The relationship between the Containing Systems (Figure 5) and the rich picture (Figure 3) assist the identification of dependences between the systems, which can determine the order of production of each one, or, if it is possible to determine if they can be produced at the same type. In addition, it provides elements to specify what must be produced.
Operation & Support	Logistic Support	Continuous training programs for operators and maintenance personnel.	Operation & Maintenance documents and installed operational system	The rich picture (Figure 3) assists the documentation of how the system should function.

Table 5. RSM Consensual Method outputs and contribution to System life cycle

5.6.1 Proposal impact

Store and make available information generated by social networks organized by the ALCUE Unit does not affect the **Storage Containing System.** Store information already was its original function.

The merge of the Containing Systems that was implemented may cause internal systems imbalances at the resulting system, because the different institutions with which the Implicit Systems are connected may demand different connection properties. However, in this phase of the System life cycle, it is too early to determine clearly this dependence scenario of connection, and "how" these connections with the different institutions will be held.

The purpose duplication of distinct systems was resolved.

5.7 Potential solution

The e-Infrastructure systems that KNOMA wishes to develop and maintain to support the ALCUE Unit activities is composed of three Containing Systems, which interact between themselves always that information is demanded or disclosed. The interaction between these systems is shown in Figure 5, in which arrows indicate the direction in which information is being sent.

5.8 Contribution to next phases of project life cycle

The process of RSM identified the symptoms and treatments of the issue on to develop and maintain an e-Infrastructure for ALCUE Unit. RSM has been chosen because according to the perspective presented earlier, it is the consensual method that provides more information for the phases that follows the requirement elicitation phase. Table 5 presents the contributions that the application of RSM brings to the phases of System Engineering life cycle model proposed by Kossiakoff and Sweet (2003).

6. Conclusion

This chapter addressed the use of Consensual Methods to assist the authors in the process of understanding a problematic situation: Design an e-Infrastructure to be used by KNOMA ALCUE Unit of VertebrALCUE Project, from ALFA III Program. According to the perspective adopted, the use of RSM provides information to all the phases of Project life cycle and was adopted. The meetings organized by the authors enabled the engagement of people with interest in the ALCUE Unit development, reduce the people dissatisfactions about the requirement elicitation process and respect the human and social dimensions. This scenario allows the development of a e-Infrastructure that minimized the difference between what is expected and what will be verified in reality. The authors decisions about the development of a TOWS Matrix was supported by VertebrALCUE Project board, which after evaluating the results obtained, demanded to all ALCUE Units the development of a TOWS Matrix.

7. Acknowledgments

The research and scholarships are partially funded by the Vertebralcue Project (http://www.vertebralcue.org). An ALFA III Program Project that aims to contribute to the development process of the regional integration among Latin American Higher Education Systems (HES´s), and the implementing process of the Common Area of Higher Education between Latin America, the Caribbean and the European Union (ALCUE in Spanish), by exploring and strengthening different levels of articulation of Latin America-Latin America and EU-Latin America academic cooperation through the design and implementation of a cooperation infrastructure at institutional, national and regional level.

8. References

Bryl, V.; Giorgini, P. & Mylopoulos, J. (2009). Designing socio-technical systems: from stakeholder goals to social networks. *Requirements Engineering*, Springer London, Vol. 14, n. 1,p. 47-70, Feb. 2009.

Chapanis, A. (1996). Human *Factors in Systems Engineering*, John Wiley & Sons, ISBN: 0471137820, New York.

Checkland, P. (2000). Soft Systems Methodology: A Thirty Year Retrospective. Systems Research and Behavioral Science Syst. Res. 17, S11–S58 , 2000

Fiadeiro, J. L. (2008). On the Challenge of Engineering Socio-Technical Systems. In: Software-Intensive Systems and New Computing Paradigms. ISBN 9783540894360. Heidelberg: Springer Berlin, v. 5380, p. 80-91, 2008.

Hitchins, D. K. (2007). Systems Engineering: A 21st Century Systems Methodology. John Wiley & Sons, ISBN 9780470058565, Chichester.

Hollnagel, E. & Woods, D.D. (2005). Joint Cognitive Systems: Foundations of Cognitive Systems Engineering, CRC Press, ISBN 9780849328213, Boca Raton.

Houwing, M; Heijnen P.W. & Bouwmans, I. (2006). Socio-Technical Complexity in Energy Infrastructures - Conceptual Framework to Study the Impact of Domestic Level Energy Generation, Storage and Exchange. In: Proceedings of the IEEE International Conference on Systems, Man, and Cybernetics. October 8-11, 2006, Taipei, Taiwan.

Kossiakoff A. & Sweet W. N. (2003), Systems Engineering Principles and Practice.: John Wiley & Sons Inc., ISBN 9780471234432. Hoboken

Nemeth, C.P. (2004), Human Factors Methods for Design: Making Systems Human-centered. CRC Press,ISBN: 0415297982. Boca Raton

Nissenbaum, H. (2001). How Computer Systems Embody Values, Computer, vol. 34, no. 3, pp. 120, 118-119, Mar. 2001, doi:10.1109/2.910905.

Ottens, M.; Franssen, M.; Kroes, P & Van de Poel, I. (2006). Modelling infrastructures as socio-technical systems. International Journal of Critical Infrastructures, 2006, Volume 2, No. 2/3, pp. 133-145.

Sandom, C. (2004), Human Factors for Engineers. Institution of Electrical Engineers, ISBN: 0863413293. London.

Soares, J. O. P. (1986), Especificação de Métodos de Desenvolvimento de Sistemas - Aplicação a Sistemas de Tempo Real. Dissertação (Mestrado) - Escola Politécnica, Universidade de São Paulo, São Paulo, Brasil.

Sommerville, I. (2007), Software Engineering. Addison-Wiley, ISBN: 9780321313799, Boston.

VERTEBRALCUE (September 2011), Project web site, presents its goals and activities. Available from http://www.vertebralcue.org

15

System Engineering Approach in Tactical Wireless RF Network Analysis

Philip Chan[2], Hong Man[1], David Nowicki[1] and Mo Mansouri[1]
[1]*Stevens Institute of Technology, Castle Point on Hudson, Hoboken, NJ,*
[2]*University of Maryland University College (UMUC), MD,*
USA

1. Introduction

Engineers dealing with different scaled and interconnected engineering systems such as tactical wireless RF communication systems have growing needs for analyzing complex adaptive systems. We propose a systemic engineering methodology based on systematic resolution of complex issues in engineering design. Issues arise which affect the success of each process. There are a number of potential solutions for these issues, which are subject to discussion based on the result assembled from a variety of sources with a range of measures. There are needs to assemble and balance the results in a success measure showing how well each solution meets the system's objectives. The uncertain arguments used by the participants and other test results are combined using a set of mathematical theory for analysis. This process-based construction helps not only in capturing the way of thinking behind design decisions, but also enables the decision-makers to assess the support for each solution. The complexity in this situation arises from the many interacting and conflicting requirements of an increasing range of possible parameters. There may not be a single 'right' solution, only a satisfactory set of resolution, which this system helps to facilitate. Applying systems engineering approaches will definitely help in measuring and analyzing tactical RF wireless networks, smart and innovative performance matrixes through tactical modeling and simulation scenarios may also be developed and enhanced. Systematic utilize of systems engineering approaches with RF electronic warfare modeling and simulation scenarios can support future research in vulnerability analysis of RF communication networks. RF electronic tactical models are used to provide a practical yet simple process for assessing and investigate the vulnerability of RF systems. The focus is also on tactical wireless network within a system of systems (SoS) context research area and to provide a comprehensive network assessment methodology. Researchers have proposed a variety of methods to build network trees with chains of exploits, and then perform normal post-graph vulnerability analysis. This chapter presents an approach to use mathematical Bayesian network to model, calculate and analyze all potential vulnerability paths in wireless RF networks.

2. Main methodology

Tactical wireless network vulnerabilities continually being reported and critically studied with many U.S. government organizations. The need for a comprehensive framework of network vulnerability assessment using systems engineering approach [24] [26] [27] [28] has been an increasing challenge to many research analysts. Researchers have proposed a more systematic way to manage wireless network nodes and trees with possible chains of events, and then perform normal post-graph vulnerability assessments with system of systems methodology. The most recent system engineering approaches are building attack trees by trying to number all potential attack paths with vulnerabilities identification, node probabilities calculations, inference analysis, and weights assignments by system experts. These are expert driven vulnerabilities analysis. Assessment and identification are one of the main key issues in making sure the property security of a given deployed tactical RF communication network. The vulnerability assessment process involves many uncertain factors reside within both the networks and the network nodes. Threat assessment or injecting threats is one of the major factors of evaluating a situation for its suitability to support decision-making and the indication of the security of a given tactical RF communication network system. One approach is using experienced decision makers database. This type of expert driven database recorded most of their decisions on vulnerability identification. The decision-makers use past experience for their decisions. The decision will be based upon previously good solutions that have worked in similar real life scenarios. The approach is to extract the most significant characteristics from the lay-down situation. Any similar situations and actions that have worked well in past cases will be considered in the assessment due to the present or the lack of certain essential characteristics. The assessment and identification is to create relevant relations between objects in the tactical RF network environment. Tactical communication RF wireless networks are best illustrated by Mr. David L. Adamy [11] in his book. Bayesian network (BN) and the related methods [17] is an effective tool for modeling uncertainty situation and knowledge. This paper discusses Bayesian's Theory [17], Bayesian networks and their ability to function in a given tactical RF communication network [11] for vulnerabilities analysis and identification. This short chapter presents an approach to use Bayesian network to model all potential vulnerabilities or attack paths in tactical RF wireless network. We will call such graph as "Bayesian network vulnerabilities graph" for a given tactical RF wireless network. It provides a more compact representation of attack paths than conventional methods. Bayesian inference methods can be used for probabilistic analysis. It is necessary to use algorithms for updating and computing optimal subsets of attack paths relative to current knowledge about attackers. Tactical RF wireless models were tested on a small example JCSS [12] network. Simulated test results demonstrate the effectiveness of approach.

2.1 Why systems engineering is used here

Systems engineering [7] [8] [27] [28] is applied here to assist the rapid design and development of complex systems such as tactical wireless communication systems. Systems engineering [29] uses engineering sciences techniques with operations research. Operations research also tackles with designing complex systems. Our goal is to utilize concurrent engineering principles in systems engineering analysis that covers our design goals and testing requirements in developing the RF communication system. The systems approach to solving complex problems are critical since integrating complex analysis and building of RF communication models requires synthesis of different methods. Systems approach is widely used and successful in fields of engineering, for example systems engineering. It is most effective in treating complex phenomena in tactical wireless RF communication networks. All this requires the use of modular views that clearly illustrate the component features of the whole system. The views may be put into different parts with proper interfaces. Extended knowledge may be gained about the parts in order to further understand the whole nature of a given tactical RF communication system. The system and its details in many levels may then be decomposed into several subsystems and into sub-subsystems, and so on, to the last details. In the same time, we can change focus to view different levels so that users are not overwhelmed by complexity. From time to time abstracts level information may be hided to gain focus on a certain task for detailed analysis. We may just simplify the system by treating some of its parts as black boxes except their interfaces. Hiding information for certain RF tactical analysis is not discarding it. The same black box can be opened at later time for other uses. Systems engineering can make a complex system more tractable and some of the parts can be studied or designed with minimal interference from other parts. All these protective measures can control defective designs and improves system level performance. The systems approach is effective not only for understanding or designing tactical RF wireless communication systems but also for abstract construction in mathematics and theories. Instead of an actual RF communication physical module, a RF wireless network "subsystem" can be a concept within a conceptual scheme and its "interfaces" can be relations to other in the scheme. Analyses and concepts are sometimes needed to approximate in the beginning. We can then refine approximations step by step towards a better answer with our method of analysis. Systems approach is not merely system-level approach but rather delving into lower-level subsystems. The system-level is powerful and appropriate in some cases, but it also misses out on most structures plus dynamics of the system and it is not employed in our systems approach, modularity study here. Systems approach is an integral part of systems engineering. Our analysis here may also call reduction, and "lessening" to yet finer information that also mean the importance of detailed analysis.

3. System of systems in tactical wireless network

In general, system of systems [9] [10] is a compilation of task-oriented or dedicated systems that bundle their resources and capabilities together to obtain a newer, more complex system that offers more functionality and performance than simply the summation of basic systems. Currently, system of systems is a critical research discipline that supplements engineering processes, quantitative analysis, tools, and design methods. The methodology to define, abstract, model, and analyze system of systems problems is typically referred to as system of systems engineering. We are going to define features for a system of systems that are unique for our study of tactical wireless communication system. The goal will be linking systems into joint system of systems allows for the interoperability and integration of Command, Control, Computers, Communications, and Information (C4I) and Intelligence, Surveillance and Reconnaissance (ISR) Systems as description in the field of information management control in modern armed forces. The system of systems integration is a method to pursue better development, integration, interoperability, and optimization of systems to enhance performance in future combat zone scenarios that related to area of information intensive integration. As one can predict that modern systems that comprise system of systems problems are not merely massive, rather they have some common characteristics: operational independence of the individual systems and managerial independence of the systems. System of systems problems are a collection of multiple domain networks of heterogeneous systems that are likely to exhibit operational and managerial independence, geographical distribution, and emergent and evolutionary behaviors that would not be apparent if the systems and their interactions are modeled separately. Taken together, all these background requirements suggest that a complete system of systems engineering framework is considered necessary to improve decision support for system of systems problems. In our case, an effective system of systems engineering framework for tactical RF communication network models are desired to help decision makers to determine whether related infrastructure, policy, and/or technology considerations are good, efficient, or deficient over time. The urgent need to solve system of systems problems is critical not only because of the growing complexity of today's technology challenges, but also because such problems require large resource commitments and investments with multi-years cost. The bird-eyes view using system-of-systems approach will allow the individual system constituting a system of systems that can be different and operate independently. The interactions expose certain important emergent properties. These emergent patterns have an evolving nature that the RF communication system stakeholders must recognize, analyze, and understand. The system of systems way of thinking promotes a new way of approach for solving grand challenges where the interactions of current technology, organization policy, and resources are the primary drivers. System of systems study is also integrated the study of designing, complexity and systems engineering with additional challenge of design. Systems of systems typically exhibit the behaviors of complex systems. However, not all complex problems fall into the area of systems of systems. System of systems by nature, are several combinations of qualities, not all of which are exhibited in the operation of heterogeneity networks of systems. Current research into effective approaches to system of systems problems includes: proper frame of reference, design architecture. Our study of RF communication network modeling, simulation, and analysis techniques will include network theory, agent-based modeling, probabilistic (Bayesian) robust design (including uncertainty modeling/management), software simulation and programming with multi-objective optimization. We have also studied and developed various numerical and visual tools for capturing the interaction of RF communication system requirements, concepts, and technologies. Systems of systems are still being employed predominantly in the defense sector and space exploration. System of Systems engineering methodology is heavily used in U.S. Department of Defense applications, but is increasingly being applied to many non-defense related problems such as commercial PDA data networks, global communication networks, space exploration and many other System of Systems application domains. System-of-Systems engineering and systems engineering are related but with slightly different fields of study. Systems engineering addresses the development and operations of one particular product like the RF communication networks. System-of-Systems engineering addresses the development and operations of evolving programs. Traditional systems engineering seeks to optimize an individual system (i.e., the target product), while System-of-Systems engineering seeks to optimize network of various interacting legacy and new systems brought together to satisfy multiple objectives of the program. It enables the decision-makers to understand the implications of various choices on technical performance, costs, extensibility and flexibility over time and the effective of methodology. It may prepare decision-makers to design informed architectural solutions for System-of-Systems context type problems. The objective in our research is to focus on tactical wireless network within a system of systems (SoS) context research area. The ultimate goal is to provide a comprehensive network assessment methodology and possible framework with systems engineering approach.

4. Approach with system engineering

Systems engineering [7] [8] [9] is employed here to look into wireless network vulnerabilities with simulation and modeling work-processes. Sets of useful tools are developed to handle the vulnerability analysis part of the RF wireless network. In the research, we have summarized a variety of methods to build network trees with chains of possible exploits, and then perform normal post-graph vulnerability assessment and analysis. Recent approaches suggest building more advanced attack trees by trying to number all potential attack paths with vulnerabilities identification, node probabilities calculations, inference analysis, weights assignments by system experts. Vulnerabilities analysis, assessment and identification are one of the key issues in making sure the security of a given tactical RF communication network. The vulnerability assessment process involves many uncertain factors. Threat assessment is one of the major factors of evaluating a situation for its suitability to support decision-making and the indication of the security of a given tactical RF communication network system. Systems engineering methodology in the research plays a critical role to help develop a distinctive set of concept and methodology for the vulnerability assessment of tactical RF communication networks. Systems engineering approaches have been developed to meet the challenges of engineering functional physical systems of tactical RF communication networks with complexity. The system engineering process employs here is a brand of holistic concept of system engineering processes. With this holistic view in mind, the systems engineering focuses are on analyzing and understanding the potential U.S. government customer needs. Re-useable RF connectivity models with requirements and functionality are implemented early in the development cycle of these RF communication network models. We then proceed with design synthesis and system validation while considering the complete problem, the system lifecycle. Based upon the concept by Oliver et al. [23], systems engineering technical process are adopted during the course of the research. Within Oliver's model [23], the technical process includes assessing available information, defining effectiveness measures, to create a behavior Bayesian vulnerabilities model, create a structure model, perform trade-off analysis, and create sequential build & test plan. At the same time, a RF communication system can become more complex due to an increase in network size as well as with an increase in the amount of vulnerabilities data, engineering variables, or the number of fields that are involved in the analysis. The developments of smarter matrices with better algorithms are the primary goals of the research. With disciplined systems engineering, it enables the use of tools and methods to better comprehend and manage complexity in wireless RF network systems for in-depth analysis. These tools are developed using modeling and simulation methodologies, optimization calculations and vulnerabilities analysis. Taking an interdisciplinary engineering systems approach to perform vulnerabilities analysis using Bayesian graph with weights calculation is inherently complex. The behavior of and interaction among RF wireless network system components can be well defined in some cases. Defining and characterizing such RF communication systems and subsystems and the interactions among them that supports vulnerabilities analysis is one of the goals of the research.

5. Insights behind research

Decision matrix is used for vulnerabilities analysis in the research. Decision matrix is an arrangement of related qualitative or quantitative values in terms of rows and columns. It allows our research to graphically identify, analyze, and rate the strength of relationships between sets of information in vulnerabilities. Elements of a decision matrix represent decisions based upon calculations and Bayesian network (BN) on certain vulnerabilities decision criteria. The matrix development is especially useful and critical for looking at large sample numbers of decision factors and assessing each factor's relative importance. Decision matrix employs in the research is used to describe a multi-criteria decision analysis (MCDA) for the tactical RF wireless network. When given a MCDA problem, where there are M alternative options and each need to be assessed on N criteria, can be described by the decision matrix which has M rows and N columns, or M × N elements. Each element, such as Xij, is either a single numerical value or a single grade, representing the performance of alternative i on criterion j. For example, if alternative i is "Wireless Node i", criterion j is "Background Noise" assessed by five grades {Excellent, Good, Average, Below Average, Poor}, and " Wireless Node i" is assessed to be "Good" on "Background Noise", then Xij = "Good". The matrix table 1 is shown below:

	Criterion 1	**Criterion 2**	**...**	**Criterion *N***
Alternative 1	x_{11}	x_{12}	...	x_{1N}
Alternative 2	x_{21}	x_{22}	...	x_{2N}
...	...	...	X_{ij} = Good	...
Alternative *M*	x_{M1}	x_{M2}	...	x_{MN}

Table 1.

5.1 Multiple criteria decision

Using a modified belief decision matrix, the research is now more refined and the matrix can describe a multiple criteria decision analysis (MCDA) problem in the Evidential Reasoning Approach. In decision theory, the evidential reasoning approach is a generic evidence-based multi-criteria decision analysis (MCDA) approach for dealing with problems having both quantitative and qualitative criteria under various uncertainties. This matrix may be used to support various decision analysis, assessment and evaluation activities such as wireless RF networks environmental impact assessment and wireless RF networks internal nodes (transceiver) assessment based on a range of quality models that are developed. For a given MCDA, there are M alternative options and each need to be assessed on N criteria, then the belief decision matrix for the problem has M rows and N columns or M X N elements. Instead of being a single numerical value or a single grade as in a decision matrix, each element in a belief decision matrix is a belief structure. For example, suppose Alternative i is "Wireless Node i", Criterion j is "Background Noise" assessed by five grades {Excellent, Good, Average, Below Average, Poor}, and "Wireless Node i" is assessed to be "Excellent" on "Message Completion Rate" with a high degree of belief (i.g. 0.6) due to its low Transmission Delay, low Propagation Delay, good Signal-to-Noise Ratio and low Bit Error

Rate. At the same time, the quality is also assessed to be only "Good" with a lower degree of confidence (i.g. 0.4 or less) because its fidelity and "Message Completion Rate (MCR) can still be improved. If this is the case, then we have Xij={ (Excellent, 0.6), (Good, 0.4)}, or Xij={ (Excellent, 0.6), (Good, 0.4), (Average, 0), (Below Average, 0), (Poor, 0)}. A conventional decision matrix is a special case of belief decision matrix when only one belief degree in a belief structure is 1 and the others are 0. The modified matrix table 2 is shown below:

	Criterion 1	Criterion 2	...	Criterion N
Alternative 1	x_{11}	x_{12}	...	x_{1N}
Alternative 2	x_{21}	x_{22}	...	x_{2N}
...	...	...	X_{ij}={ (Excellent, 0.6), (Good, 0.4)}	...
Alternative M	x_{M1}	x_{M2}	...	x_{MN}

Table 2.

5.2 Probability distributions

The research may help to develop a more systematic and automated approach for building "Bayesian network vulnerabilities graph" with weights assignment for vulnerability study in tactical wireless RF networks [11]. Bayesian network [17] is designed in vulnerabilities graph and models all potential attack steps in a given network. As describe by T. Leonard and J. Hsu [17], using Bayesian's rule as a special case involving continuous prior and posterior probability distributions and discrete probability distributions of data, but in its simplest setting involving only discrete distributions, the theorem relates the conditional and marginal probabilities of events A and B, where B has a certain (non-zero) probability as in (1):

$$P(A \mid B) = (P(B \mid A)P(A))/P(B) \tag{1}$$

Each term in the theorem has a conventional name: P(A) is the prior probability or marginal probability of A. It is "prior" in the sense that it does not take into account any information about B. P(A | B) is the conditional probability of A, given B. It is also called the posterior probability because it is derived from or depends upon the specified value of B. P(B | A) is the conditional probability of B given A. P(B) is the prior or marginal probability of B, and acts as a normalizing constant. The theorem in this form gives a mathematical representation of how the conditional probability of even A given even B is related to the converse conditional probability of even B when given even A. In our research, each wireless network node represents a single security and vulnerability point and contains property violation mode; each link edge corresponds to an exploitation of one or more possible vulnerabilities and each network path represents a series of exploits that can signify a potential vulnerability for attack within the RF wireless network. The communication model takes on characteristics of a tactical wireless RF network, and we consider an integrated posterior probability of Bayesian networks (BN) [17] with well-defined security metric represents a more comprehensive quantitative vulnerability assessment of a given tactical RF networks which contain different communication stages. Posterior probability is a revised probability that takes into account new available information. For example, let there be two stages within a given wireless transceiver. Wireless stage A having vulnerability or 0.35 accuracy due to noise factor and 0.85 accuracy due to jamming factor and wireless stage B having vulnerability or 0.75 accuracy due to noise factor and 0.45 accuracy due to jamming. Now if wireless stage is selected at random, the probability that wireless stage A is chosen is 0.5 (50% chance, one out of two stage). This is the a priori probability for the vulnerability of wireless communication stage. If we are given an additional piece of information that a wireless stage was chosen at random from the wireless network, and that the factor is noise, what is the probability that the chosen wireless stage is A? Posterior probability takes into account this additional information and revises the probability downward from 0.5 to 0.35 according to Bayesian's theorem. Also, the noise factor effect is more probable from stage B (0.75) than stage A (0.35). When the factor is jamming instead, the probability that the chosen wireless stage is A will be revised upward from 0.5 to 0.85 instead. Then, the vulnerability related jamming factor now is definitely less probable from stage B (0.45) than stage A (0.85). With conditional independence relationship encoded in a Bayesian network (BN) can be stated as follows: a wireless node is independent of its ancestors given its parents, where the ancestor/parent relationship is with respect to some fixed topological ordering of the wireless nodes. Using figure 1 below to demonstrate the outcomes, by the chain rule of probability with stages C, S, R & W, the joint probability of all the nodes in the vulnerabilities graph is now become: $P(C, S, R, W) = P(C) * P(S|C) * P(R|C,S) * P(W|C,S,R)$. By using conditional independence relationships, we can rewrite this as: $P(C, S, R, W) = P(C) * P(S|C) * P(R|C) * P(W|S,R)$ where we are allowed to simplify the third term because R is independent of S given its parent C, and the last term because W is independent of C given its parents S and R. We can see that the conditional independence relationships allow us to represent the joint more compactly. Here the savings are minimal, but in general, if we had n binary nodes, the full joint would require $\mathbf{O}({}_2n\ N)$ space to represent, but the factored form would require $\mathbf{O}(n\ {}_2k)$ space to represent, where k is the maximum fan-in of a node with fewer overall parameters.

5.3 Wireless communication models

In the model, we concern about the vulnerability of the wireless network caused by the failure of various communication stages in the wireless RF communication network. Figure 2 clearly presents the logical communication block diagram of our RF model. Each stage in a RF network is profiled with network and system configurations with exhibited vulnerabilities. They are identified through the breaking down of a given transceiver into transmitter and receiver with different stages. The purpose of our modeling and simulation goals is to make use the DISA JCSS Transceiver Pipeline stages [12]. All vulnerabilities data may be collected and the following information may be collected at run-time: (1) Effect of the transmission on nodes in the vicinity. (2) Set of nodes will attempt to receive the packet. (3) Determine a node attempting to receive a packet successfully. (4) Time it take for a packet to be transferred to the receiver. To start with the transmitter, we break down the transceiver into different radio pipeline stages. On the transmitter side, the transmitter has a

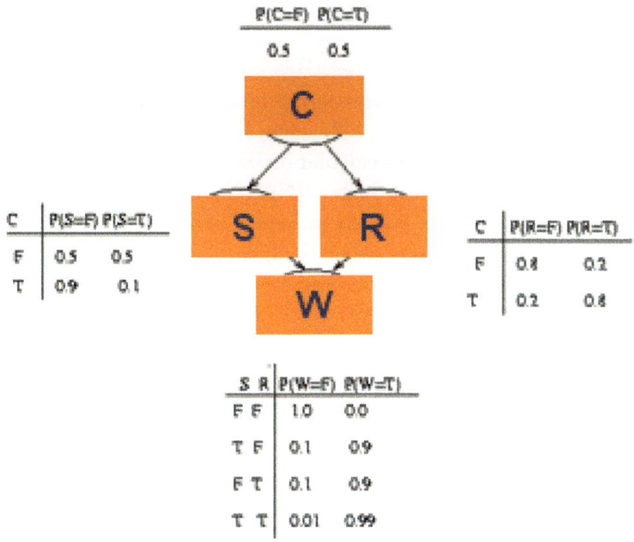

Fig. 1. Vulnerabilities graph (simple stage within a wireless node)

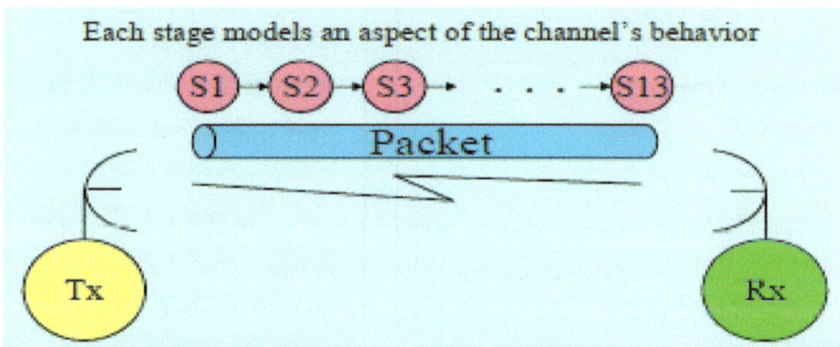

Fig. 2. JCCS pipeline stages are defined for a wireless communication model

"Group Receiver" start with the index "Group 0". The transmitter executed once at the start of simulation for each pair of transmitter and receiver channels or dynamically by OPNET JCSS's [12] Kernel Procedure (KP) calls. Inside the radio pipeline stages of the receiver side, for every receiver channel which "passed" the transmission checks, the simulated RF packet will "flow" through the pipe. Using JCSS [12] and OPNET Modeler, it is very critical to make sure the JCSS Radio Pipeline Model [12] attributes are being configured correctly. This is particular important for military RF radios like EPLRS [12] during a lay-down of network nodes in different scenarios. In all cases, configuration should be retained and saved in the node model. In summary, for Radio Transmitter, there are six (6) different stages (stage 0 to stage 5) associated with each Radio Transmitter. The following are six of the stages for a give Radio Transmitter (RT): Receiver Group, Transmission Delay, Link Closure, Channel Match, Transmitter (Tx) Antenna Gain and Propagation Delay. As for the Radio Receiver, there are altogether eight (8) stages (stage 6 to stage 13) that associated with a Radio Receiver (RR): Rx Antenna Gain, Received Power, Interference Noise, Background Noise, Signal-to-Noise Ratio,

Bit Error Rate, Error Allocation and Error Correction. In JCSS [12] and OPNET Modeler, there are altogether 14 Pipeline Stages (PS) that have implemented vulnerabilities graph for Bayesian networks (BN) [17] analysis. These are customized collections sequence of 'C' or 'C++' procedures (code & routines) with external Java subroutines and portable applications written for research purposes. In figure 2, each 14 different stages that comprised in a transceiver network perform a different calculation. For example in (1) Line-of-sight, (2) Signal strength & (3) Bit errors rates, Pipeline Stages (PS) code & routines are written in C, C++ and with external subroutine interfaces written in Java. Each procedure has a defined interface (prototype) with arguments typically a packet. Unlike most available vulnerability bulletins on public domains, we classify tactical wireless networks with vulnerabilities inside the 14 different stages of a given tactical wireless RF communication transceiver. So the vulnerabilities graph for a given tactical transceiver may be classified as vulnerabilities in Radio Transmitter are: (Vt1) Receiver Group, (Vt2) Transmission Delay, (Vt3) Link Closure, (Vt4) Channel Match, (Vt5) Transmitter Antenna Gain and (Vt6) Propagation Delay. On the hand the vulnerabilities for the Radio Receiver are: (Vr1) Rx Antenna Gain, (Vr2) Received Power, (Vr3) Interference Noise, (Vr4) Background Noise, (Vr5) Signal-to-Noise Ratio, (Vr6) Bit Error Rate, (Vr7) Error Allocation and (Vr8) Error Correction.

Vulnerabilities

precond:

Radio Transmitter:

(Vt1) Receiver Group

(Vt2) Transmission Delay

(Vt3) Link Closure

(Vt4) Channel Match

(Vt5) Transmitter Antenna Gain

(Vt6) Propagation Delay

prostcond:

(Vt1) Receiver Group = 0.99

(Vt2) Transmission Delay = 0.55

(Vt3) Link Closure = 0.65

(Vt4) Channel Match = 0.85

(Vt5) Transmitter Antenna Gain = 0.15

(Vt6) Propagation Delay = 0.25

Vulnerabilities

precond:

Radio Receiver:

(Vr1) Rx Antenna Gain

(Vr2) Received Power

(Vr3) Interference Noise

(Vr4) Background Noise

(Vr5) Signal-to-Noise Ratio

(Vr6) Bit Error Rate

(Vr7) Error Allocation

(Vr8) Error Correction

prostcond:

(Vr1) Rx Antenna Gain = 0.75

(Vr2) Received Power = 0.65

(Vr3) Interference Noise = 0.85

(Vr4) Background Noise = 0.10

(Vr5) Signal-to-Noise Ratio = 0.90

(Vr6) Bit Error Rate = 0.25

(Vr7) Error Allocation = 0.35

(Vr8) Error Correction = 0.40

Fig. 3. An example of vulnerabilities template for JCSS (transmitter / receiver pair) and related simulations.

Using the existing JCSS tactical RF hosts configuration and profile editors with wireless networking analysis tools [13] [14], we can construct generic, vulnerabilities graph and templates to describe possible exploitations conditions with certain vulnerabilities in a given transceiver and then on to a larger scale, a given tactical communication network's overall situation. Each template contains some pre-conditions and post-conditions of an atomic event related to the communication stage along with some security metric(s) information. A successful JCSS simulation will lead to better understanding for a more secure tactical RF communication model. Since we build vulnerability graphs using Bayesian networks (BN), we also assign probability of success after a failure in a pipeline stage's link-edge weight.

5.4 Algorithm within vulnerabilities graph

Specifying valid probability of communication in different stages requires domain expert knowledge. Most existing vulnerabilities scanning tools report those vulnerabilities with a standard set of categorical security measurements, such as severity level and vulnerability consequences. Therefore, considering the nature of a wireless network, one can define a more than one dimension security or vulnerabilities matrix using these categorical information and quantify levels of each category into numerical values for computation and comparison basis. Our approach is to make each matrix entry value related to each stage in a given transceiver. The result can then be computed and derived by a mathematical function that receives contributions from various dimensions like a normal linear addictive function $f(x + y) = f(x) + f(y)$ or multiplicative function $f(ab) = f(a)\, f(b)$. Then, it can be converted to a value within range [0,1] by applying a special scalar function. A function of one or more variables whose range is one-dimensional, this scalar function can be applied to the matrix. Such value may be represented the probability of a given vulnerability with respect to the transceiver. For example, One can define a two dimension m × n security matrix W = (wij), with one dimension wi to denote severity levels and another dimension wj to denote ranges of exploits. A 3-scale severity level may be specified as {high = 0.95, medium = 0.65, low = 0.35}, and 2-scale exploit ranges may be specified as {remote = 0.55, local = 0.95}. If applying a multiplicative function to the matrix, then each entry value is given by wij = wi × wj. Our research constructs Bayesian vulnerabilities graphs with our graph generation and mapping routine by matching a list of stages in a given transceiver on a wireless network with profile information against a library of computed vulnerabilities specified node characteristic templates. For any vulnerability, if all pre-conditions are met, values of post-condition attributes are updated with an edge that is assigned with weight. It is then added to the vulnerability graph. The most common task we wish to solve using Bayesian networks (BN) is probabilistic inference. For example, consider the network G with a current vulnerability status W, and suppose we observe the fact that G with a status of W. There are two possible causes for this: either it is due to factor R, or the due to factor S is on. Which is more likely? We can use Bayesian's rule to compute the posterior probability of each explanation (where 0==false and 1==true).

$$\Pr(S=1 \mid W=1) = \frac{\Pr(S=1, W=1)}{\Pr(W=1)} = \frac{\sum_{c,r} \Pr(C=c, S=1, R=r, W=1)}{\Pr(W=1)} = 0.2781 / 0.6471 = 0.430$$

$$\Pr(R=1 \mid W=1) = \frac{\Pr(R=1, W=1)}{\Pr(W=1)} = \frac{\sum_{c,s} \Pr(C=c, S=s, R=1, W=1)}{\Pr(W=1)} = 0.4581 / 0.6471 = 0.708$$

where

$$\Pr(W=1)=\sum_{c,r,s}^{c,r,s}\Pr(C=c,S=s,R=r,W=1)=0.6471$$

Pr(W = 1) is a normalizing constant, equal to the probability (likelihood) of the data. So we see that it is more likely that the network G will have a status of W, because of the weight in factor R is more than factor S: i.e. the likelihood ratio is 0.7079/0.4298 = 1.647. With variable elimination techniques illustrated below and using vulnerabilities graph in figure 4, we use Bayesian networks (BN) with Bucket Elimination Algorithm implementation in the models with belief updating in our scenarios, to the most probable explanation. We need to provide vulnerability values in each communication stage within each transceiver plus the network scores on the entire tactical network. Finding a maximum probability assignment to each and the rest of variables is a challenge. We may really need to maximizing a posteriori hypothesis with given evidence values, finding an assignment to a subset of hypothesis variables that maximize their probability. On the other hand we may need to maximize the expected utility of the problem with given evidence and utility function, finding a subset of decision variables that maximize the expected utility. Any other consideration is Bucket Elimination Algorithm. It may be used as a framework for various probabilistic inferences on Bayesian Networks (BN) in the experiment. Finally, a RF Vulnerability Scoring System (RF-VSS) analysis is in development. It is based upon the Common Vulnerability Scoring System [22] and associates with additional features of Bayesian networks [17] (also known as belief network) that in turn yields a more refined belief decision matrix and the matrix can then describes a multiple criteria decision analysis (MCDA) with evidential reasoning approach for vulnerabilities analysis of a given tactical wireless RF network.

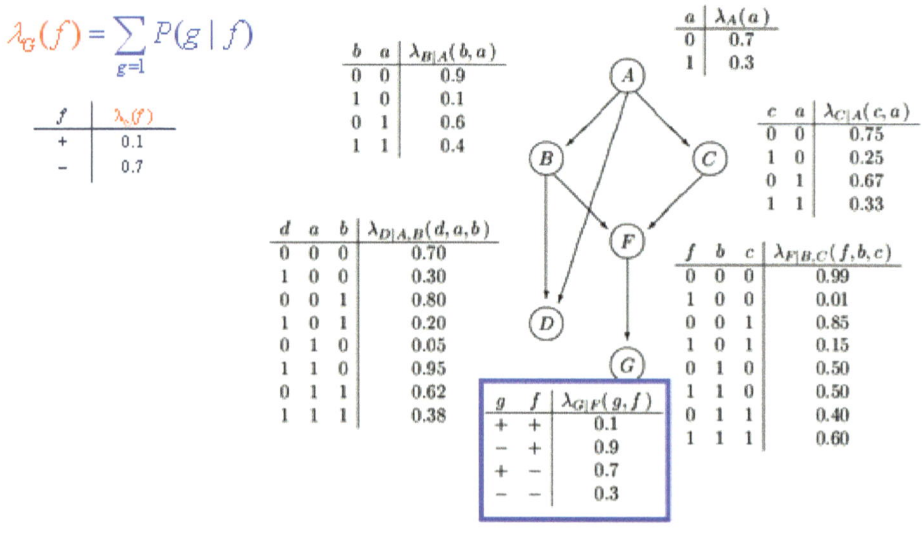

Fig. 4. Use of Bucket Elimination Algorithm within vulnerabilities graph

6. Result generated from sample experiments

For simplicity in terms of network radio analysis, we provide here a rather simple two (2) nodes wireless RF network scenarios that are communicating with each other via UDP protocol. A more complex one is illustrated in figure 5b. Using some of the available wireless networking analysis toolkits [13] [14] as in figure 5a, a set of JCSS EPLRS Scenarios with a link being jammed. Packets were being captured and exported into Microsoft EXCEL spreadsheet. Jamming occurs between 2 wireless links for this network: EPLRS_6004 and EPLRS_6013. EPLRS_6013 transceiver model was changed to a special EPLRS EW network vulnerability model as in figure 5c. The receiver link was intentionally jammed (by increase the noise level to an extremely high value, i.e. the vulnerabilities within one of the wireless stage are increased by many fold) so that no more simulated packet will be "successful" in getting through from EPLRS_6004 to EPLRS_6013 and the results are listed and illustrated in figure 5d with some sample data.

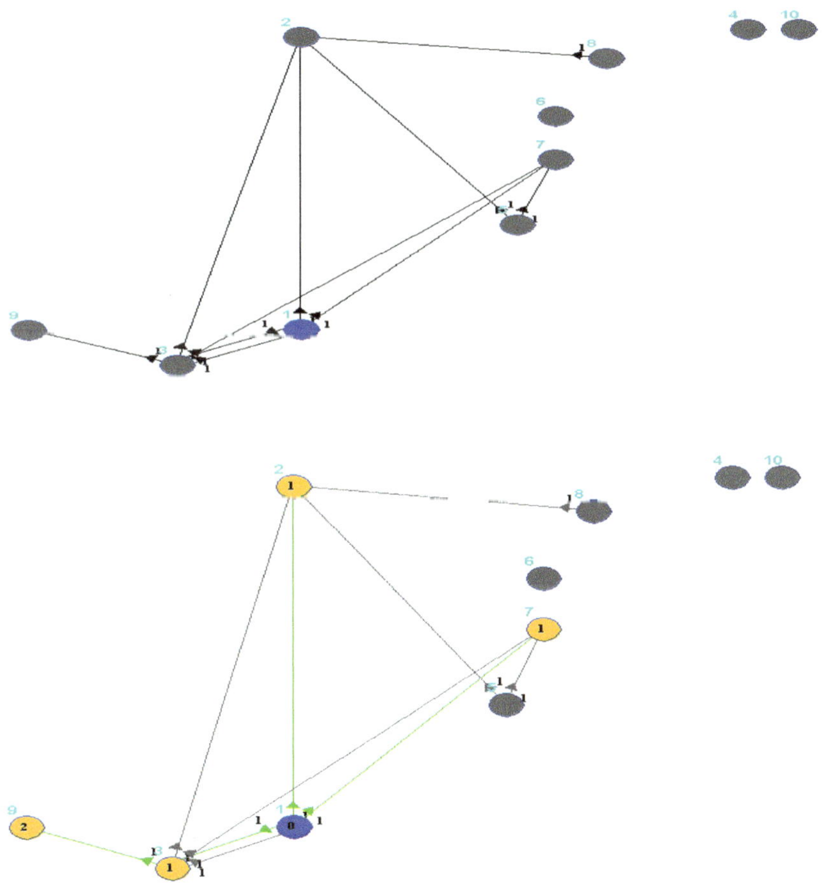

Fig. 5a. Before and after scenarios using wireless networking analysis toolkits in Java

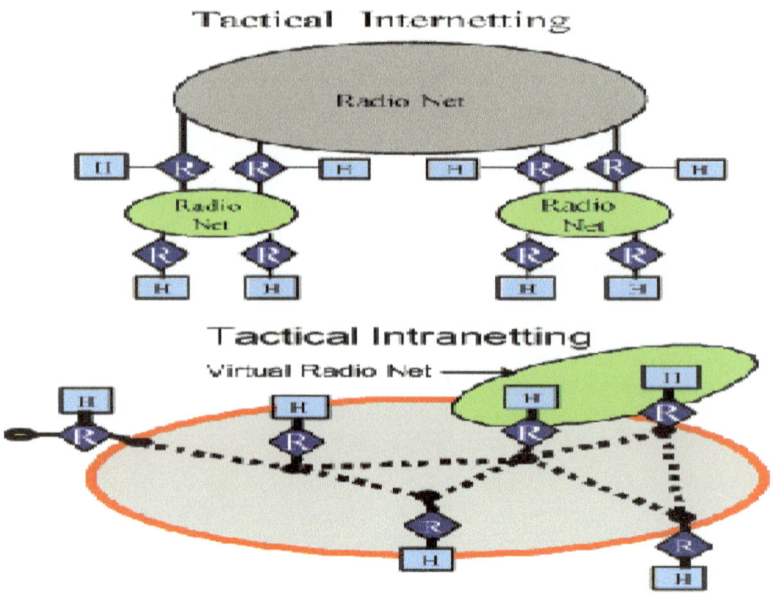

Fig. 5b. Wireless RF Networks

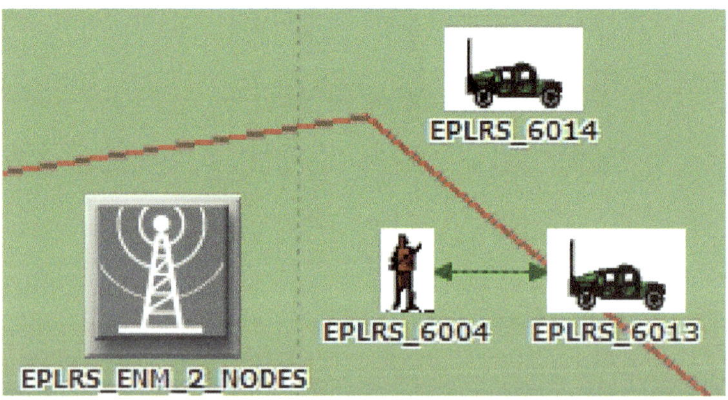

Fig. 5c. Two wireless nodes network

7. Future possibilities

Bayesian Analysis [17] – the Bayesian's Theorem looks at probability as a measure of a state of knowledge, whereas traditional probability theory looks at the frequency of an event happening. In other words, Bayesian probability looks at past events and prior knowledge and tests the likelihood that an observed outcome came from a specific probability distribution. With some sample field data the Bayesian's Theorem can be applied including wireless RF communications & computer networking science in tactical military applications. The research presented here is for building a set of "Bayesian network vulnerabilities graph" for vulnerability study in tactical wireless RF networks. Bayesian network is designed in vulnerabilities graph and model all potential attack steps in a given network. Each wireless network node represents a single security property violation mode; each link edge corresponds to an exploitation of one or more possible vulnerabilities and each network path represents a series of exploits that can signify a potential vulnerability for attack within a tactical RF wireless communication network. Inference is played a major part in our vulnerability calculations. Future research work will involve looking into different kinds of Bayesian's network (BN) with advanced topological arrangements as in figure 6 below with multiple experts and multiple factors analysis for our more advanced JCSS wireless RF vulnerabilities analysis.

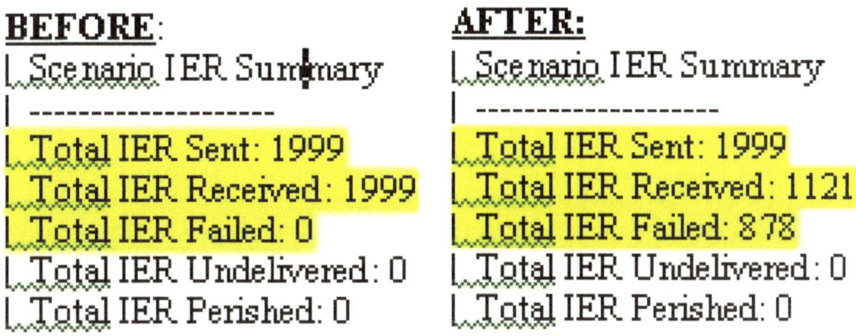

BEFORE:
| Scenario IER Summary
| --------------------
| Total IER Sent: 1999
| Total IER Received: 1999
| Total IER Failed: 0
| Total IER Undelivered: 0
| Total IER Perished: 0

AFTER:
| Scenario IER Summary
| --------------------
| Total IER Sent: 1999
| Total IER Received: 1121
| Total IER Failed: 878
| Total IER Undelivered: 0
| Total IER Perished: 0

Fig. 5d. Sample results generated by JCSS scenarios

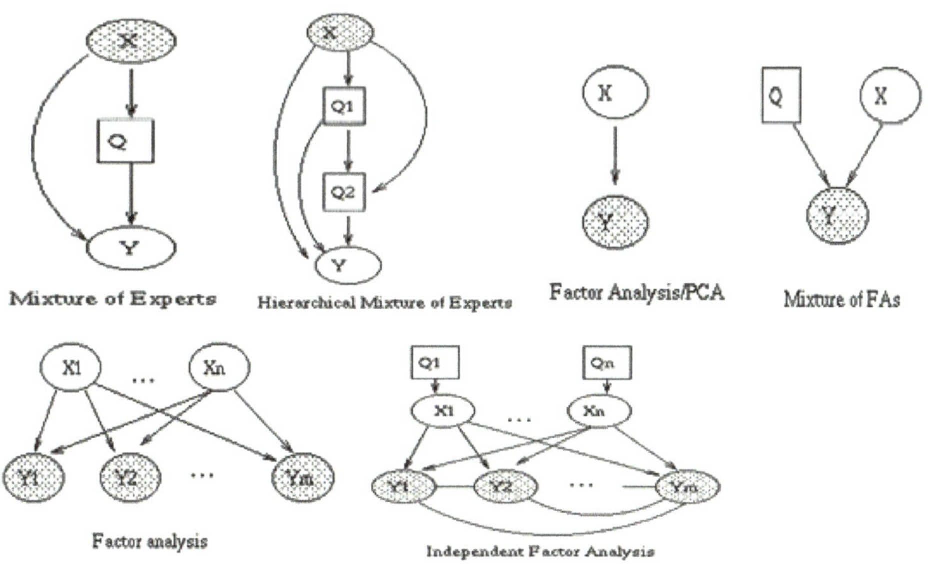

Fig. 6. Multiple experts and multiple factors analysis

7.1 Adaptive Bayesian network and scoring system

Finally, we may consider an adapted Bayesian network (BN) of wireless tactical network analysis with a RF Vulnerability Scoring System (RF-VSS) that can generate weighted scores in the research. The Common Vulnerability Scoring System (CVSS), a NIAC research project from U.S. Department of Homeland Security. This rating system is designed to provide open and universally standard severity ratings of vulnerabilities in certain specific systems. It creates a global framework for disclosing information about security vulnerabilities. The CVSS may be recognized and generally accepted by the public in support, international coordination and communication to ensure successful implementation, education and on-going development of the scoring system. It serves a critical need to help organizations appropriately prioritize security vulnerabilities across different domains. A common scoring system has the advantages of solving the similar problems with better coordination. Based upon the Common Vulnerability Scoring System develops by Peter Mell et al. [22], we think this is a very valuable, useful tool and scoring system for quickly assessing wireless RF security and vulnerabilities. RF-VSS scores are derived from three scores: a "base network" score, an "adversaries impact" score, and an "environmental impact" score. These can better be described as "fixed" score, "external variable" score, and "wireless RF network experts" assigned score. The base network system score is fixed at the time the vulnerability is found and its properties do not change. The base assigned score includes numerous scoring metrics. Each of these metrics will then be chosen from a pre-determined list of options. Each option has a value. The values are then fed into a formula to produce the base network score. Next comes the temporal or adversaries impact score. The adversaries impact score changes and revises the base network score up or down. The temporal or adversaries impact score can also change over time (thus it is "time sensitive"). For example, one of the component metrics of the adversaries impact score is System Remediation Level (SRL). This means, there exists a possible common defense fixes out there, maybe from a contractor or vendor or an emergency research workaround. If, when the detected vulnerability is first encountered, there may be no possible fix, then the temporal or adversaries impact score will be much higher. But when a solution or fix is possible, then the score will go down dramatically. Again, it was temporary and a changing factor. There are three possible vulnerabilities metrics that make up the temporal or adversaries impact score. This score is then multiplied by the base network score to produce a new score. This first computed new score will be produced based upon the current operating wireless RF network scenarios set up via background expert diagnostic. The final part is the environmental impact score. This is how the final vulnerability will affect the wireless RF network. The researchers get to determine how the combined vulnerabilities might affect the overall wireless RF network in field deployment. If the vulnerability has very little risk or to do with all the listed factors then this computed score will be very, very low (like zero). There are five metrics that affect the environmental impact score. This portion is combined with the base network and temporal adversaries impact score to produce a final score. The score will be on a scale of 1-10. If it is a low 2, then don't be too worried. However, a rather higher score like 6 or above might indicate major security issues in terms of security. We will provide a vulnerabilities smart index by constructing a novel calculator with a set of RF Vulnerability Scoring System (RF-VSS) for final system vulnerability analysis. For an example: For a given wireless RF radio network, according to expert released analysis and advisory, there are a set of "RF wireless network vulnerabilities" being assigned. The example metrics for the given wireless RF network scenarios with vulnerabilities are: (1) base network impact, (2) temporal or adversaries' impact and (3) Environmental impact.

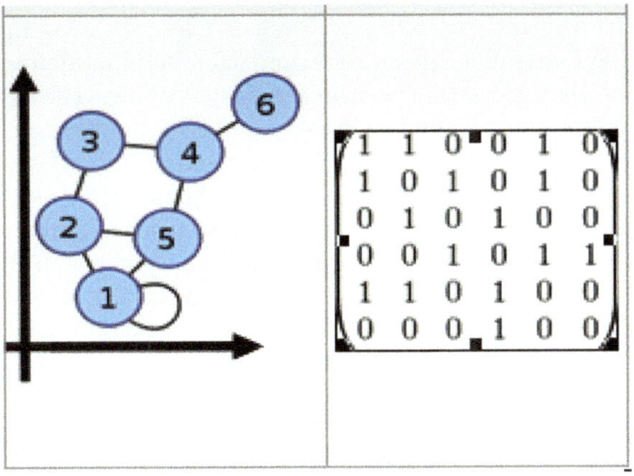

Fig. 7. Transposing the vulnerabilities graph into a matrix for analysis

So, overall a base RF wireless network vulnerability score of 8.8 (very bad) that is slightly mitigated to 7.9 by the temporal or adversaries metrics. Still, 7.9 is not a great score and still has considerable amount of risk. Now, this is where the final environmental impact score comes in to alter the landscape. The negative impact may be bad for the overall wireless RF network when we look at the environmental impact metrics calculated before for certain wireless network scenarios as illustrated above. We gather all those factors into the RF Vulnerability Scoring System (RF-VSS) calculator and it produces an environmental score of 6.5 which translates into high vulnerabilities. This is a relatively good approach to determine what the overall risk is for a give wireless RF network and the RF Vulnerability Scoring System (RF-VSS) analysis is based upon the Common Vulnerability Scoring System develops by Peter Mell [22] and associates with additional features of Bayesian networks [17] (also known as belief network). Using adjacency-matrix as a starting point, a more quantitative wireless RF network vulnerability assessment may be achieved. An adjacent edge counts as 1 unit in the matrix for an undirected graph as illustrated in figure 7. (For example a given X, Y coordinates that are numbered below from #1 to #6 may be transposed into a 6x6 matrix.)

8. Conclusion

A possible framework with systems engineering approach [7] [8] is utilized. The ultimate goal is now partially achieved by providing a comprehensive network assessment methodology. Our study illustrates using system engineering thinking, Bayesian networks [17] can be applied during the analysis as a powerful tool for calculating security metrics regarding information system networks. The use of our modified Bayesian network model with the mechanisms from CVSS is in our opinion an effective and sound methodology contributing towards improving the research into the development of security metrics by constructing a novel calculator with a set of RF Vulnerability Scoring System (RF-VSS) for final system vulnerability analysis. We will continue to refine our approach using more dynamic Bayesian Networks to encompass the temporal domain measurements established in the CVSS. This short paper demonstrated an approach to model all potential vulnerabilities in a given tactical RF network with Bayesian graphical model. In addition, using a modified belief decision matrix, the research can describe a multiple criteria decision analysis (MCDA) using Evidential Reasoning Approach [3] [4] [5] [6]. It was used to support various decision analysis, assessment and evaluation activities such as impact and self assessments [1] [2] based on a range of quality models. In decision theory, evidential reasoning approach (ER) is generally a evidence-based multi-criteria decision analysis (MCDA) for dealing with some problems having both quantitative and qualitative criteria with various uncertainties including ignorance and randomness. With evidential reasoning approach, a generic evidence-based multi-criteria decision analysis (MCDA) approach is chosen for dealing with problems having both quantitative and qualitative criteria with variables. This matrix may be used to support various decision analysis, assessment and evaluation activities such as wireless RF networks environmental impact assessment and wireless RF networks internal nodes (transceiver) assessment based on a range of quality models that are developed. Bayesian vulnerabilities graphs provide comprehensive graphical representations with conventional spanning tree structures. The Bayesian vulnerabilities graph model is implemented in Java, and it is deployed along with JCSS software. JCSS is the Joint Net-Centric Modeling & Simulation Tool used to assess end-to-end communication network capabilities and performance. It is the Joint Chiefs of Staff standard for modeling military communications systems. JCSS is a desktop software application that provides modeling and simulation capabilities for measuring and assessing the information flow through the strategic, operational, and tactical military communications networks. Our new tool can generate implement vulnerabilities network graph with link edges and weights. All these may be transposed into an adjacency-matrix as illustrated before for a more quantitative wireless RF network vulnerability assessment. The convention followed here is that an adjacent edge counts as one in a matrix for an undirected graph as illustrated before in figure 7. For a given X, Y coordinates, for instant; they can be numbered from one to six and may also be transposed into a 6x6 matrix. The vulnerabilities analysis with the help of system engineering approach [25] [26] [29] of a wireless RF network is then achieved by assigning corresponding measurement metrics with posterior conditional probabilities of Bayesian network [17]. The Bucket Elimination algorithm is adapted and modified for probabilistic inference in our approach. The most common approximate inference algorithms are stochastic MCMC simulation, bucket algorithm and related elimination steps which generalizes looping and aggregated belief propagation, and variation methods. A better approximate inference mechanism may be deployed in the near future for more complex vulnerabilities graph. Our method is very applicable to tactical wireless RF networks by picking, implementing each model's communication stages and states. The result when using with OPNET JCSS [12] simulation and modeling will provide both graphical quantitative and real assessment of RF network vulnerabilities at a network topology state and during time of actual deployment.

9. Acknowledgment

The authors thank Dr. John V. Farr, Dr. Ali Mostashari, Dr. Jose E. Ramirez-Marquez of the Department of Systems Engineering, Stevens Institute of Technology for many fruitful discussions in the early stages, the referees for helpful comments, for providing outstanding intellectual environments and significant guidance. We appreciate the assistance of Dr. Jessie J. Zhao who proof read some of the technical materials. Finally, we also thank the countless effort and insight contributed to this research by Dr. Mung Chiang, Professor of Electrical Engineering Department from Princeton University for his excellent technical skills, strategies and valuable advice in applying conventional and distributed convex optimization in the area of wireless communication network.

10. References

[1] Wang Y.M., Yang J.B. and Xu D.L. (2006). "Environmental Impact Assessment Using the Evidential Reasoning Approach". *European Journal of Operational Research,* 174 (3): 1885–1913.

[2] Siow C.H.R., Yang J.B. and Dale B.G. (2001). "A new modeling framework for organizational self-assessment: development and application". Quality Management Journal 8 (4): 34–47.

[3] Keeney, R. & Raiffa, H. (1976). Decisions with Multiple Objectives. Cambridge University Press. ISBN 0521438837.

[4] Shafer, G.A. (1976). Mathematical Theory of Evidence. Princeton University Press. ISBN 0691081751.

[5] Yang J.B. & Xu D.L. (2002). "On the evidential reasoning algorithm for multiple attribute decision analysis under uncertainty". IEEE Transactions on Systems, Man and Cybernetics Part A: Systems and Humans 32 (3): 289–304.

[6] Xu D.L., Yang J.B. and Wang Y.M. (2006). "The ER approach for multi-attribute decision analysis under interval uncertainties". European Journal of Operational Research 174 (3): 1914–43.

[7] Popper, S., Bankes, S., Callaway, R., and DeLaurentis, D., System-of-Systems Symposium: Report on a Summer Conversation, July 21-22, 2004, Potomac Institute for Policy Studies, Arlington, VA.

[8] Manthorpe Jr., W.H., "The Emerging Joint System-of-Systems: A Systems Engineering Challenge and Opportunity for APL," Johns Hopkins APL Technical Digest, Vol. 17, No. 3 (1996), pp. 305–310.

[9] Li, B.; Xu, Y. & Choi, J. (1996). Applying Machine Learning Techniques, *Proceedings of ASME 2010 4th International Conference on Energy Sustainability*, pp. 14-17, ISBN 842-6508-23-3, Phoenix, Arizona, USA, May 17-22, 2010

[10] Kotov, V. "Systems-of-Systems as Communicating Structures," Hewlett Packard Computer Systems Laboratory Paper HPL-97-124, (1997), pp. 1–15.

[11] Luskasik, S. J. "Systems, Systems-of-Systems, and the Education of Engineers," Artificial Intelligence for Engineering Design, Analysis, and Manufacturing, Vol. 12, No. 1 (1998), pp. 55-60

[12] Adamy, D. L. "EW103: Tactical Battlefield Communications Electronic Warfare", Artech House, ISBN-13: 978-1-59693-387-3, 2009.

[13] JCSS. The Joint Net-Centric Modeling & Simulation Tool. JCSS Project Manager, JCSS@disa.mil Commercial: (703) 681-2558.

[14] Chan P., U.S. Army, ARL patent (pending) - ARL Docket No. ARL 06-37. "Network Security and Vulnerability Modeling & Simulation Libraries".

[15] Chan P., U.S. Army, ARL patent (pending) - ARL Docket No. ARL 10-09. "Wireless RF Network Security and Vulnerability Modeling & Simulation Toolkit - Electronic Warfare Simulation & Modeling of RF Link Analysis with Modified Dijstrka Algorithm".

[16] Swiler, Phillips, Ellis and Chakerian, "Computer-attack graph generation tool," in Proceedings of the DARPA Information Survivability Conference & Exposition II (DISCEX'01), vol. 2.

[17] Liu Yu & Man Hong, "Network vulnerability assessment using Bayesian networks," Data Mining, Intrusion Detection, Information Assurance, and Data Networks Security 2005.Proceedings of the SPIE, Volume 5812, pp. 61-71 (2005).

[18] Leonard T. & Hsu J., "Bayesian Methods: An Analysis for Statisticians and Interdisciplinary Researchers," Cambridge University Press, ISBN 0-521-00414-4, 1997.

[19] Sheyner, Lippmann and J. Wing, "Automated generation and analysis of attack graphs," in Proceedings of the 2002 IEEE Symposium on Security and Privacy (Oakland 2002), pp. 254–265, May 2002.

[20] Dijkstra E., Dijkstra's algorithm. Dutch scientist Dr. Edsger Dijkstra network algorithm. http://en.wikipedia.org/wiki/Dijkstra's_algorithm

[21] Phillips & Swiler, "A graph-based system for network-vulnerability analysis," in Proceedings of the 1998 workshop on New security paradigms, pp. 71–79, January 1999.

[22] Ammann, Wijesekera and S. Kaushik, "Scalable, graph-based network vulnerability analysis," in Proceedings of 9th ACM conference on Computer and communications security, pp. 217–224, November 2002.

[23] Mell & Scarfone, "A Complete Guide to the Common Vulnerability Scoring System Version 2.0", National Institute of Standards and Technology. http://www.first.org/cvss/cvss-guide.html#n3.

[24] Systems Engineering Fundamentals. Defense Acquisition University Press, 2001.

[25] Chan P., Mansouri M. and Hong M., "Applying Systems Engineering in Tactical Wireless Network Analysis with Bayesian Networks", Computational Intelligence, Communication Systems and Networks (CICSyN), 2010 Second International Conference, Publication Year: 2010 , Page(s): 208 - 215.

[26] Defense Acquisition Guidebook (2004). Chapter 4: Systems Engineering.

[27] Bahill, T. & Briggs, C. (2001). "The Systems Engineering Started in the Middle Process: A Consensus of Systems Engineers and Project Managers". in: Systems Engineering, Vol. 4, No. 2 (2001)

[28] Bahill, T. & Dean, F. (2005). What Is Systems Engineering?

[29] Boehm, B. (2005). "Some Future Trends and Implications for Systems and Software Engineering Processes". In: Systems Engineering, Vol. 9, No. 1 (2006)

[30] Vasquez, J. (2003). Guide to the Systems Engineering Body of Knowledge - G2SEBoK, International Council on Systems Engineering.

[31] Lima, P.; Bonarini, A. & Mataric, M. (2004). *Application of Machine Learning,* InTech, ISBN 978-953-7619-34-3, Vienna, Austria

16

Creating Synergies for Systems Engineering: Bridging Cross-Disciplinary Standards

Oroitz Elgezabal and Holger Schumann
Institute of Flight Systems, German Aerospace Center (DLR)
Germany

1. Introduction

The increasing complexity of technical systems can only be managed by a multi-disciplinary and holistic approach. Besides technical disciplines like aerodynamics, kinematics, etc. cross-disciplines like safety and project management play an immanent role in the Systems Engineering approach. In this chapter, standards from different cross-disciplines are discussed and merged together to elaborate synergies which enable a more holistic Systems Engineering view.

After this introductory section, definitions of the terms *system* and *complexity* are given and the problems associated with the development of complex systems are introduced. The third section presents existing development philosophies and procedures. Additionally the mentioned cross-disciplines are introduced together with international standards widely established in the respective fields. Because the selected standards are not only complementary but also overlapping, the fourth section describes the harmonization approach carried out, together with the resulting holistic view. This combination of the standards enhances the benefits of the "traditional" Systems Engineering approach and solves many of the mentioned problems associated to the development of complex systems by taking also project management and safety aspects into a deeper and therefore, more holistic, account.

2. Background

The concept *system* has been defined in multiple ways since Nicolas Carnot introduced it in the modern sciences during the first quarter of the 19th century. Most of the definitions assigned to it are based on the Greek concept of *"σύστημα systēma"*, which means: *a whole compounded of several parts or members, literally "composition"*. An example of the remanent influence of the original *system* concept on the modern one is the definition provided by Gibson et al. (Gibson et al., 2007) which defines a system as *a set of elements so interconnected as to aid driving toward a defined goal.*

As an extension to the concept *system*, the term *complex system* is interpreted very broadly and includes both physical (mostly hardware and software) groupings of equipment to serve a purpose, and sets of procedures that are carried out by people and/or machines (Eisner, 2005). In complex systems, characteristics and aspects belonging to different fields of expertise interact with each other. The factors which make a system to be complex are the interactions and interdependencies between the different components of a system. Those dependencies are not always obvious, intuitive or identifiable in a straightforward way. Especially, keeping a perspective of the whole system, together with all its implications in big projects, is complicated if not almost impossible at all. Even if the size is not a determinant factor for complexity, complex systems tend to be relatively large, with lots of internal and external interfaces. Additionally, in complex systems other kinds of considerations than those purely technical come frequently into play like political interests, international regulations, social demands, etc.

2.1 Problems associated with complex systems development

The development of complex systems implies other kinds of problems apart from those directly related with the different technical fields involved in it. Eisner summarizes in (Eisner, 2005) some of the problems associated with the design and development of complex systems. Eisner further classifies those problems into four different categories: Systems-, Human-, Software- and Management-related problems. Fig. 1 lists the mentioned problem categories together with their respective problems associated with the development of complex systems.

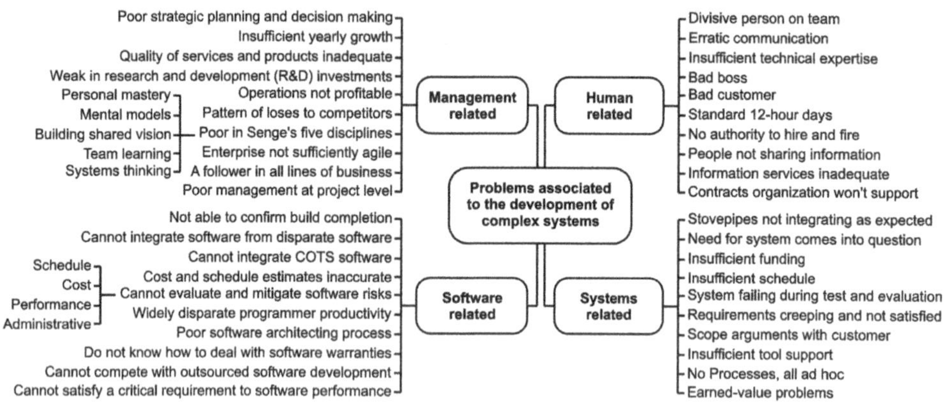

Fig. 1. Problems associated with the development of complex systems

As a consequence of all those problems, the efficiency during system development process decreases, which in fact can lead to a loss of money or project cancellation, both due to lower productivities. Besides, this efficiency decrease can result in higher project risks i.e. violation of deadlines or project failure during system verification phase due to poor system quality.

Another critical point associated with problems which belong to the previous classification like: *Erratic communication, People not sharing information, Requirements creeping and not validated, No processes, all ad hoc* and *Poor software architecting process* is the fact that they make the achievement and maintenance of traceability very difficult. Traceability is a key source of know-how in every company since it condensates the rationale behind every decision made during the system design process. Traceability is also vital for finding the location of design and production failures in case they are detected internally or reclamations from customers take place. Finally, in the case of safety-related systems, it is a mandatory requirement for system certification as well as for failure and accident investigation.

All these problems result in a poor execution of system development processes which, in case they get established in the every-day working methodology of a company, could even threat the profitability and continuity of the company itself.

3. Existing development philosophies and procedures

3.1 System development philosophies

The development of systems in general, and of technical systems in particular, has been carried out since the foundation of engineering sciences, or even earlier. During that time, many different terms like systems analysis and systems integration have been used to make reference to the concept represented by the modern Systems Engineering approach. Currently two philosophies with different focuses are applied in the development of technical systems, the analytic and the holistic approach (Jackson, 2010).

On the one hand, the traditional approach taken for the development of systems is the *analytic approach,* which concentrates on the development of each system's element independently, without paying any attention neither to the system as a whole, nor to the interactions among the different elements conforming the system once they are assembled together. This design process is carried out according to the problem solving methodology stated by Descartes, which consists on dividing the complex problems into smaller and simpler problems. Once the top problem has been decomposed into a collection of atomic entities, the problems are solved hierarchically in an ascent way until a solution for the complex problem on the top is achieved. This kind of methodology, applied in the conventional engineering design, is suitable and valuable for the design of systems where the technological environment is subject to minor changes, system's goals are clear, and the amount of uncertainties is low.

On the other hand, the *holistic approach* is based on the *Systems thinking* philosophy which considers a system as a whole rather than as simply the sum of its parts, and tries to understand how the different parts of a system influence each other inside the whole. This approach also takes into consideration the boundaries and environment of the system-of-interest by determining which entities are inside the system and which are not, as well as by analysing the influence of the operating environment on the system to be developed. The *holistic approach* has also been considered as a problem solving method in which the different aspects of a problem can most effectively be understood if they are considered in the context of interactions among them and with other systems rather than in isolation. This problem solving nature has been also stated by Sage and Armstrong in (Sage & Armstrong, 2000). According to them, the *holistic approach* stresses that *there is not a single correct answer or solution to a large-scale problem or design issue. Instead, there are many different alternatives that can be developed and implemented depending on the objectives the system is to serve and the values of the people and organizations with a stake in the solution.*

The principles of *Systems thinking* state that events can act as catalysts which can heavily influence complex systems. Thereby, the events as well as the systems can be completely different. The events can have a technical, natural or timely source amongst others, while the systems can be from technical, political, social, or any other kind. In fact, identifying the so-called *emergent* properties of a system that cannot be predicted by examining its individual parts is an exclusive feature of the *holistic approach* not provided by the *analytical approach*. This kind of methodology is suitable and valuable for the design of systems where the technological environment is subject to significant changes, system's goals are not clear, and the amount of uncertainties is high.

According to the provided definition of *complex system* and the description of its characteristics, it can be stated that the features of the *holistic approach* make it to be best suited to the characteristics required for the process of developing this kind of systems. Table 1 maps the specific challenges associated with the development of complex systems to the characteristics and features provided by the holistic system design approach. It shows how *holistic approach* provides measures to manage all the concerns present in a typical development process of complex systems.

The argument of the *holistic approach* being more suitable for developing complex systems is supported by the statement made by Gibson et al. in (Gibson et al., 2007) in which *system team members are supposed to be able to work across disciplinary boundaries toward a common goal when their disciplinary methodologies are different not only in detail but in kind*. A design process based on the *analytical approach* cannot fulfill this requirement since the system team members work exclusively in their own disciplines and they do not have access neither to a vision in perspective of the whole system nor to the context information related to the other elements in the system. The former is necessary for identifying the interacting elements while the latter is necessary for assessing the way the different elements interact with each other.

The characteristics of the *holistic approach* described above may propitiate the assumption that this approach remains pretty much superficial and that it does not get very detailed or specific. This assumption is incorrect in the sense that, inside the *holistic approach*, there is much effort devoted to in-scoping, high-fidelity modeling, and specification of system requirements and architecture (Sage & Armstrong, 2000).

Mapping of characteristics	
Complex systems	**Holistic approach**
• Difficulty to maintain whole system under perspective	• Systems considered as a whole, not as a sum of parts
• Big amount of internal and external interfaces	• Focus on understanding how the different parts of a system influence each other inside the whole • System aspects considered in the context of interactions among components and with other systems rather than in isolation
• Implication of different technical fields	• Identification of *emergent* properties that cannot be predicted by examining individual parts of a system • Analysis of unexpected interactions and cause-effect events
• Broad and heterogeneous stakeholders	• Consideration of system boundaries and operating environment

Table 1. Mapping of characteristics of complex systems and holistic approach

3.2 Standardized procedures as a means for managing complexity

As in any other field of life, the experience and knowledge acquired with the time plays a vital role in the design of complex systems. Past experience provides the system engineer with a set of rules of thumb, intuition and sense of proportion and magnitude, which combined together, result in a very valuable toolbox to be applied for proposing solutions, supporting judgements and making decisions during the development of complex systems. Those design principles, guidelines, or rules that have been learned from experience, especially with respect to the definition of the architecture of a system, have been considered by Jackson to constitute which is called heuristics (Jackson, 2010).

It is common that companies rely on heuristics-dominant system teams for the development of systems in areas considered as sensitive for the companies. However, this is a very individual-centred approach, in which system's or even company's know-how is concentrated in specific people and thus dependent on them. This kind of know-how is critical, since in the case of one key person leaving the team or the company, the know-how it possesses leaves with him or her, thus creating a loss of knowledge with two different consequences: On one side, the company loses all the existing information, creating a regression of company's know-how in the field. On the other side, it takes a lot of time to determine exactly which specific know-how has been lost and to assess which part of the know-how still remains in the company.

Another aspect of heuristics to be considered is that human beings unconsciously make use of the knowledge they possess in a specific situation in order to interpret the reality they confront. In other words, heuristics provide background information and helps to put the facts and figures in context and to interpret them. This means that two different members of the same system team might interpret in a different way and derive different conclusions from the same information just because they possess different background knowledge.

A standardized know-how management system can help making company's dependency on individuals' heuristics unnecessary or at least, less critical. The generation of standard documentation with predefined structure and contents allows condensing the most important information about projects and its transmission. A key piece of information that must be included in the standard documentation is the rationale behind the different decisions made in the project, in order to provide traceability. Standardized documentation means that anyone working in a company knows exactly which documents are available inside a project and which information do they contain. This makes possible to minimize the consequences of a key person leaving the team, since its successor ideally would be able to achieve the same knowledge status about the project in a fast and efficient way thanks to the traceability of decisions made. For the same reasons stated before, the information contained in the standardized documentation can be transmitted to every other member of the team or the company in a transparent way, thus enabling the achievement of homogeneous background information about the project that can be shared by all team members.

In the modern and globalized industrial market, where trends, products and technologies change very rapidly and companies worldwide compete fiercely for the same business niche, the reputation of a company frequently plays a determinant role. This reputation basically depends on the quality of the products they produce or the services they provide, which at the same time, greatly depends on the quality of the processes used during the whole product's life-cycle. The definition of efficient and high-quality working methodologies and best practices takes place as a result of an iterative learning process which refines itself making use of the lessons learned during the development of past projects. All this know-how is considered as a strategic business active of every company and therefore it is condensed in standard practices and regulations that become mandatory for every employee of the company. Every time a new employee joins the entity, he or she must get started with those internal regulations and assimilate them.

Nowadays, the system development strategies based on the black box approach, which uses in-house developed proprietary technologies, has been substituted by a white box approach based on Commercial-of-the-Shelf technologies, where most of the system development workload is subcontracted to external entities. This subcontracting strategy has many associated advantages like the reduction of development costs and risks (derived from delegating the development of specific system parts to companies with more experience in that type of elements) among others. However, this strategy has also associated risks that must be correctly managed in order not to become drawbacks with highly negative effects. One of those risky factors is a higher communication flow between at least two different entities, which in general possess different working methodologies and tools. A standardized system development process, makes the exchange of information effective and efficient, since on one side, there is no risk of misinterpretation of the transmitted information and on the other side, the number of required transactions decreases due to the fact that every part knows which documents with which specific content must be delivered in every phase of the development process.

All these aspects have been also considered by Sage and Armstrong (Sage & Armstrong, 2000) who stated that the development process of any system in general, and of complex systems in particular, should fulfil amongst others, the following requirements:

- Systems engineering processes should be supportive of appropriate standards and management approaches that result in trustworthy systems.
- Systems engineering processes should support the use of automated aids for the engineering of systems, such as to result in production of high-quality trustworthy systems.
- Systems engineering processes should be based upon methodologies that are teachable and transferable and that make the process visible and controllable at all life-cycle phases.
- Systems engineering processes should be associated with appropriate procedures to enable definition and documentation of all relevant factors at each phase in the system life cycle.

In summary, standardized processes help to increase the productivity in system development activities by improving the transparency of all team members' work, which eases and advances communication and collaboration. They also help to increase the quality of working methods and products, as well as to manage company's know-how by enabling traceability of requirements, decision, rationales and deliverables. This traceability makes all working steps reproducible and improves consistency and integrity of all deliverables, contributing to the management of knowledge created during the process. Additionally, standardized processes help to mitigate risks by enabling comparability with previous development projects, amongst others, which supports monitoring and controlling of cost and schedule.

3.3 Fundamental development disciplines

The fundamentals of building and managing complex systems at the top level have been identified by Eisner in (Eisner, 2005). According to him, there are three areas which are critically important in building and managing complex systems: Systems engineering, project management and general management. The importance of these three areas has also been identified by Sage and Armstrong in (Sage & Armstrong, 2000) in which they state that, *Systems engineering processes should enable an appropriate mix of design, development and systems management approaches.*

Additionally, in the special case of developing systems whose failure could imply catastrophic consequences like big economic losses or human casualties, the concepts, methods and tools belonging to the safety engineering discipline must also be considered as fundamental.

The area of general management is an extremely broad topic, which is out of the scope of the current chapter and therefore the chapter's contents will concentrate on the other disciplines mentioned, i.e. Systems engineering, project management and safety engineering.

3.3.1 Systems engineering

Systems engineering is an interdisciplinary approach and means to enable the realization of successful systems (Haskins, 2010). It is based on well-defined processes considering customer needs and all other stakeholders' requirements and it always profits from providing a holistic view on all problems across the whole development life-cycle. It has progressively attracted the attention in different fields of industry, as a methodology for managing the design and development of complex systems in a successful, efficient and straightforward way. According to (Gibson et al., 2007), *it is a logical, objective procedure for applying in an efficient, timely manner new and /or expanded performance requirements to the design, procurement, installation , and operation of an operational configuration consisting of distinct modules (or subsystems), each of which may embody inherent constraints or limitations.*

This conceptual definition of Systems engineering, states implicitly that the development process is defendable against external critics and that all the decisions made inside are objective and traceable. As it has been reasoned previously, traceability is a fundamental characteristic that must be present in every development process because of the multiple benefits it has associated with it, i.e. project reproducibility or the creation of know-how by means of stating the rationale behind the design decisions made, or listing and describing the risks found out and resolved during the development process.

Additionally, previous definition of Systems engineering also describes implicitly its holistic nature, by taking in consideration all the phases of a system's life-cycle and the interfaces and interactions between the system of interest and the systems related to it.

The field of Systems Engineering has published an international standard called *ISO/IEC 15288 – Systems and software engineering* (ISO 15288, 2008). It provides a *common framework for describing the life-cycle of systems* from conception up to retirement and defines associated processes and terminology. Processes related to project management are specified therein, but because of standard's scope focusing on Systems engineering, those processes do not cover the complementary domain of project management. The last update of the ISO 15288

Standard was released in 2008[1] which points to it as an active standard which is still in an iterative improvement status. Nevertheless, the standard has been consolidated with the INCOSE Handbook (Haskins, 2010) which is broadly established worldwide.

3.3.2 Project management

Project Management is the application of knowledge, skills, tools, and techniques to project activities to meet the project requirements (PMI, 2008). It is also based on well-defined processes regarding planning, executing, monitoring, and controlling of all working activities and the effective application of all assigned project resources. Project management profits from an always transparent status of all activities and deliverables and from the early identification of any risks.

It must be remarked that project management consists not only on applying the specific skills necessary for carrying out a project once it has been accepted, but also on managing the systems team itself on an effective manner.

Gibson et al. identify in (Gibson et al., 2007) some requirements for building an effective systems team. Aspects like having a leader, defining a goal and using a common working methodology with a well-balanced set of skills among members who pull together towards the goals have been identified as critical for achieving project's goal on schedule.

Sage and Armstrong (Sage & Armstrong, 2000) state in addition to this that systems engineering processes should possess following characteristics from the point of view of project management: 1) *they should support the quality assurance of both the product and the process that leads to the product*, 2) *they should be associated with appropriate metrics and management controls* and 3) *they should support quality, total quality management, system design for human interaction, and other attributes associated with trustworthiness and integrity*. These statements support the idea of a holistic design process for developing complex systems.

The Project Management Institute (PMI) has published the guide to the *Project Management Body of Knowledge (PMBOK)* (PMI, 2008). This document is recognized as a standard by classical standardization entities like ANSI and IEEE. It covers all management topics completely, without taking engineering aspects into scope.

3.3.3 Safety engineering

Safety engineering can be seen as *a set of well-defined processes aiming at achieving freedom from unacceptable risk* (ISO 61508, 2009), together with the application of methodologies in order to quantify and to prove it. Due to the fact that not only the complexity of modern systems increases, but also their capabilities, the amount of functions performed by a system also raises. Inside those functions, there are safety-related functions performed by specific systems included whose failure would lead to important economical and material damages, severe injuries, or even fatalities. The increase of capabilities in systems, together with the growing humankind's dependency on them, leads to the fact that more often the safety depend directly on a fail-safe operation of systems. Furthermore, the more safety-related

[1]The references made to ISO 15288 relate to the 2008 version of the standard, if not explicitly mentioned otherwise.

equipment is integrated into a system, the bigger is the probability that one system element fails. This, in turn, increases the concern about safety among the society. Additionally, due to the overall complexity of systems, assessing the impact of single failures on them and setting up preventive or corrective actions is a very challenging task. All these facts mentioned above have contributed to making safety considerations become more and more essential to modern development processes.

In the field of Safety engineering, a widely considered international standard is the *ISO/IEC 61508: Functional Safety of Electrical/Electronic/Programmable Electronic Safety-Related Systems* (ISO 61508, 2008) which sets out a generic approach for all safety life-cycle activities for systems whose elements perform safety functions. This standard is field-independent and sets the basis framework in which additional, branch-specific industrial safety standards are based, e.g. ISO 26262, the new safety standard for the automotive domain, or EN 50128 for railway systems.

3.4 Drawbacks of standards involved in complex-systems development

The historical evolution of every of the standards presented above has grown independently from each other. This fact implies that, even if the participation of the three cross-disciplines and their combined use has been recognized as critical for the development of complex systems by the industry, the different standards are poorly connected or not connected at all among them. This leads to a situation in which the standards overlap with each other in many processes and activities, and in the worst case they even could contain conflicting directives. Additionally, there is a lack of consolidated set of terms used inside the standards. Every standard makes its own definition of terms which creates confusion and misunderstandings and makes the cross-disciplinary communication difficult.

Besides, the standards themselves possess some deficiencies that difficult their interpretation and understanding, and consequently, their implementation. On one side, the ISO 15288 standard does not provide any sequence diagrams showing the relationships between the processes and activities contained in it. On the other side, the ISO 61508 standard lacks of a detailed description of the inputs and outputs associated with the different activities it describes.

4. Systems engineering approach based on international standards

4.1 General description

The holistic Systems engineering view described in this work takes the ISO 15288 standard as its core and tries to combine it with the other two standards introduced above. Some of the technical processes contained in the ISO 15288 are also addressed by the safety and project management standards respectively, providing interfaces where information can be exchanged among them or even where processes can be merged together. This combination of standards can be noticed in the case of the project related processes of ISO 15288, which are completely replaced by those defined inside the PMBOK standard, due to the fact that this standard considers them in a much more detailed way. The agreement processes defined by the ISO 15288 standard are also considered by the PMBOK standard inside the procurement area, but in this work, merging the agreement processes of both standards has been considered out of scope.

From the five organizational project-enabling processes defined by the ISO 15288 standard, only the *Human resource management* and *Quality management* processes are explicitly addressed by the PMBOK standard. The remaining three processes are not explicitly treated by the project management standard and therefore they are not considered inside the present work. Fig. 2 shows the process groups defined by the systems engineering standard together with an overview of the process groups also addressed by the project management and safety standards.

4.2 Harmonization process

The analysis and comparison of different items like the standards mentioned above, is logically impossible without a common reference framework in which all the items to be compared can be represented.

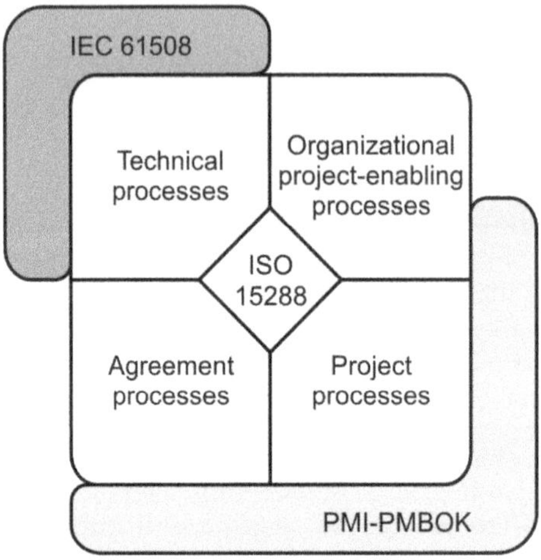

Fig. 2. Overlapping of considered standards regarding process groups

A detailed analysis of the three international standards has revealed that no common reference framework exists among them. This fact implies that before any task of the merging process can be carried out, e.g. comparison and identification of interfaces among the standards, a reference framework must be defined. The PMBOK standard provides a clear overview of its management processes structured in a two-dimensional matrix, representing different process groups in its columns against specific knowledge areas in its rows. This kind of representation based on a matrix has been considered by the authors as a clear and valuable means for analysing, comparing and merging the different international standards and consequently, it has been selected as the reference framework for the merging process.

None of the ISO standards analysed defines process groups or knowledge areas in the way that PMBOK does. The PMBOK standard defines process groups according to a temporal sequence while the ISO 15288 standard defines the process groups on a purpose basis. As a consequence, their respective reference matrices of both ISO standards need to be created from the scratch. Instead of the process groups used by PMBOK standard, the different life-cycle stages named by the ISO 15288:2002 standard have been taken. In the case of the knowledge areas, if the ones from the PMBOK standard were not appropriate, new ones have been defined.

This approach showed that the matrices of both ISO standards can be merged to one unique matrix while the mapping of management process groups of the PMBOK standard into the life-cycle stages of the ISO standards is not possible. This is due to the fact that project management activities are carried out during the whole life-cycle of the system-of-interest and not just during a specific stage. Besides, there are also several knowledge areas regarding management, e.g. procurement, which cannot be considered together with technical processes. In consequence, the management and life-cycle stages have to be considered as parallel stages and two different process matrices have been created; one for management processes and another one for technical processes, respectively.

Finally, the processes being assigned to the same stage and knowledge area inside the technical processes' matrix are good candidates for interfacing or merging. After the description of the two matrices, a detailed analysis of the processes follows based on the matrices.

4.2.1 Management processes

The matrix shown in Fig. 3 is taken from the PMBOK standard. The columns represent process groups which can also be seen as project management stages starting with *Initiation* and ending with *Closing*. Each of the rows represents a typical project management topic, which is further called knowledge area. All of the forty two management processes specified by the PMBOK standard are classified into the cells resulting from the crossing of five process groups' columns with the nine knowledge areas' rows.

4.2.2 Technical processes

In the case of technical processes, the ISO 15288 standard does not define stages for the life-cycle of systems. However, a division of the life-cycle in various stages was provided in its previous version, ISO 15288:2002. These life-cycle stages have been assigned to the columns of the respective matrix. For the rows, ISO 15288 standard defines four knowledge areas (as shown in Fig. 2), in which the life-cycle processes are grouped by their purpose. However, these knowledge areas are not useful for comparing the processes with those contained in the other standards. Therefore, those used in the project management matrix were considered. Only two knowledge areas, *Scope* and *Quality*, were found to be also relevant for technical processes. Two further knowledge areas have been defined by the authors. On one hand, *Realisation* represents all activities which elaborate the outputs of the *Scope* area, which then can be quality-checked. On the other hand, *Service* describes all the activities to be carried out during the operating life of a system.

The ISO 15288 standard does not explicitly assign any processes to any life-cycle stages. In fact, the processes are initiated in one or more stages and some can be executed sequentially or in parallel. In this work, an interpretation process has been carried out in which the processes of the standard have been assigned to the cells of the matrix described above. The aim of this interpretation work was to enable the comparison and analysis of the processes and activities of the three standards in order to facilitate the identification of possible interfaces and overlapping areas between the different standards.

Knowledge areas	Project management process groups: Initiating	Planning	Executing	Monitoring & Controlling	Closing
Integration	Develop project charter	Develop project management plan	Direct & manage project execution	Monitor & ctrl. project work Perform integrated change ctrl.	Close project or phase
Scope		Collect requirements Define scope Create WBS		Verify scope Control scope	
Time		Define activities Sequence activities Estimate activity resources Estimate activity durations Develop schedule		Control schedule	
Cost		Estimate costs Determine budget		Control costs	
Quality		Plan quality	Perform quality assurance	Perform quality control	
Human resource		Develop human resource plan	Acquire project team Develop project team Manage project team		
Communications	Identify stakeholders	Plan communications	Distribute information Manage stakeholder expectations	Report performance	
Risk		Plan risk management Identify risks Perform qual. risk analysis Perform quan. risk analysis Plan risk responses		Monitor and control risks	
Procurement		Plan procurements	Conduct procurements	Administer procurements	Close procurements

Fig. 3. Project management processes assigned to process groups and knowledge areas

The eleven technical processes specified inside ISO 15288 have been spread over the matrix using a black font, as depicted in Fig. 4. Inside the *Conception* stage, three different processes have been assigned to two different knowledge areas. The two processes dealing with requirements have been assigned to the *Scope* area, while the *Architectural design* process has been assigned to the *Realisation* area. In the first case, requirements specify the scope of the system. In the second case, the process was assigned to that specific area because one of the process' activities is to evaluate different design candidates, which cannot be done in the development stage or later ones. Besides, the process generates a system design based on the requirements elicited in the scope area, which supports its assignment to the *Realisation* row.

The *Production* stage contains *Transition* and *Validation* processes in two different knowledge areas. *Transition* process has been assigned to the *Production* stage because the development ends before the transition of the system (ISO 24748-1, 2010). In the same way as with *Verification, Validation* has also been seen in the *Quality* area. It must be remarked that the *Validation* process has been considered by the authors to take place at the end of the transition, in which at the end, the customer accepts the system delivered and installed in its operational environment. *Operation* and *Maintenance* belong to *Utilization* and *Support,* while *Disposal* can be found in the *Retirement* stage. All of them are assigned to the *Service* area. The activities of the Disposal process can also be seen as a service in the widest sense.

The ISO 61508 standard defines sixteen so-called life-cycle phases. In this work, they are interpreted as activities because for each of them, its inputs and outputs are defined and in the corresponding standard's chapters, tasks are indicated that have to be carried out. The standard neither defines any superior life-cycle stages comparable to ISO 15288:2002 nor defines any knowledge areas. For this reason, and because the activities are also of a technical kind like the processes of ISO 15288, they have been assigned to the same matrix shown in Fig. 4.

The matrix contains all of the sixteen activities defined by ISO 61508, illustrated by a grey font. Six activities are assigned to the *Conception* stage divided in two different knowledge areas. *Concept, Overall scope definition, Hazard and risk analysis* and *Overall safety requirements* have been assigned to the *Scope* area because they contribute to defining the scope and safety related requirements for the design. The *Overall safety requirements allocation* and *System safety requirements specification* have been assigned to the *Realisation* area. This is due to the fact that both processes specify and allocate safety requirements to designed system elements during the *Architectural design* process.

Inside the development stage five different processes have been assigned into three different knowledge areas. First, *Realisation,* Other *risk reduction measures* and *Overall installation and commissioning planning* have been assigned to the *Realisation* area because they address questions related to the physical implementation of the system. The two remaining planning activities, i.e. *Overall safety validation planning* and *Overall operation and maintenance planning* have been assigned to the *Quality* and *Service* knowledge areas respectively. The planning activities typically take place in parallel to the implementation and they must be carried out before the system is installed and validated in its operational environment.

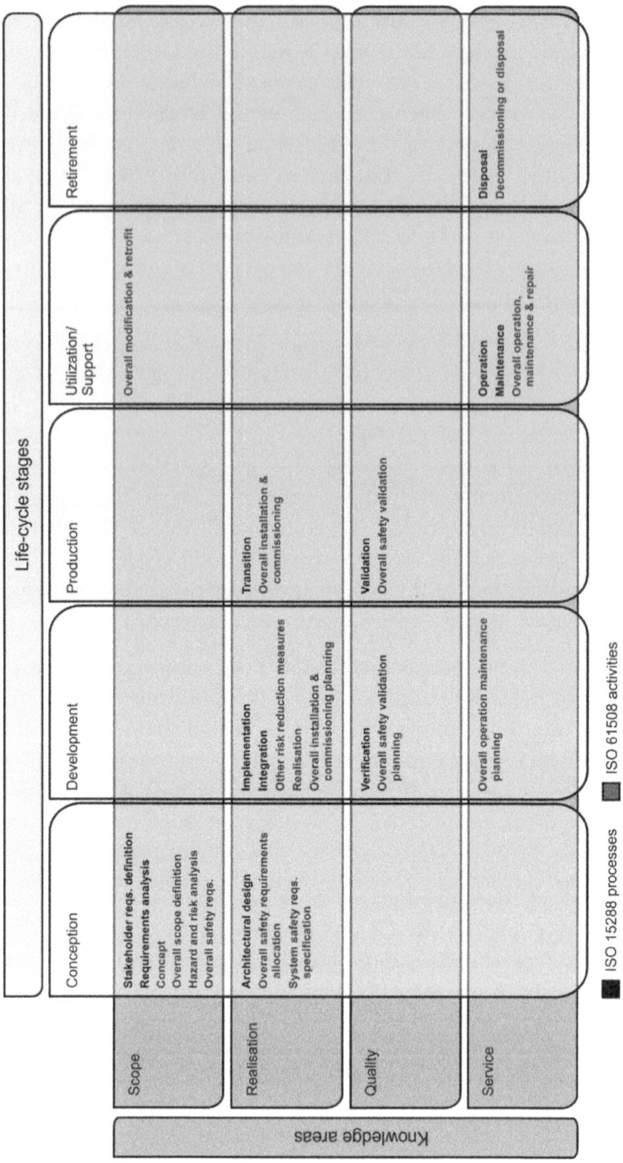

Fig. 4. ISO 15288 technical processes and ISO 61508 activities assigned to life-cycle stages and knowledge areas.

Inside the *Development* stage, another three processes have been assigned to the *Realisation* and *quality* areas. On one hand, the processes *Implementation* and *Integration* have been assigned to the *Development* and *Realisation* area because the physical creation of the real system takes place inside them. On the other hand, the *Verification* process is part of the *Quality* area because it contributes to guarantee the quality of the system-of-interest under development.

The *Overall modification and retrofit* activity has been assigned to the *Support* stage because this activity is typically initiated when the system is under operation and its support is active. Due to the fact that the output of this activity can affect all knowledge areas including the scope, it has been assigned to this overall area. The last two activities of the ISO 61508 standard, i.e. *Overall operation, maintenance and repair* and *Decommissioning or disposal* can be found in the *Service* area, assigned to the corresponding life-cycle stage.

4.3 Detailed standards interfacing and merging process

Those processes which are in the same life-cycle stage or knowledge area or both, bear potential for being harmonized. After an in-depth analysis of the three standards, eleven information and twelve process interfaces have been respectively identified. On one hand, information interfaces represent some kind of information generated by any of the standards, which is provided to the other standards for its use. E.g. safety requirements provided by the ISO 61508 are merged into the *System requirements* document generated by the ISO 15288 standard. On the other hand, process interfaces represent similar activities that are carried out in at least two of the standards, which in consequence, can be put together in order to avoid duplicities that constitute a waste of resources.

Because processes basically describe a sequence of activities, they are typically represented by some kind of flow diagram. For this reason, a standardized graphical notation for process diagrams has been selected to represent the relevant process parts and the outcome of their merging.

4.3.1 Business Process Model and Notation (BPMN) specification

The Object Management Group (OMG), a non-profit consortium dedicated to developing open computer industry specifications, took over the development of the BPMN specification in 2005. BPMN's primary goal is to provide a notation that is readily understandable by all business users, from the business analysts, to technical developers, and to managers who will manage and monitor those processes. (OMG, 2011)

The notation used in Fig. 5 to Fig. 8 corresponds to BPMN. The processes defined in the different standards (*activities* in BPMN) are represented as boxes, their outputs (*data objects*) are depicted by the leaf symbol, and the arrows illustrate the sequence flow. Circle symbols represent either the start or end event, or they describe an incoming or outgoing link to another diagram or (not depicted) process. In BPMN, a diamond symbol illustrates a gateway control type which marks the point where sequence flow paths are joined or divided. Gateways that initiate or merge a parallel sequence flow are expressed by a diamond containing a *plus* symbol. In the following diagrams, those gateways have been mostly omitted for the sake of simplicity and size. Gateways that introduce a conditional sequence flow are expressed by an empty diamond. Horizontal *pool lanes* represent a categorization of activities.

4.4 Harmonization result: The Holistic Systems Engineering view (HoSE)

Fig. 5 to Fig. 8 represent the product life-cycle stages defined in ISO 15288:2002 respectively. Every figure contains the project management as well as technical processes corresponding to the specific life-cycle stage. Due to length constraints a complete in depth representation of all standard's levels is not possible, thus only the top level view has been provided. The processes of every standard are contained in a pool lane. The ISO 15288 standard is depicted in the middle pool lane of each figure. In case of the PMBOK standard, only the processes related to technical activities have been considered. Every sequence flow arrow crossing a lane represents an information interface between the corresponding standards.

4.4.1 HoSE conception stage

In Fig. 5, the corresponding processes of the three international standards for the *Conception* stage are shown. This includes eight of the technical processes already assigned to the conception stage as depicted in Fig. 4 as well as three related management processes.

In every standard, one initiating process is defined. Regarding the PMBOK, the first process is the *Identify stakeholders* process, for the ISO 15288 it is the *Stakeholder requirements definition* process, and so on. In this case, the *Identify stakeholders* process has been selected. Looking at the activities of the ISO 15288 *Stakeholder requirements definition* process and its outputs, it shows that it includes a sub-activity called *Identify the individual stakeholders*. This activity matches exactly with the *Identify stakeholders* process from PMBOK which identifies the related stakeholders and which lists them in an output document called *Stakeholder register*. As a consequence, the ISO 15288 sub-activity has been merged together with the PMBOK process and the *Stakeholder register* document it produces has been provided as an input to the remaining activities inside the *Stakeholder requirements definition* process of ISO 15288.

In the PMBOK lane in Fig. 5, the next process is the *Collect requirements* process. This can be merged with the activity *Elicit stakeholder requirements* of the *Stakeholder requirements definition* process from ISO 15288. At this point, a distinction between product and project requirements, as explicitly recommended by the PMBOK, helps to differentiate between project's progress and system-of-interest's advancements. In this way, PMBOK's activity of eliciting product requirements is merged into the ISO 15288 process, which also includes merging the techniques of facilitated workshops and prototypes into the ISO standard. In consequence, the output documents of the *Collect requirements* process are changed to project (only) requirements, project (only) requirements traceability matrix, and an (unchanged) requirements management plan. The sequence flow of the documents is kept as defined in the PMBOK, as illustrated by the grey lines.

The separation of the requirements into stakeholder and system requirements, as explicitly recommended by ISO 15288, enables the consideration of different views on the requirements. Stakeholder requirements define high-level functions from the point of view of client's expectations, while system requirements define functions in more detail from a technical perspective. Both kinds of requirements belong to the problem domain and not the solution domain. In other words, they try to specify what should be developed and not how it should be done. The stakeholder requirements constitute the input for the *Concept* activity of ISO 61508 and provide the level of understanding of the system-of-interest and its environment, required by this task. The *Concept* activity includes performing a *Functional hazard analysis (FHA)* which contributes together with safety-related requirements to the stakeholder requirements by identifying the likely sources of top-level hazards for the system. Those enhanced stakeholder requirements complement the requirements flowing to further PMBOK or ISO 15288 processes.

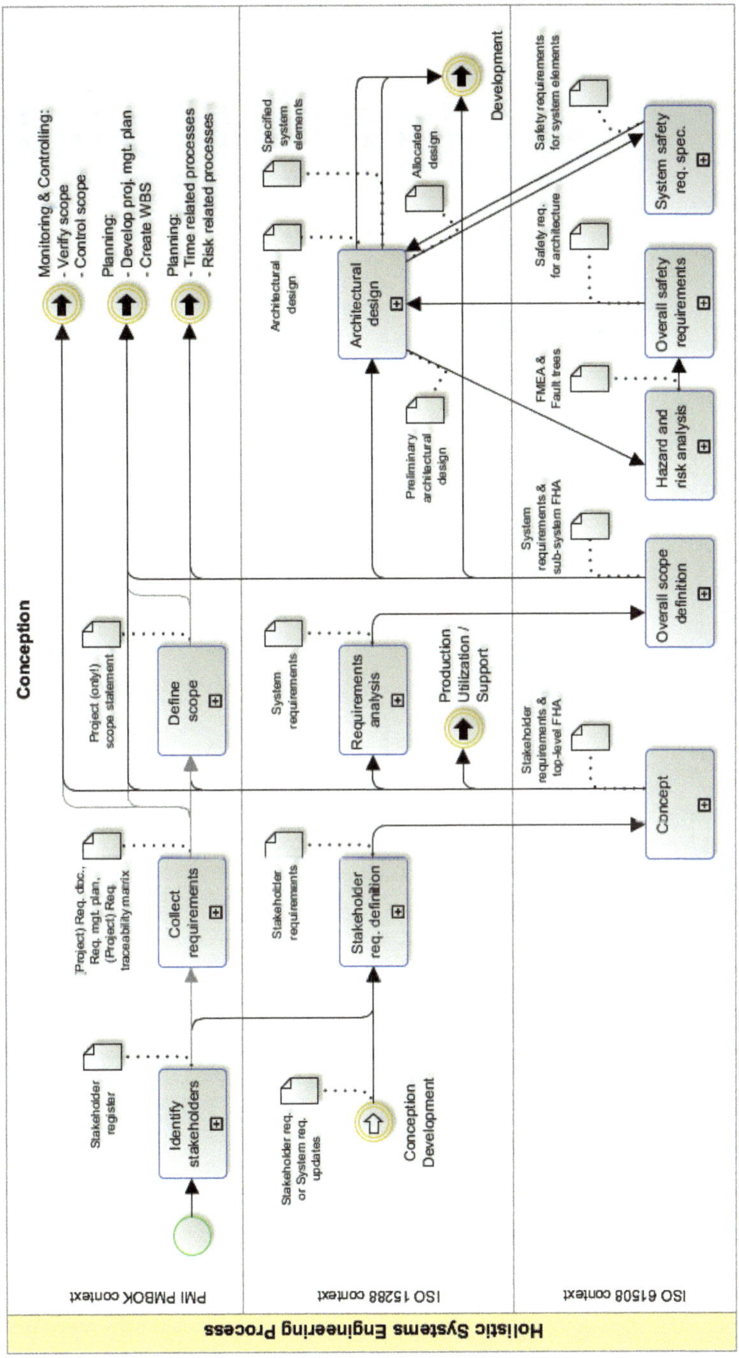

Fig. 5. Conception stage of the holistic systems engineering view

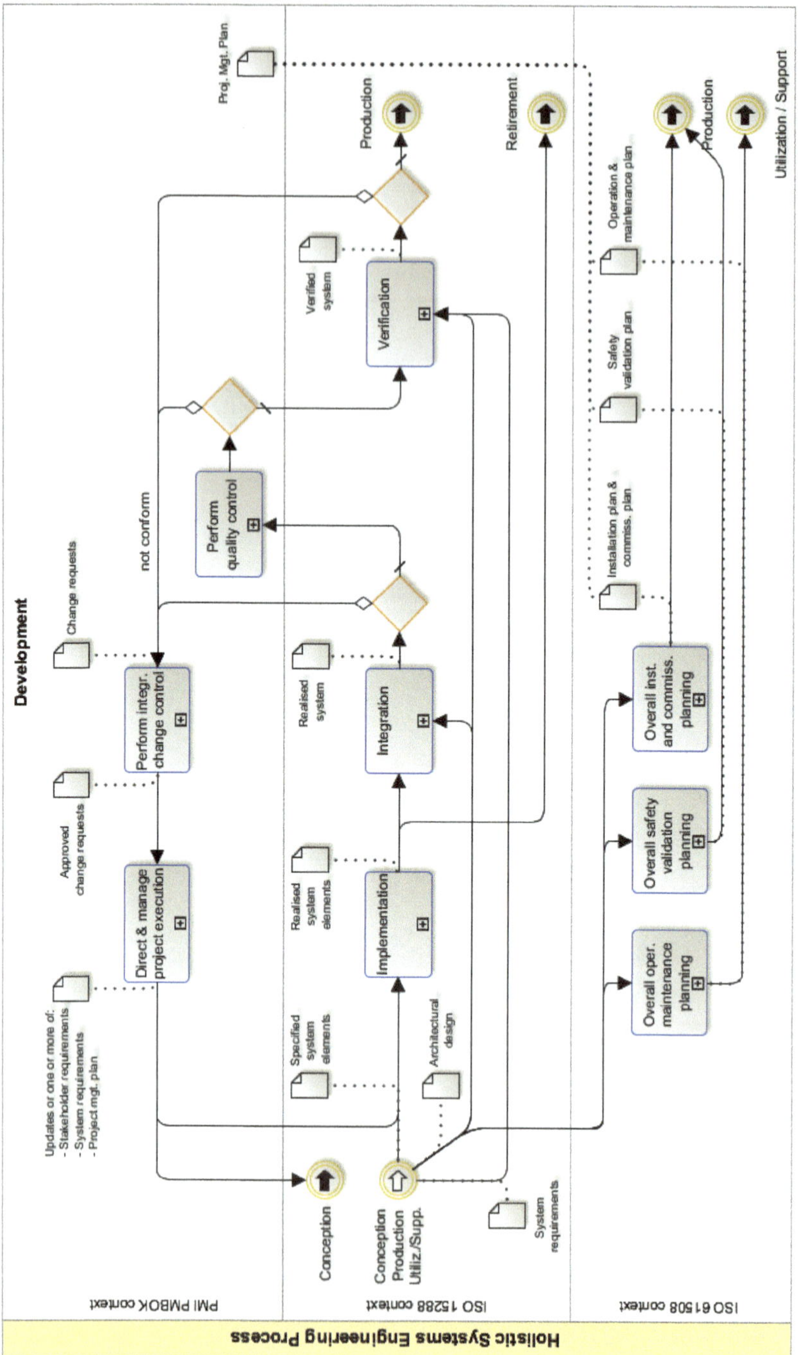

Fig. 6. Development stage of the holistic systems engineering view

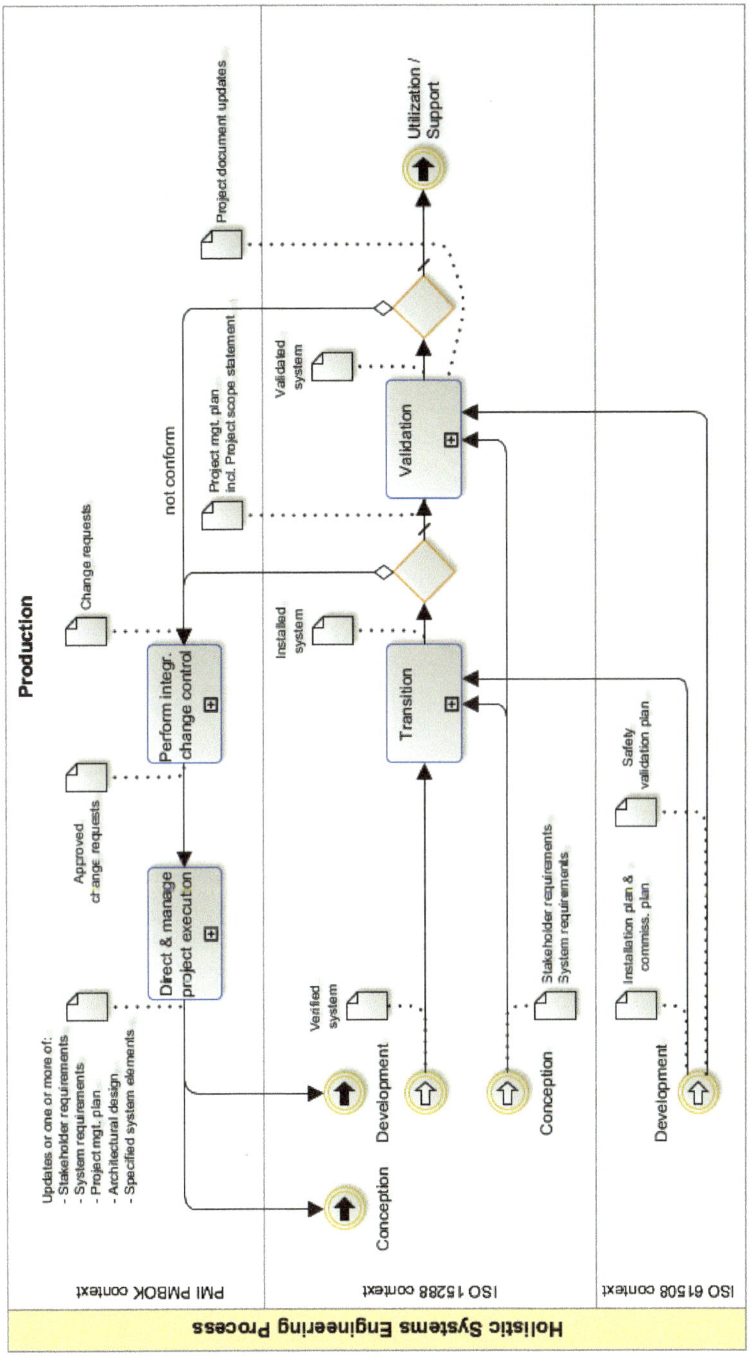

Fig. 7. Production stage of the holistic systems engineering view

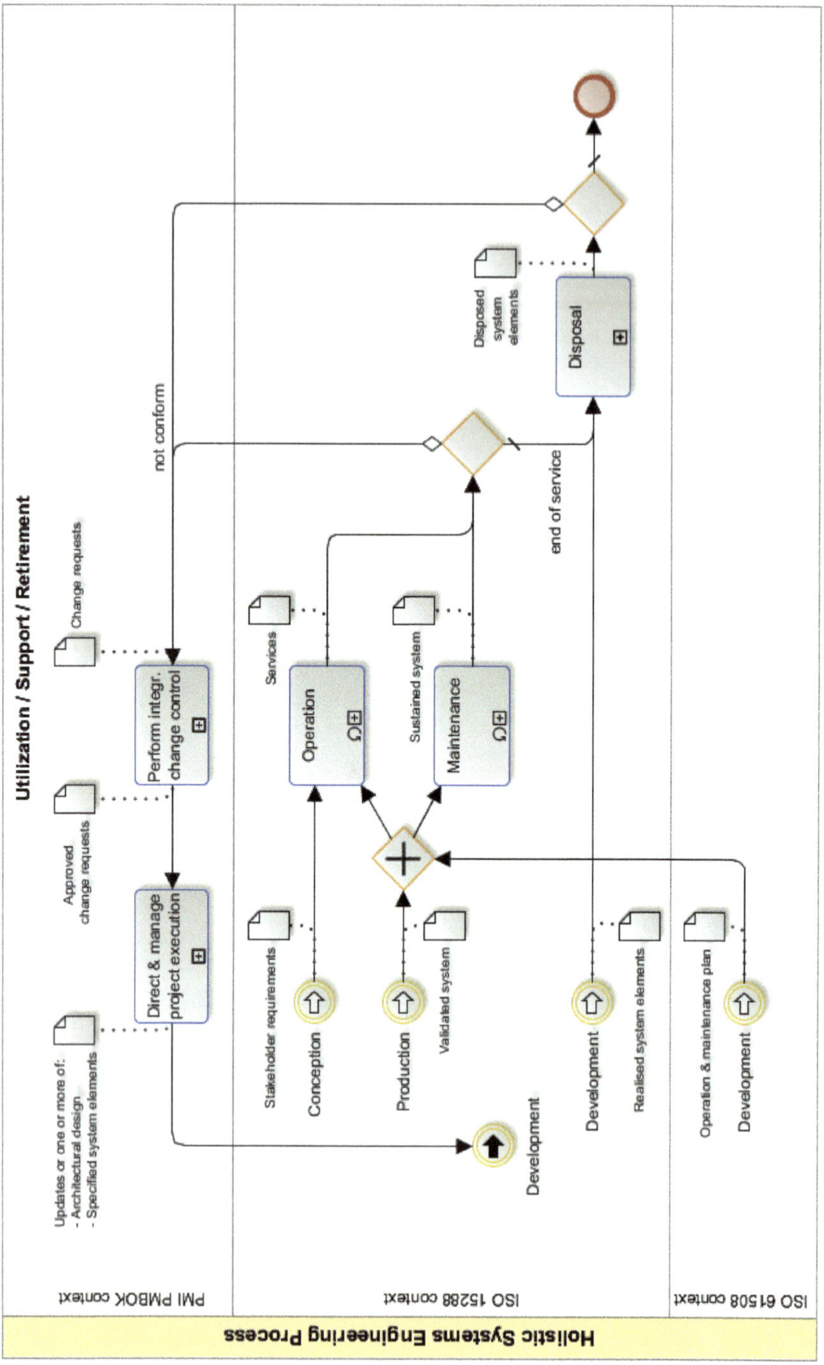

Fig. 8. Utilization, support, and retirement stages of the holistic systems engineering view

The *Requirements analysis* process of the ISO 15288 refines stakeholder requirements into technical system requirements. In the holistic view, the technique of *Product analysis*, specified in PMBOK's *Define scope* task, and the product related *Scope statement* are moved into this process. The complete system requirements are used by the *Overall scope definition* activity of ISO 61508 to refine the identified hazards and to specify the boundary and scope of the system-of-interest from the safety perspective. Both, *Requirements analysis* and *Overall scope definition* processes, could disclose weaknesses in the stakeholder requirements, which enforce the revision of the requirements (not depicted in the figure for the sake of clearness). The resulting enhanced system requirements flow into related PMBOK processes and into the *Architectural design* process of ISO 15288.

As shown in Fig. 5, the *Architectural design* process of ISO 15288 is split into several parts for being able to accommodate the safety assessment related activities. First, a preliminary architectural design is created and passed to the *Hazard and risk analysis* activity of ISO 61508. In this process, a *Failure Modes and Effects Analysis (FMEA)* together with a *Fault Tree Analysis (FTA)* is performed based on the provided design. The FMEA table and the fault trees are used in the *Overall safety requirements* activity to create safety related requirements for the architecture like required reliability and redundancy levels.

Those requirements are fed back into the *Architectural design* process which provides a refined design where system elements are identified and all requirements are allocated to the related elements (*Allocated design*). The allocation activity also includes the allocation of safety requirements which means that the *Overall safety requirements allocation* activity of ISO 61508 standard can be merged into the *Architectural design* process. In the *System safety requirements specification* activity, safety requirements for the system elements are identified which again influence the design refined in the *Architectural design* process. Finally, an architectural design is created representing the whole system, its decomposition, and its interfaces. Additionally, all system elements are specified in detail to enable their realization in the next stage: the development stage.

4.4.2 HoSE development stage

Fig. 6 shows the six technical processes assigned to the development stage in Fig. 4 as well as three related management processes. The specified system elements created in the *Conception* stage are realized inside the *Implementation* process of ISO 15288. *Realization* and *Other risk reduction measures* activities of ISO 61508 have been merged into this process since both of them are related with the physical implementation of the system-of-interest. The realized system elements resulting from the *Implementation* process are passed to the *Integration* process for further development or to the *Disposal* process, in case that the production of the system-of-interest has been cancelled. On the sub-contractor side, verification, quality control, and validation tasks may also follow directly after or within the *Implementation* process.

During the Integration process, the physical system elements are assembled together according to the architectural design. This process ends with the physical implementation of the system-of-interest including its configuration. During system integration, problems or non-conformances may arise, which lead to change requests.

Those requests are explicitly managed by PMBOK's *Perform integrated change control* process. Approved change requests enforce corrective actions to be carried out within the *Direct and manage project execution* process of the same standard. This may include revising the corresponding requirements, updating the project management plan, implementing an improved system element, or cancellation of the project, in the worst case. *Overall modification and retrofit* activity of the ISO 61508 standard is also responsible for managing change requests with regard to safety aspects, thus it has been merged into the change control process of the PMBOK standard.

As shown in Fig. 6, a PMBOK process called *Perform quality control* follows a successful integration, but it can also be carried out after *Implementation* and/or *Verification* processes. The goal is to check the quality of the output provided by the related process. Any non-conformances are managed like described in the previous paragraph. The *Verification* process of ISO 15288 checks if the realized system meets the architectural design and the system requirements which can also include quality requirements. Again, non-conformances may arise in this process; otherwise, the verified system can be transferred into the *Production* stage.

During the implementation of the system or its elements, safety related planning must be performed according to ISO 61508. The corresponding outputs are plans regarding installation, commissioning, safety validation, operation, and maintenance. Those plans have to be integrated into the project management plan.

4.4.3 HoSE production stage

In the *Transition* process of ISO 15288, the verified system is set up in its operational environment. This is done under consideration of stakeholder and system requirements and the installation plan provided by ISO 61508 which contains a description of the operational environment. The *Overall installation and commission* activity of ISO 61508 also deals with the installation aspects of safety-critical systems. Therefore, it has been merged into the *Transition* process of ISO15288 standard.

After the transition, during the ISO 15288 *Validation* process, the installed system is validated against the requirements and the safety validation plan. PMBOK's *Verify scope* process and the *Overall safety validation* activity of ISO 61508 have been merged into this process due to their common goals. To enable the verification of project's scope as required by PMBOK, the *Validation* process is enhanced by the project validation task from PMBOK which requires the project scope statement as an input document. This additional task may lead to project document updates regarding the current state of the project or product.

Non-conformances during *Transition* or *Validation* are managed as already described. They can affect any requirements, designs, plans, or realized system elements which leads to a reiteration of the corresponding process. After a successful *Validation* process, the system, including its operational configuration, can be passed to the *Utilization* and *Support* stage.

4.4.4 HoSE utilization, support, and retirement stages

The validated system and the safety related operation and maintenance plan are the inputs for the next processes of ISO 15288. During the *Operation* process, the system is used to deliver the expected services meeting the stakeholder requirements. The *Maintenance* process is typically applied in parallel to *Operation*. It enables a sustained system. The *Overall operation, maintenance and repair* activity of ISO 61508 is split in two and the corresponding parts are merged into the respective processes. *Operation* and *Maintenance* are carried out uninterruptedly until non-conformances arise or the end of service is reached.

During system operation and/or maintenance, change requests regarding the system or the services it delivers may arise. These must be evaluated through PMBOK's *Perform integrated change control* process. The *Overall modification and retrofit* activity of ISO 61508, responsible for guaranteeing the safe operation of the system, has been merged into this process. If the intended modification is unfeasible or system's end of service is reached, the *Disposal* process organizes the system's retiring and disposing. The *Decommissioning or disposal* activity of ISO 61508 has the same function, thus they have been merged together.

4.5 Harmonization summary

Fig. 9 illustrates a general overview of the harmonization work done. It shows the considered disciplines of project management, systems engineering and safety engineering together with their identified interfaces. There are two kinds of interfaces: On one side, *Information interfaces* express a dependency between information as well as documents of different standards. An information interface results in a merge or change of the information, or document flow. On the other side, *Process interfaces* represent a merge of whole processes or process parts of different standards.

It must be remarked that interfaces between the three standards are present in every of the life cycle stages. This reinforces the usefulness of consolidating the processes of those three standards into a holistic view.

4.6 Benefits of the holistic systems engineering view

The use of standardized procedures during the development of complex systems has many associated advantages. As previously stated in section 3.2, these advantages arise in different aspects of a company. From a commercial point of view, standardized procedures contribute to increase the efficiency of company's processes, to improve the communication with subcontractors and clients, and as a result of those, to increase the quality of the products or services a company offers. From a corporate point of view, standardized procedures provide the basis for traceability and storing of decisions' rationale, which constitute the fundamental factors for generating and managing company's know-how.

Most of the systems development problems mentioned in section 2 can be solved or at least reduced by applying the mentioned standards. However, some of the problems can be solved more effectively by applying the presented harmonized view on the standards. This is especially true for those problems which address the topics knowledge management, risk management, communication and systems thinking.

Using the classification of problems provided in Fig. 1, it can be stated that the use of the HoSE view contributes to solve problems in all the problem areas homogenously, thus reinforcing its holistic character.

In the case of *Human-related* problems, the *Bad customer* and *Erratic communication* problems are solved. On one side, in the bad customer case, a holistic approach based on standardized processes generates standardized documentation. One of those documents is the stakeholder requirements document which must be approved by all the stakeholders. Using this document, later discussions about uncovered topics or not fulfilled objectives can be rejected. The *Systems-related* problem of *Scope arguments with customer* is solved in the same way. On the other side, the problem of erratic communications is solved more effectively in the case that the project manager and the systems engineer are different people. Following the holistic view presented, the project manager and the systems engineer follow the same processes now, e.g. in the field of requirements definition, which avoid any misunderstandings.

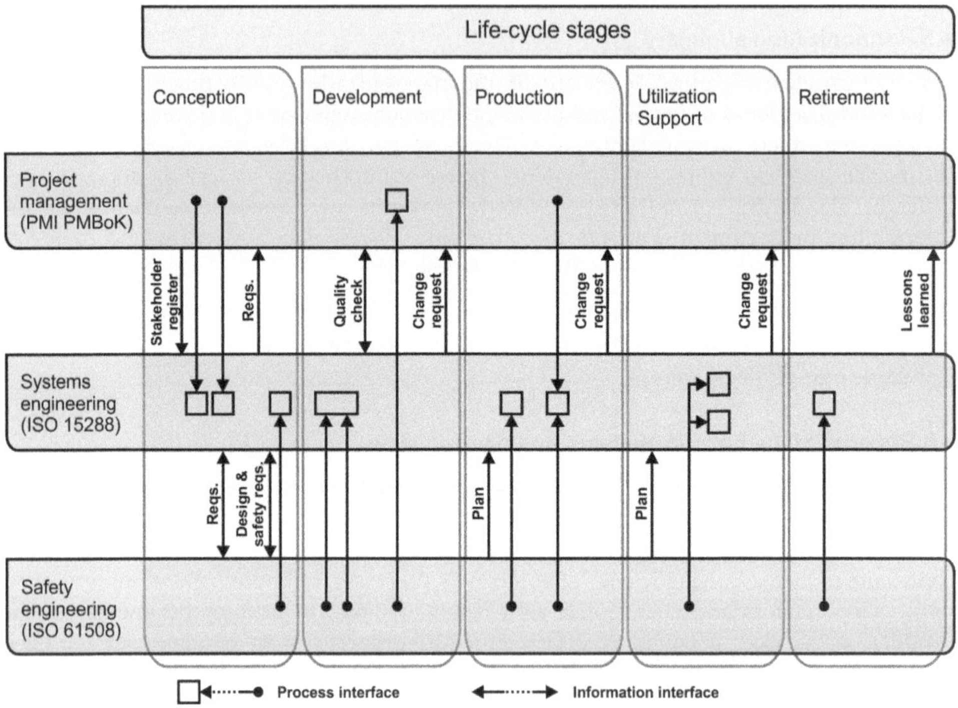

Fig. 9. General overview of the holistic Systems Engineering view

In the case of remaining *Systems-related* problems, *Insufficient funding* and *Insufficient schedule* problems are solved. All the different standards generate and store information during the whole life-cycle of previous projects. A holistic view condensates information from many different sources, thus providing an extremely valuable information source for the planning of further projects. This cumulated information supports an accurate and realistic calculation of resources during project planning.

In the case of *Software-related* problems, problems associated with risk management, performance and quality management like, *Cannot evaluate and mitigate software risks, Do not know how to deal with software warranties* and *Cannot satisfy a critical customer requirement to software performance* respectively, are solved due to the advantages provided by the HoSE view. In this case, the cross-discipline of safety engineering provides means for assessing risks, assessing the proper operation of the system and guaranteeing the satisfaction of critical requirements, which are all not present in a non-holistic approach. The same argument is applicable to the *Management-related* problem of *Quality of services and products inadequate*.

Finally, inside the *Management-related* problems, the HoSE view contributes to achieve one of the most important disciplines of a learning organization as stated by Senge in (Senge, 1994), Systems Thinking.

5. Conclusions

Increasing complexity of contemporary technical systems has led to several problems, inefficiencies and safety threats during their whole life-cycle. The system thinking philosophy, initiated as a consequence of the common need for a better understanding of multidisciplinary dependencies, surfaced the need of a holistic approach for the development of complex systems.

Standardized processes support the management of complexity in a critical way. Additionally, they improve risk mitigation, productivity and quality, and they serve as a basis for generating and managing the knowledge of a company.

Two different disciplines are considered to be essential in the development of modern complex systems: systems engineering and project management. In a reality were more and more responsibilities are being delegated to technical systems, the safety engineering discipline has become substantial also. For each of the three cross-disciplines, one internationally accepted standard has been chosen. ISO 15288 has been widely recognized as means for managing complexity and coping with uncertainties. The PMI PMBOK standard is comprised of detailed project management processes and activities and has gained the biggest support in the industry world-wide. Finally, ISO 61508 is a basic industrial standard which sets out a generic approach for developing safety-critical functions. This standard has been used as a reference for domain-specific safety standards.

Despite of the existing interdependencies regarding systems engineering, all three cross-disciplines have developed their corresponding standards with minimal consideration in form of referencing each other. This leads to a situation in which the standards overlap with each other in many processes and activities, and in the worst case, they even could contain conflicting directives. Additionally, some deficiencies like missing sequence diagrams or a clear description of inputs and outputs of the associated activities have been identified.

A unique kind of representation has been conceived in order to enable the comparison of the different standards. The processes belonging to different cross-disciplines have been arranged together in a matrix form, representing life-cycle stages and knowledge areas. Processes being assigned to the same stage and knowledge area were identified as possible candidates for being harmonized. Interacting processes and activities were either merged together or their information flows were adapted into a holistic view. The resulting view, called HoSE view, has been illustrated using the standardized *Business Process Model and Notation (BPMN)*.

The results of the work carried out disclose that several interfaces and synergies do exist between the three standards. The holistic view arisen from this work aims to provide a good basis for further harmonization and consolidation within standardisation activities. Furthermore, it also makes a contribution to enhance the systems engineering approach by further improving its capabilities regarding productivity, quality and risk mitigation.

6. References

Eisner, H. (2005). *Managing Complex Systems: Thinking Outside the Box,* John Wiley & Sons, Inc., ISBN 978-0-471-69006-14, Hoboken, USA.

Gibson, J. E., Scherer W. T., & Gibson, W. F. (2007). *How to Do Systems Analysis,* John Wiley & Sons, Inc., ISBN 978-0-470-00765-5, Hoboken, USA.

Haskins, C. (Ed.). (2010). *Systems Engineering Handbook: A Guide for System Life-Cycle Processes and Activities v. 3.2,* International Council on Systems Engineering (INCOSE), San Diego, USA.

*ISO/IEC 51, Safety Aspects: Guidelines for their Inclusion in Standards (*1999*),* International Organization for Standardization *(ISO),* Geneva, Switzerland.

ISO/IEC 15288:2002, Systems and Software Engineering: System Life Cycle Processes (2002), International Organization for Standardization (ISO), Geneva, Switzerland.

ISO/IEC 15288:2008 Systems and Software Engineering: System Life Cycle Processes (2008), International Organization for Standardization (ISO), ISBN 0-7381-5666-3, Geneva, Switzerland.

ISO/IEC TR 24748-1, Systems and Software Engineering: Life cycle management -- Part 1: Guide for life cycle management (2010), International Organization for Standardization (ISO), ISBN 978-0-7381-6603-2, Geneva, Switzerland.

ISO/IEC 61508:2010, Functional Safety of Electrical/Electronic/Programmable Electronic Safety-Related Systems (2010), International Organization for Standardization (ISO), ISBN 978-2-88910-524-3, Geneva, Switzerland.

Jackson, S. (2010). *Architecting Resilient Systems: Accident Avoidance and Survival and Recovery from Disruptions,* John Wiley & Sons, Inc. ISBN 978-0-470-40503-1, Hoboken, USA.

Object Management Group (2011). *Business Process Model and Notation (BPMN) v. 2.0,* Needham, USA.

Project Management Institute, Inc. (2008). *A Guide to the Project Management Body of Knowledge (PMBoK Guide), 4th ed.,* ISBN 978-1-933890-51-7, Newtown Square, USA.

Sage, A. P. & Armstrong, J. E. Jr. (2000). Introduction to Systems Engineering, John Wiley & Sons, Inc., ISBN 0-471-02766-9, Hoboken, USA.

Senge, P. M. (1994). The Fifth Discipline: The Art & Practice of the Learning Organization, Doubleday Business, ISBN 0-385-26095-4, New York, USA.

Permissions

The contributors of this book come from diverse backgrounds, making this book a truly international effort. This book will bring forth new frontiers with its revolutionizing research information and detailed analysis of the nascent developments around the world.

We would like to thank Prof Boris Cogan, MSc, PhD, DSc, for lending his expertise to make the book truly unique. He has played a crucial role in the development of this book. Without his invaluable contribution this book wouldn't have been possible. He has made vital efforts to compile up to date information on the varied aspects of this subject to make this book a valuable addition to the collection of many professionals and students.

This book was conceptualized with the vision of imparting up-to-date information and advanced data in this field. To ensure the same, a matchless editorial board was set up. Every individual on the board went through rigorous rounds of assessment to prove their worth. After which they invested a large part of their time researching and compiling the most relevant data for our readers. Conferences and sessions were held from time to time between the editorial board and the contributing authors to present the data in the most comprehensible form. The editorial team has worked tirelessly to provide valuable and valid information to help people across the globe.

Every chapter published in this book has been scrutinized by our experts. Their significance has been extensively debated. The topics covered herein carry significant findings which will fuel the growth of the discipline. They may even be implemented as practical applications or may be referred to as a beginning point for another development. Chapters in this book were first published by InTech; hereby published with permission under the Creative Commons Attribution License or equivalent.

The editorial board has been involved in producing this book since its inception. They have spent rigorous hours researching and exploring the diverse topics which have resulted in the successful publishing of this book. They have passed on their knowledge of decades through this book. To expedite this challenging task, the publisher supported the team at every step. A small team of assistant editors was also appointed to further simplify the editing procedure and attain best results for the readers.

Our editorial team has been hand-picked from every corner of the world. Their multi-ethnicity adds dynamic inputs to the discussions which result in innovative outcomes. These outcomes are then further discussed with the researchers and contributors who give their valuable feedback and opinion regarding the same. The feedback is then collaborated with the researches and they are edited in a comprehensive manner to aid the understanding of the subject.

Apart from the editorial board, the designing team has also invested a significant amount of their time in understanding the subject and creating the most relevant covers. They scrutinized every image to scout for the most suitable representation of the subject and create an appropriate cover for the book.

The publishing team has been involved in this book since its early stages. They were actively engaged in every process, be it collecting the data, connecting with the contributors or procuring relevant information. The team has been an ardent support to the editorial, designing and production team. Their endless efforts to recruit the best for this project, has resulted in the accomplishment of this book. They are a veteran in the field of academics and their pool of knowledge is as vast as their experience in printing. Their expertise and guidance has proved useful at every step. Their uncompromising quality standards have made this book an exceptional effort. Their encouragement from time to time has been an inspiration for everyone.

The publisher and the editorial board hope that this book will prove to be a valuable piece of knowledge for researchers, students, practitioners and scholars across the globe.

List of Contributors

Boris Cogan
Faculty of Computing, London Metropolitan University, UK

Sergio Chiesa, Marco Fioriti and Nicole Viola
Politecnico di Torino, Italy

Sabrina Corpino and Fabrizio Stesina
Politecnico di Torino, Italy

Guido Ridolfi
Politecnico di Torino, Italy
Delft University of Technology, The Netherlands

Erwin Mooij
Delft University of Technology, The Netherlands

John V. Farr
United States Military Academy, USA

Sofia Lingegård, Tomohiko Sakao and Mattias Lindahl
Department of Management and Engineering, Linköping University, Sweden

Derek Fowler and Ronald Pierce
JDF Consultancy LLP, UK

Jason Sherwin
Columbia University in the City of New York, USA

Dimitri Mavris
Georgia Institute of Technology, USA

Guillaume Auriol, Claude Baron, Vikas Shukla and Jean-Yves Fourniols
CNRS, LAAS, Université de Toulouse, UPS, INSA, INP, ISA, UT1, UTM, LAAS, France

Moti Frank
HIT-Holon Institute of Technology, Israel

Joseph Kasser
NUS-National University of Singapore, Singapore

Joongyoon Lee and Youngwon Park
Ajou University/SE Technology Ltd., Republic of Korea

Didier Fass
ICN Business School and MOSEL - LORIA, Lorraine University, Nancy, France

Marko E. Leinonen
Nokia Oyj, Finland

Alper Pahsa
Havelsan Inc., Turkey

Marcel J. Simonette and Edison Spina
University of São Paulo, Escola Politécnica, Computer and Digital Systems Department, Brazil

Hong Man, David Nowicki and Mo Mansouri
Stevens Institute of Technology, Castle Point on Hudson, Hoboken, NJ, USA

Philip Chan
University of Maryland University College (UMUC), MD, USA

Oroitz Elgezabal and Holger Schumann
Institute of Flight Systems, German Aerospace Center (DLR), Germany